Lecture Notes in Mathematics 1656

Editors:
A. Dold, Heidelberg
F. Takens, Groningen

Subseries: Fondazione C. I. M. E., Firenze
Advisor: Roberto Conti

T0254762

Springer
Berlin
Heidelberg
New York
Barcelona
Budapest
Hong Kong
London
Milan
Paris
Santa Clara
Singapore
Tokyo

B. Biais T. Björk J. Cvitanić
N. El Karoui E. Jouini J. C. Rochet

Financial Mathematics

Lectures given at the 3rd Session of the
Centro Internazionale Matematico Estivo
(C.I.M.E.) held in Bressanone, Italy,
July 8–13, 1996

Editor: W. J. Runggaldier

Fondazione
C.I.M.E.

Springer

Authors

Bruno Biais
Jean Charles Rochet
Université des Sciences Sociales
de Toulouse
Institut d'Economie Industrielle
F-31000 Toulouse, France

Thomas Björk
Stockholm School of Economics
Department of Finance
Box 6501
S-11383 Stockholm, Sweden

Jakša Cvitanić
Columbia University
Department of Statistics
New York, NY 10027, USA

Nicole El Karoui
Université de Paris VI
Laboratoire de Probabilités
4, Place Jussieu
F-75005 Paris, France

Elyés Jouini
Ecole Nationale de la Statistique
et de l'Administraion Economique (ENSAE)
3, av. Pierre Larousse
F-92245 Malakoff Cedex, France

Editor

Wolfgang J. Runggaldier
Dipartimento di Matematica Pura e Applicata
Università di Padova
Via Belzoni, 7
I-35131 Padova, Italy

Cataloging-in-Publication Data applied for
Die Deutsche Bibliothek – CIP-Einheitsaufnahme

Centro Internazionale Matematico Estivo <Firenze>: Lectures given at the ... session of the Centro
Internazionale Matematico Estivo (CIME) ... – Berlin; Heidelberg; New York; London; Paris; Tokyo;
Hong Kong: Springer
Früher Schriftenreihe. – Früher angezeigt u.d.T.: Centro Internazionale Matematico Estivo: Proceedings
of the ... session of the Centro Internazionale Matematico Estivo (CIME)
NE: HST 1996,3. Financial mathematics. – 1997
Financial mathematics: held in Bressanone, Italy, July 8–13, 1996 / B. Biais ... Ed.: W. Runggaldier.
– Berlin; Heidelberg; New York; Barcelona; Budapest; Hong Kong; London; Milan; Paris; Santa Clara;
Singapore; Tokyo: Springer, 1997 (Lectures given at the ... session of the Centro Internazionale
Matematico Estivo (CIME) ... ; 1996,3)
(Lecture notes in mathematics; Vol. 1656 : Subseries: Fondazione CIME)
ISBN 3-540-62642-5
NE: Biais, Bruno; Runggaldier, Wolfgang [Hrsg.]; 2. GT

Mathematics Subject Classification (1991): 90A09, 90A10, 90A12, 90A60, 60H30,
60H10, 60G40, 60G44, 93E20, 49N15

ISSN 0075-8434
ISBN 3-540-62642-5 Springer-Verlag Berlin Heidelberg New York

Typesetting: Camera-ready TEX output by the authors
SPIN: 10520353   46/3142-543210 - Printed on acid-free paper

Preface

Financial Mathematics has become a growing field of interest for mathematicians and economists alike, and so it was a very welcome decision by the CIME Scientific Committee to devote one of the three 1996 Sessions to the field. This Session/Summer School on "Financial Mathematics" has attracted a surprisingly large number of interested participants from 16 countries, ranging from mathematicians to economists to people operating in the finance industry.

This volume collects the texts of the five series of lectures presented at the Summer School. They are arranged in alphabetic order according to the name of the (first) lecturer.

The lectures were given by six outstanding scientists in the field and reflect the state of the art of a broad spectrum of topics within the general area of Financial Mathematics, ranging from more mathematical to more economic issues, where also the latter are treated with mathematical formalism and rigour. It should allow economists to familiarize themselves with advanced mathematical techniques and results relevant to the field and, on the other hand, stimulate mathematicians to attack important issues raised by the underlying economic problems with appropriate mathematical tools.

As editor of these Lecture Notes it is my pleasure to thank various persons and Institutions that played a major role for the success of the School. First of all my thanks go to the Director and the Members of the CIME Scientific Committe, in particular to Prof. Arrigo Cellina, for the invitation to organize the School and their support during the organization; to the CIME staff, lead by Prof. Pietro Zecca, for the really efficient job they did, and to the CIME as such for the financial support. I would also like to mention here Professors Ivar Ekeland and Jean Michel Lasry : Prof. Ekeland from the Université de Paris Dauphine was originally appointed as a Scientific Director for this CIME Session; an important governemental appointment prevented him however from continuing his role in the organization of the School. Still, the program of the School reflects in part his organizational contribution. After it became impossible for Prof. Ekeland to continue his role, Prof. Lasry, now President of the Caisse Autonome de Refinancement, offered his support and his contribution is also partly reflected in the program. My very sincere thanks go to the Lecturers for their excellent job of preparing and teaching the Course and for having made available already before the start of the Course a preliminary version of their lectures to be distributed among the participants. Particular thanks go to all the participants for having created an extraordinarily friendly and stimulating athmosphere.

Finally I would like to thank the Director and staff of the Cusanus Academy in Bressanone/Brixen for the kind hospitality and efficiency; the Town Council of Bressanone/Brixen and the "Associazione Amici dell'Universitá di Padova in Bressanone" for additional financial and organizational support as well as all those who have contributed to make the stay in Bressanone/Brixen more enjoyable.

Padova, December 1996

Wolfgang J. Runggaldier

TABLE OF CONTENTS

RISK SHARING, ADVERSE SELECTION AND MARKET STRUCTURE*

Bruno Biais and Jean Charles Rochet

Gremaq, Idei, Toulouse University

Abstract

The objective of this essay is to bring to the attention of mathematicians of finance the field of market micro-structure and the issues raised by the design of trading markets in presence of asymmetric information. In the first part of the essay we present a synthesis of several papers which analyse trading volume and price formation in presence of informed and uninformed traders as well as noise traders : Grossman and Stiglitz (1980) where agents are competitive, Kyle (1985) where the informed agent is strategic, and Rochet and Vila (1994) where convex analysis techniques are used to show that a variant of the Kyle (1985) equilibrium is the solution of a market mechanism design problem. In the second part of this essay, we present the case where there is no noise trading and trading endogenously stems from risk sharing as well as informational motivations as in Glosten (1989). Our reformulation of the analysis of Glosten (1989) illustrates that variation calculus and convex analysis provide a simple and powerful way to deal with the rather difficult problems raised by the design of market mechanisms.

Paper prepared for the third 1996 session of the Centro Internazionale Matematico Estivo, on Financial Mathematics. Many thanks to Rose Anne Dana, Nicole El Karoui, Vincent Lacoste and participants at the CIME session for helpful comments. We were also influenced by numerous discussions with David Martimort. The usual disclaimer applies.

Contents

Introduction

So far, the main focus of financial mathematics has been arbitrage pricing and general equilibrium. Recent work in this field has focused on extending the analysis to markets with frictions such as transaction costs, bid–ask spread, market impact or leverage constraints.[1] In those analyses however, these frictions are exogenously specified, rather than the consequence of the optimal actions taken by economic agents.

On the other hand, a considerable body of research in economics has been devoted to the consequences of asymmetric information on the workings of markets. This strand of literature shows that asymmetric information endogenously generates the above mentioned market frictions. For example Stiglitz and Weiss (1981) show how it leads to leverage constraints, while Glosten and Milgrom (1985) and Easley and O'Hara (1987) show that it can lead to bid–ask spreads and Kyle (1985) shows that it generates market impact.[2]

---

[1] Early contributions include Magill and Constantinides (1976), Leland (1985), Constantinides (1986), Davis and Norman (1990), Grossman and Laroque (1990), and Dumas and Luciano (1991). Recent advances are presented in the contributions of Cvitanic (1996), El Karoui and Quenez (1996), and Jouini (1996) to the present book.

[2] Another strand of literature in the analysis of the bid–ask spread relates it to inventory control and risk–sharing, see Stoll (1978), Amihud and Mendelson (1980), Ho and Stoll (1981, 1983) and Biais (1993). O'Hara (1996) provides an insightful survey of the recent literature in these areas.

These analyses are carried within the context of given market structures, or micro–structures, inspired by the observation of the actual workings of financial markets. In Grossman and Stiglitz (1980) and Kyle (1989), agents submit demand or supply functions, and a Walrasian auctioneer sets the price such that supply equals demand. In Kyle (1985) or Glosten and Milgrom (1985) the informed agent places market orders and competitive market makers are assumed to quote prices equal to the conditional expectation of the value of the asset given the order flow. Note that when agents are strategic their interaction corresponds to a game (in the technical sense of game theory) and the structure of the market can be seen as the rule of this game.

A third strand of economic analysis, referred to as "mechanism design", has studied general abstract mechanisms designed to deal with information asymmetries (see Fudenberg and Tirole (1991) chapter 7 for a synthesis). The idea is that, instead of considering a given game structure, it is interesting to analyze the optimal rules of the game, which will best enable the agents to cope with the asymmetric information problem. In the case of financial markets, this line of approach suggests to solve for the optimal market structure, rather than reasoning within a given market structure. Glosten (1989) and Rochet and Vila (1994) take this approach.

The objective of the present essay is to bring to the attention of financial mathematicians this field of study, dealing with asymmetric information, endogenous market frictions, market microstructure and mechanism design. While standard market microstructure models have not, so far, made use of sophisticated mathematical methods, the mathematical methods involved in mechanism design theory are non trivial, and quite different from those used in arbitrage pricing or general equilibrium analysis.

The three major ingredients of market microstructure models are i) the motivations for trade of the agents : private information or liquidity, ii) the nature of their behaviour: rational or irrational, competitive or strategic, and iii) the structure of the market. The organisation of the present essay can be described in terms of these ingredients:

- In the next section we present some notations and assumptions which will be useful throughout the analysis.

- Then we analyze the case where the liquidity traders are irrational "noise traders". In this context we examine in turn :

  1. the case where the informed agents are competitive and the market structure is Walrasian (Grossman and Stiglitz (1980)),

  2. the case where the informed agents are strategic and trading takes place in a dealer market (Kyle (1985)), and

  3. the case where the informed agent is strategic and the market structure is endogenously derived as an optimal mechanism (Rochet and Vila (1994)).

- In the last part of this essay we dispense with the "noise traders" assumption and turn to the case where one strategic agent, with both informational and liquidity motivations to trade, submits buy or sell orders to market makers. In this context we examine in turn :

  1. a benchmark case with a benevolent social planner,
  2. the case where there is a single monopolistic market maker,
  3. and the case where there is a fringe of competitive market makers.

This second part is directly inspired by the analysis of Glosten (1989). The main difference is that we rely on different mathematical techniques (namely variation calculus and convex analysis). We hope that this will illustrate how one can use the sophisticated mathematical tools of mechanism design theory to analyze financial markets.

1   Some notations and assumptions

Consider the market for one risky and one riskless assets. Trading in these assets takes place at time $t = 0$. At time $t = 1$ the liquidation value of the assets is realized and consumption takes place. For simplicity the riskfree rate is normalized to 0. The liquidation value of the risky asset at time $t = 1$ is denoted v. We assume :

$$v = s + \epsilon, \tag{1.1}$$

where :

$$s \perp \epsilon,$$

(where the symbol $\perp$ denotes independence of random variables), and

$$\epsilon \rightsquigarrow N(0, \sigma_\epsilon^2).$$

s is the private information received by the informed agents on v. ϵ is a noise term.

An important fraction of the market microstructure literature makes the assumption that all random variables are normally distributed. This often greatly enhances the tractability of the models. In the present essay we will make this assumption when presenting the seminal contributions of Grossman and Stiglitz (1980) and Kyle (1985). In the presentation of Rochet and Vila (1994), as well as in our analysis of the case where liquidity trading is endogenous, we will dispense with the normality assumption, however, and consider general distributions for s (and hence v) and the non-informational motivations to trade of the agents.

Another difference between our assumptions on distributions and those often met in the literature is the following. In Grossman and Stiglitz (1980) or Glosten (1989) it is assumed that the informed agents observe $z = v + e$ where e and v are jointly normal, $E(e) = 0$ and $e \perp v$. Hence, while in Grossman and Stiglitz (1980) or Glosten (1989) the difference between what the informed agent observes and the value of the security is independent of v, in our case it is independent to the private signal. In spite of this difference, our analysis in the next section is identical to that of Grossman and Stiglitz (1980). However, this slight change in assumptions will generate some differences between our analysis and that of Glosten (1989).

2 Price formation and market mechanism with exogenous "noise" trading

2.1 The competitive case : Grossman and Stiglitz (1980)

Agents, information and market structure

Demand and supply in the market are determined as follows:

- There is a continuum of rational competitive traders, indexed by a number j between 0 and 1. Without loss of generality, the variable j that indexes traders is taken to be uniformly distributed on $[0, 1]$. These agents have exponential negative utility functions with equal CARA coefficient ρ, that is the utility they derive from $x \in \mathcal{R}$ is $:-e^{-\rho x}$. This functional form is chosen because, when combined with normal distributions (such as the distribution of ϵ), it leads to quadratic objective functions. Each of these agents has initial wealth W_0. They can choose to acquire private information (that is observe s) at cost c. They place demand schedules $X_j(p)$ to maximize their expected utility. If agent j decides to become informed, and buys $X_j(p)$ shares of the risky asset at price p, his final wealth (and therefore consumption) is:

$$W = W_0 + X_j(p)(v - p) - c.$$

The demand of this agent at price p, $X_j(p)$ solves:

$$Max_{X_j(p)} E[-e^{-\rho(W_0 + X_j(p)(v-p)-c)}|I_j],$$

where I_j denotes the information set of agent j, which includes s. If agent j does not decide to acquire the private information and buys $X_j(p)$ shares of the risky asset at price p his final wealth (and therefore consumption) is:

$$W = W_0 + X_j(p)(v - p).$$

The demand of this agent at price p, $X_j(p)$ solves:

$$Max_{X_j(p)} E[-e^{-\rho(W_0 + X_j(p)(v-p)-c)}|I_j],$$

where I_j denotes the information set of agent j, which does not include s, but includes the equilibrium market price p.

- There are also noise traders, with demand: $u \rightsquigarrow N(0, \sigma_u^2)$.

- The total supply of the security is normalized to one.

- p is set to equate demand and supply:

$$\int X_j(p)dj + u = 1. \tag{2.1}$$

In this subsection we assume that u, s and ϵ are jointly normal. A fraction λ of the rational competitive agents observe the private information, s, and, of course the price p. However s contains more information than p (s is a sufficient statistic for (s, p)) therefore the informed agent's demand solves:

$$Max_{X_j(p)} E[-e^{-\rho(W_0 + X_j(p)(v-p))}|s]$$

that is:

$$Max_{X_j(p)} X_j(p)(s - p) - \frac{1}{2}\rho\sigma_\epsilon^2 X_j(p)^2$$

The first order condition is:

$$X_j(p) = \frac{s - p}{\rho\sigma_\epsilon^2}. \tag{2.2}$$

The complementary fraction of the population of rational competitive agents do not observe the private signal.

Characterization of the equilibrium price function

First we solve for the rational expectations price function, for a given value of λ. We look for a random variable p that realizes almost surely the equality between supply and demand for the security. This random variable will obviously depend on the signal s received by informed traders and on the amount u of noise trading. We will denote by $P(s, u)$ the value of p as a function of the realizations of s and u. The rational agents are assumed to be able to compute this price function. They use it to infer information on s from the observed realization of p.

Definition :

A Rational Expectations Equilibrium of the Grossman and Stiglitz (1980) model is a price function $P(.,.)$ mapping the pair (s, u) into the price p such that:

- *demand equals supply for $p = P(s, u)$,*
- *agents place demand functions maximizing their expected utility,*
- *and they correctly use this price function $P(s, u)$ to compute expectations and make inferences on s using p.*

Note that this equilibrium can also be viewed as a Bayesian Nash equilibrium,[3] whereby each agent rationally anticipates the strategy of the others, i.e., the mapping from their information to their trades, and plays his best response to these strategies. In the determination of his best response, the agent takes into account the information content of the price p, as in the Rational Expectations Equilibrium.

However, there are in general two important differences between the concept of rational expectations equilibrium and the concept of Bayesian Nash equilibrium.

- First, if there was a finite number of agents, the rational expectations equilibrium whereby the agents are competitive would differ from the Bayesian Nash equilibrium, where they are strategic. This difference does not arise here because there is a continuum of infinitesimal agents who have no impact on the price.

- Second, the rational expectations equilibrium is less demanding than the Bayesian Nash equilibrium as regards the rationality of the agents. In the former the agents only need to rationally anticipate one function (the equilibrium price function) while in the latter they must rationally anticipate many functions (the strategies of each of their opponents). Analyses of the foundations of Nash equilibrium in terms of common knowledge have shown that this difference can matter (for an analysis in the case of financial markets see Guesnerie and Rochet (1993)).

In this context, Grossman and Stiglitz (1980) obtain the following proposition.

Proposition 1 *There exists a linear equilibrium price function, given by:*

$$P(s, u) = a + bw \tag{2.3}$$

where

$$w = s + (\rho\sigma_\epsilon^2/\lambda)u \tag{2.4}$$

is a sufficient statistic for the information revealed by the price.

Sketch of proof :
For the informed agents the liquidation value is normal with expectation s and variance σ_ϵ^2. The uninformed agents must compute the distribution of v given p. If they anticipate that the price p is $P(s, u)$ linear in s and u, as in equations (2.3) and (2.4), then they believe that v is normal with expectation: $E(v|p)$, and variance: $Var(v|p)$. Hence their demand solves:

$$Max_{X_j(p)} X_j(p)(E(v|p) - p) - \frac{1}{2}\rho Var(v|p)X_j(p)^2.$$

---

[3] See for instance Fudenberg and Tirole (1991) for the definition and properties of Bayesian Nash equilibria.

Consequently:

$$X_j(p) = \frac{E(v|p) - p}{\rho Var(v|p)}.$$

(2.5)

Hence the market clearing equation is:

$$\lambda \frac{s-p}{\rho \sigma_\epsilon^2} + (1-\lambda) \frac{E(v|p) - p}{\rho Var(v|p)} + u = 1.$$

(2.6)

Now, since p is a linear function of s and u, $E(v|p)$ is a linear function of p. Also, since p and v are jointly normal, $Var(v|p)$ is a constant. Hence the left-hand-side of equation (2.6) is linear in s and u. Substituting (2.3) and the explicit form of $E(v|p)$ and $Var(v|p)$ the market clearing equation can be rewritten:

$$A + Bs + Cu = 1,$$

(2.7)

where A, B and C are coefficients to be determined. In the rational expectations equilibrium, equation (2.7) must hold for almost every realization of u and s, which means that $B = C = 0$ and $A = 1$. Now, these coefficients depend on the parameters of the model, such as $\sigma_\epsilon, \sigma_u, E(s), \lambda$, and ρ as well as on the coefficients of the rational expectations equilibrium price function (defined in equation (2.3)). There exists a rational expectations price function if there exists values of these parameters such that $B = C = 0$ and $A = 1$. Grossman and Stiglitz (1980) prove that this is the case.

∎

Note that the random variable w can be interpreted as a noisy version of the signal observed by informed agents. The importance of this noise can be measured by the conditional variance of w: $var(w|s) = \sigma_u^2(\rho \sigma_\epsilon^2/\lambda)^2$. It is small when:

- the variance of noise trading σ_u^2 is small,

- or the traders'risk aversion ρ is small,

- or the precision of the signal is large(i.e. σ_ϵ^2 is small),

- or the proportion λ of informed traders is large.

In each of these cases, observing the price gives precise information on v. This makes private information acquisition less attractive.

Note also that the case $\lambda = 0$ is degenerate : when nobody observes the signal, p can only reflect information on u, which is useless. In this case, the only equilibrium is $p = E(v)$.

Endogenizing λ

We now solve for the equilibrium proportion λ of informed traders. To carry out this analysis we consider the following sequence of events:

1. First, each agent decides whether or not to acquire information. This gives rise to a certain value of λ.

2. Second, for this given value of λ trading takes place in the financial market.

We look for a Bayesian Nash equilibrium whereby:

- each agent decides to acquire the private information or not based on his rational expectations about the decisions of the others and the subsequent trading game,

- and once λ is determined trading takes place as described in Proposition 1.

There can be two types of equilibrium outcomes : either nobody buys the information ($\lambda = 0$), or a positive fraction of the population purchases the private signal ($\lambda > 0$). The former is an equilibrium if for $\lambda = 0$ the expected utility of the agents who do not acquire information is larger than the expected utility of the agents who would acquire it. The latter is an equilibrium for $\lambda^* < 1$ if for this value of λ the expected utility of the agents who do not acquire information is equal to the expected utility of the agents who acquire it. $\lambda^* = 1$ is an equilibrium if for this value of λ the expected utility of the agents who acquire information is larger than that of the agents who do not acquire it.

To simplify matters we will carry out the analysis in the simple case without noise trading : $u = 0$. Also for simplicity we will only consider pure strategy equilibria, whereby the agents deterministically decide whether to buy the information or not. Our third simplification is to focus on the case where $\lambda^* = 1$ is not an equilibrium.

Consider as an equilibrium candidate the case where $\lambda > 0$. In this case, because there is no noise trading, we know from Proposition 1 that the price perfectly reveals v. Hence the information set in the expected utility of the informed agent and the expected utility of the uninformed agent are the same. In this case, however, the two expected utilities cannot be equal since they differ only because of c. Hence $\lambda > 0$ cannot be an equilibrium.

Now consider the alternative equilibrium candidate, for which : $\lambda = 0$, i.e., there are no informed traders and p has no informational content. In this case, since there is no noise trading and no risk sharing to be undertaken, there is no trading in the market, and the equilibrium market clearing price is $p = E(v)$. Hence the expected utility of the uninformed agents is:

$$-e^{-\rho W_n}.$$

For this to be an equilibrium, it must be that no agent has any incentive to deviate from the no information acquisition strategy. If agent j did acquire information, she would demand:

$$X_j(p) \doteq \frac{s-p}{\rho\sigma_\epsilon^2} = \frac{s - E(v)}{\rho\sigma_\epsilon^2}.$$

Thus her expected utility would be :

$$E - e^{-\rho[W_0 + \frac{1}{2}\frac{(s-p)^2}{\rho\sigma_\epsilon^2} - c]} = -e^{-\rho W_0} E e^{-\rho[\frac{1}{2}\frac{(s-p)^2}{\rho\sigma_\epsilon^2} - c]}. \tag{2.8}$$

Therefore, no agent has any incentives to deviate from the no-information acquisition equilibrium if:

$$1 \le E e^{-\rho[\frac{1}{2}\frac{(s-p)^2}{\rho\sigma_\epsilon^2} - c]}, \tag{2.9}$$

or:

$$e^{-\rho c} \le E e^{-\rho[\frac{1}{2}\frac{(s-p)^2}{\rho\sigma_\epsilon^2}]}, \tag{2.10}$$

finally:

$$c \ge \frac{-1}{\rho} ln\{E e^{-\rho[\frac{1}{2}\frac{(s-p)^2}{\rho\sigma_\epsilon^2}]}\}. \tag{2.11}$$

Hence the following proposition obtains:[4]

Proposition 2 *When there is no noise trading, the only possible equilibrium is when nobody acquires information ($\lambda = 0$). It is indeed an equilibrium when the cost of information is large enough, i.e., inequality (2.11) holds. If this condition is not satisfied, there is no pure strategy Bayesian Nash equilibrium.*

This result is known as the "Grossman and Stiglitz paradox" : agents anticipate that the equilibrium price will reflect the private signal, hence they do not pay the cost of acquiring it, but this results in uninformative prices... This is why Grossman and Stiglitz (1980) titled their paper "On the impossibility of informationally efficient markets." Note however that an equilibrium will exist if there is noise trading, or if we allow for mixed strategies at the information acquisition stage.

2.2   The case of a monopolistic insider : Kyle (1985)

Agents, information and market structure

We now turn to a trading game model which exhibits the following differences with the model of Grossman and Stiglitz (1980) analysed above:

- While the rational uninformed agents are competitive as in Grossman and Stiglitz (1980), the informed agents are strategic. In fact there is a single (monopolistic) informed agent, referred to as "the insider."

---

[4]Since $s - p$ is normally distributed, its square follows a χ^2 distribution, hence the right hand side of inequality (2.11) can be computed more explicitly. This is performed in Grossman and Stiglitz (1980).

- The signal observed by the informed agent is perfect:

$$\sigma_\epsilon = 0.$$

- All agents are risk neutral:

$$\rho = 0.$$

- The market structure is different from the Walrasian market assumed in Grossman and Stiglitz (1980). The agents move sequentially.

  1. First the insider chooses the quantity he desires to trade: x. Also the noise trading u is determined, but the insider does not observe it. In the next subsection we examine the case where the insider observes the realization of u.

  2. Second, the fringe of competitive, rational uninformed agents observe the sum of the informed order and the noise trade:

$$z = x + u,$$

  and they accomodate this liquidity demand, i.e., they trade $1 - z$. They set the price p at which this trade occurs, conditionally on the information they infer from z. Because they are risk neutral and competitive they set:

$$P = E(v|z). \tag{2.12}$$

  Because they set the price, the rational competitive uninformed agents are referred to as the "market makers."

Remark : Because the price is the expectation of the liquidation value conditional on the market public information, it is said to be "weakly informationally efficient", and condition (2.12) is referred to as the efficiency condition. Note that this informational efficiency bears no relation with the allocative efficiency studied in welfare economics. Indeed because all rational agents are risk neutral and because noise traders have no utility function, there is no well defined notion of welfare in this model.

Definition :

A perfect Bayesian equilibrium [5] *of the Kyle (1985) trading game is a pair of functions:*

$$P(.) : \quad \mathbb{R} \to \mathbb{R}$$
$$z \mapsto p$$

---

[5]The notion of perfect Bayesian equilibrium is a refinement of the previous notion of Bayesian Nash equilibrium, in the context of sequential games. See Fudenberg and Tirole (1991) for the general definition and properties of perfect Bayesian equilibrium.

$$X(.): \quad \mathbb{R} \to \mathbb{R}$$
$$v \mapsto x$$

such that $X(.)$ maximizes the expected profit of the insider based on his rational expectation of the pricing function $P(.)$:

$$X(v) \, solves \quad Max_x(v - E_u P(x + u))x, \tag{2.13}$$

and $P(.)$ sets the price to the conditional expectation of the value of the asset, computed using rational expectations about the insider's trading strategy $X(.)$:

$$P = E(v|X(v) + u = z). \tag{2.14}$$

Characterization of the linear equilibrium in the Gaussian case

When v and u are jointly Gaussian, there exists a unique *linear* equilibrium :

$$P(z) = a + bz, \; X(v) = c + dv.$$

Let us determine under what conditions on the coefficients this is indeed an equilibrium. By the standard conditional expectation formula for normal variables we have:

$$a = E(v) - bE(z) = E(v) - b(c + dE(v)),$$

$$\text{and} \quad b = \frac{d\sigma_v^2}{d^2\sigma_v^2 + \sigma_u^2}.$$

Similarly by solving the optimization program of the insider, we get:

$$c = \frac{-a}{2b}, \; d = \frac{1}{2b}.$$

Therefore, by identification :

$$a = E(v) \; , \; b = \tfrac{1}{2}\sigma_v/\sigma_u \; , \; c = -E(v)\sigma_u/\sigma_v, \; d = \sigma_u/\sigma_v.$$

Proposition 3 When v and u are jointly gaussian, there exists a unique linear equilibrium given by:

$$P(z) = E(v) + (\tfrac{1}{2}\sigma_v/\sigma_u)z \; , \; X(v) = (\sigma_u/\sigma_v)(v - E(v)).$$

Properties of the linear equilibrium

The sensitivity of the insider's demand to information, measured by the derivative $X'(v)$, is proportional to σ_u, (the standard deviation of noise trading), and inversely proportional to σ_v, (the standard deviation of his information). Similarly the market depth, defined as the inverse of the derivative of P with respect to z, has the same properties. Note also that the informational content of prices, measured as the proportion of the variance of v that is explained by p, is constant:

$$var(v|p) = \frac{1}{2}var(v). \tag{2.15}$$

This means that the insider chooses his strategy in such a way that he only reveals half of his information. This has to be contrasted with the Grossman-Stiglitz case, where competitive informed traders do not take into account the information revealed by their trades.

The profit of the insider is:

$$E(W|v) - W_0 = (v - E(v) - bX(v))X(v) = (\sigma_u/2\sigma_v)(v - E(v))^2.$$

Hence, before observing $v = s$ the expected profit of the informed agent is :

$$\frac{1}{2}\sigma_u\sigma_v.$$

The expected profit of the insider is thus proportional to σ_v,which measures the quality of his information, and σ_u, which measures the importance of noise trading, behind which the insider is able to "hide" his orders. For instance when the aggregate demand z is high , the market makers do not know whether this is because the insider knows that the liquidation value of the security is high, or if this is just because noise trading is important,which does not bring any information about the value of the security.

Equilibrium multiplicity

The trading game analysed above falls within the general class of signalling games. It is well known that such games usually admit multiple equilibria.[6] One source of multiplicity is related to the "out of equilibrium beliefs" (i.e., the beliefs that would be entertained by the agents in the game if they observed actions which in equilibrium will never occur) which are not restricted by the concept of perfect Bayesian equilibrium. We now present a variant of the above trading game, drawn from Rochet-Vila (1994), in which there is a large multiplicity (actually, a non countable infinity) of equilibria. The trading game is as above except that we make different distributional assumptions :[7]

---

[6] For an analysis of this problem see Cho and Kreps (1987).

[7] We keep the assumption that u and v are independent.

- the distribution of u is discrete and takes the values $+1$ and -1 with equal probability $1/2$.

- v takes the values -2, -1, 1 and 2 with equal probability $1/4$.

We are going to construct a non-countable family of equilibria, indexed by $a\epsilon]0, 2/3[$ and described in the next proposition.

Proposition 4 *Take a non-gaussian specification of the Kyle (1985) model for which the distribution of u is discrete and takes the values $+1$ and -1 with equal probability $1/2$, v takes the values -2, -1, 1 and 2 with equal probability $1/4$. Then, there is a non-countable family of perfect Bayesian equilibria, indexed by $a\epsilon]0, 2/3[$ such that,*

- *The insider's demand is given by :*

$$\begin{cases} X_a(2) = & -X_a(-2) = & 1+a \\ X_a(1) = & -X_a(-1) = & 1-a. \end{cases}$$

- *For the set of equilibrium trades, P_a is even, i.e.,*

$$\forall z \in \{-2-a, -2+a, -a, a, 2-a, 2+a\}, P_a(z) = -P_a(-z).$$

and :

$$P_a(a) = \frac{1}{2},$$

$$P_a(2+a) = 2,$$

$$P_a(2-a) = 1.$$

Proof of Proposition 4 :

Consider the candidate equilibrium a whereby the trading strategy of the insider is $X_a(\cdot)$ such that :

$$\begin{cases} X_a(2) = & -X_a(-2) = & 1+a \\ X_a(1) = & -X_a(-1) = & 1-a. \end{cases}$$

The proof consists in two steps. First we compute the prices that would arise for this strategy, using the property that prices are conditional expectations. Second, we show that for these prices the strategy $X_a(.)$ does maximize the expected profit of the insider.

Step 1 :

For the trading strategy $X_a(.)$ the aggregate demand $Z = Z(u,v)$ satisfies :

$$Z(-1,2) = Z(1,-1) = a$$
$$Z(1,-2) = Z(-1,1) = -a.$$

Thus when observing an aggregate purchase of a the market makers do not know if it stems from $v = 2$ (and $u = -1$) or from $v = -1$ (and $u = 1$). In fact they put equal probability on these two events. Hence the conditional expectation price function corresponding to the trading strategy $X_a(.)$ satisfies :

$$P_a(a) = \frac{2 + (-1)}{2} = \frac{1}{2}.$$

Using symmetric arguments :

$$P_a(-a) = \frac{-2 + 1}{2} = -\frac{1}{2}.$$

Note further that the aggregate demand is $2 + a$ if and only if $v = 2$ and $u = 1$. Hence :

$$P_a(2 + a) = 2.$$

Similarly, the aggregate demand is $2 - a$ if and only if $v = 1$ and $u = 1$. Hence :

$$P_a(2 - a) = 1.$$

Finally it is easy to check that for the set of equilibrium trades, P_a is even, i.e.,

$$\forall z \in \{-2 - a, -2 + a, -a, a, 2 - a, 2 + a\}, P_a(z) = -P_a(-z).$$

To finish the construction of the price function in the candidate equilibrium a, we need to construct the price quoted by the market makers facing trades that are inconsistent with the equilibrium, i.e., trades that do not belong to: $\{-2 - a, -2 + a, -a, a, 2 - a, 2 + a\}$. Since prices are conditional expectations, we need to compute the probabilities of the different possible values of v conditional on the observation of :

$$z \notin \{-2 - a, -2 + a, -a, a, 2 - a, 2 + a\}$$

It is impossible however to use Bayes rule to compute such probabilities since the conditioning set (z) has zero prior probability. The simplest approach to this problem is to specify these "updated" probabilities in an arbitrary way, such that they support the equilibrium, that is such that they deter the agents from engaging in out-of-equilibrium actions. In the present case we will use the following :

$$\forall z > 0, z \neq 2 + a, 2 - a, a, Prob(v = 2|z) = 1$$

and

$$\forall z < 0, z \neq -2 - a, -2 + a, -a, Prob(v = -2|z) = 1$$

We now turn to the second step of the proof.

Step 2 :

We now establish that for the above defined price function it is indeed optimal for the insider to follow strategy $X_a(.)$. Denote $B_a(.)$ the function mapping the couple (value v, trade x) of the insider into his expected profit :

$$B_a(v, x) = x(v - E[P_a(x + u)]).$$

Given the out of equilibrium beliefs we have specified above, the insider never finds it optimal to trade outside the set $\{1 + a, 1 - a, -1 + a, -1 - a\}$. Also because of the symmetry of the problem (reflected for example in the fact that P_a is even) we can restrict our attention to the buy side of the problem. Therefore to establish the optimality of $X_a(.)$ we only need to prove that :

$$\begin{aligned} B_a(2, 1 + a) &\geq B_a(2, 1 - a), \text{and} \\ B_a(1, 1 - a) &\geq B_a(1, 1 + a). \end{aligned}$$

An immediate computation shows that these inequalities are satisfied when $a \geq \frac{2}{3}$.

■

Proposition 4 illustrates our earlier statement that the indeterminacy of out–of–equilibrium beliefs in signalling games generates equilibrium multiplicity. Indeed, in the present case, for each value of $a \leq 2/3$ it is possible to specify out–of–equilibrium beliefs supporting X_a.

2.3   A variant of Kyle (1985) where there is a unique equilibrium : Rochet and Vila (1994)

The trading game

We now consider a variant of the above analyzed trading game which exhibits the following two differences with the Kyle (1985) model:

1. To simplify the analysis, we assume that the insider can observe the noise trade u.

2. Instead of assuming that the random variables are jointly normal, we allow for any distribution of (u, v) on a bounded support, say the rectangle :

$$\Omega = [u_0, u_1] \times [v_0, v_1].$$

In this context, the insider chooses his demand $X(u, v)$ after observing the volume u of noise trading to maximize his profits :

$$x = X(u, v) \quad \text{solves} \quad Max_x(v - P(x + u))x. \tag{2.16}$$

It is more convenient to use $z = x + u$ as a strategic variable:

$$z = Z \quad \text{solves} \quad Max_z(v - P(z))(z - u) \stackrel{d}{=} \pi(u, v). \tag{2.17}$$

The weak efficiency condition remains the same as in the previous subsection :

$$P(z) = E(v|z = Z(u, v)). \tag{2.18}$$

It is easy to see that this mild modification of the model does not alter the intuitive properties of the linear equilibrium obtained in the original formulation when the distribution of (u, v) is gaussian (see Rochet-Vila (1994) for details). However, we are now going to prove that equilibrium is unique in this variant of the Kyle model.

A mechanism design problem

So far, we have considered given market structures. We now adopt a different approach, and study *how the exchange should be organized.* To analyze this problem we take a mechanism design approach. We consider the situation where the organizer of the market (whom we might also call the "central planner") offers to the insider a menu of contracts specifying, for each level of the aggregate demand z, the price $P(z)$ and a monetary transfer to the insider : $T(z)$. This includes as a particular case the above described trading game, for which $P(z) = E(v|z)$ and $T(z) = 0$. It is more general however, since it raises the possibility that other trading mechanisms would be used, whereby the monetary transfer would not be 0, and whereby the price would not be the conditional expectation of the value. Among this large class of possible market mechanisms the market organizer chooses the optimal one. This is the sense in which, in this approach, as opposed to the above described models, the market structure is endogenous. We need to define, however, the criterion with respect to which the market structure is optimal. This is a difficult task in the present context, because noise traders have no well defined utility, and as a result the gains from trade are not properly defined. Since, in other contexts, asymmetric information, and the profits it generates prevents exploiting the gains from trade, it is sensible to take as optimal the market mechanism which minimizes the profit of the insider.[8]

First, we characterize the set of possible profits for the insider which the central planner could implement. The strategy $Z(u, v)$ of the insider solves:

$$Max_z \{(v - P(z))(z - u) + T(z)\} \stackrel{d}{=} \Pi(u, v). \tag{3.10}$$

---

[8] In the next subsection of this essay we will provide a more in depth analysis of the welfare issues raised in these market games by dispensing with the noise trading assumption and considering endogenously the non–informational motivations to trade.

Proposition 5 *Given* $\Pi : \Omega \longrightarrow I\!R$, *there exists a mechanism* $T(z)$, $P(z)$ *which implements* Π *if and only if* $(u, v) \longrightarrow \Pi(u, v) + uv$ *is convex and lower semi continuous.*

Proof of Proposition 5 :

By definition of $\Pi(u, v)$, we have:

$$\Pi(u, v) + uv = Max_z \{vz - P(z)(z - u) + T(z)\},$$

therefore it is convex lower semi continuous, as a supremum of affine functions of (u, v). Conversely, by Fenchel's duality theorem (see Rockafellar (1972)), any convex lower semi continuous function can be obtained in such a way.   ∎

By Proposition 5 finding the market mechanism which minimizes the insider's profit is equivalent to solving the following infinite dimensional optimization problem:

$$(\mathcal{P}_1) \begin{cases} \min E[\Pi(u, v)] \\ \text{under the constraints :} \\ \Pi(u, v) + uv \text{ convex} \\ \Pi(u, v) \geq 0 \end{cases}$$

This infinite dimensional optimization problem is at first sight rather difficult to solve. In fact, we will show below that any equilibrium of the modified Kyle model is associated to a solution of the above problem. Since this solution is typically unique, this will also establish the (essential) uniqueness of the equilibrium. Even if there are several solutions (which happens non generically) the value of the problem is unique. Therefore the expected profit of the insider at (any) equilibrium is uniquely determined.[9]

Proposition 6 *Let* $(P^*(\cdot), Z^*(\cdot, \cdot))$ *be an equilibrium of the modified Kyle model, i.e., it satisfies :*

$$P^*(Z) = E\left[v|Z^*(u, v) = Z\right] \tag{2.20}$$

and

$$Z^*(u, v) \rightarrow \max_Z (v - P^*(Z))(Z - u). \tag{2.21}$$

---

[9] A similar method (although in a very different context) has been used by Negishi for proving existence (and uniqueness) of competitive equilibrium in the presence of gross substitutability. It has been extended to infinite-dimensional spaces, with applications to continuous time finance (see for instance Mas - Colell (1986)).

Let $\Pi^*(u, v)$ be the profit function associated to P^* :

$$\Pi^*(u, v) = Max_z(v - P^*(z))(z - u). \tag{2.22}$$

Then Π^* solves :

$$(\mathcal{P}_1) \begin{cases} \min E[\Pi(u, v)] \\ \Pi(u, v) + uv \quad convex \\ \Pi(u, v) \geq 0 \end{cases}$$

Before presenting the proof of this proposition we rapidly comment on its economic consequences. Proposition (6) states that even if there is no benevolent central planner, economic agents, simply pursuing their own interests (therefore playing a Nash equilibrium) will set the price and allocations in the socially "optimal" way, as long as they are competitive. This is reminiscent of Adam Smith analysis of the "invisible hand" establishing that, in competitive markets, the spontaneous forces of individual egoism result in the social optimum. However we term this result "weak invisible hand" property since, as we already remarked, there is no clear measure of social welfare in this model.

Proof of Proposition 6 :

Before establishing the proposition we will first establish 3 preliminary lemmas.

Lemma 1 Π^* *is admissible, i.e.,* $\Pi^* \geq 0$ *and* $\Pi^* + uv$ *is convex.*

Proof of lemma 1 :
$\Pi^* \geq 0$ since the insider can always choose $Z = u$ in (2.22).
By adding uv to both sides of (2.22), we get :

$$\Pi^* + uv = \max_Z(vZ + uP^*(Z) - ZP^*(Z))$$

which is convex l.s.c. since it is the supremum of continuous affine functions. ∎

Lemma 2 *For a.e.* (u, v), P^* *is differentiable at* $Z = Z^*(u, v)$.

Proof of Lemma 2 :

The proof goes in three steps :

- **step 1** : $\{(u,v), \ \Pi^*(u,v) = 0\}$ is Lebesgue negligible (left to the reader).

- **step 2** : For a.e. (u,v) there is $(u',v') \neq (u,v)$ such that $Z^*(u,v) = Z^*(u',v')$. Indeed if it was not the case then the price would be a.s. fully revealing : $p(Z^*(u,v)) = v$ and $\Pi^*(u,v) = 0$.

  One can also prove by contradiction that every buyer $(Z^*(u,v) > u)$ is pooled with at least one seller $(Z^*(u',v') < u')$. If for instance $Z^*(u,v) > u$ for all (u,v) such that $Z^*(u,v) = Z$, then $P^*(Z) < E \ [v|Z]$ (since $Z^*(u,v) = Z \Rightarrow u > P^*(Z)$).

- **step 3** : Let (u,v) and (u',v') be such that:

$$u < Z^*(u,v) = Z^*(u',v') < u'.$$

This means that the insider buys in state (u,v), while he sells in state (u',v'). Thus the two functions

$$(z,p) \mapsto (v-p)(z-u),$$

and

$$(z,p) \mapsto (v'-p)(z'-u),$$

attain their maximum on the curve $p = P(z)$ at the same point (z^*,p^*). Therefore P is differentiable in z^* and $P'(z^*)$ equals the common tangent to the two indifference curves. This is illustrated graphically in Figure 1. This implies in particular that for a.e. (u,v) we can write the first order condition :

$$v - P^*(Z) = (P^*)'(Z)(Z-u) \ \text{ for } \ Z = Z^*(u,v).$$

■

Lemma 3

$$E(u|Z^*(u,v) = Z) = Z.$$

Proof of lemma 3 :
By assumption,

$$E \ [v - P^*(Z)|Z] = 0$$

Now if we apply the first order condition just derived above and we use the fact that :

$$(P^*)'(Z) > 0$$

we obtain the desired property.

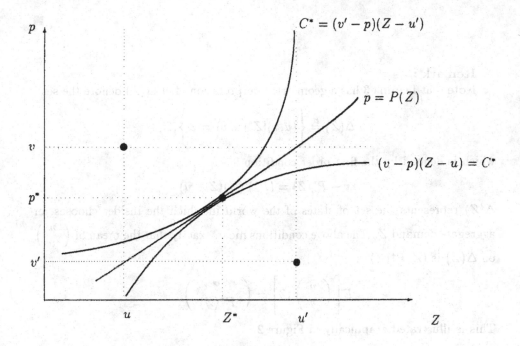

Figure 1 : Graphical illustration of Step 3 in the proof
of Lemma 2.

■

Remark :

Note that Lemma 3 has a geometric interpretation. Let $\Delta(Z)$ denote the set :

$$\Delta(Z) \stackrel{d}{=} \left\{ (u,v) | Z^*(u,v) = Z \right\},$$

which is defined by the first order condition :

$$v - P^*(Z) = (P^*)'(Z)(Z - u)$$

$\Delta(Z)$ represents the set of states of the world in which the insider chooses an aggregate demand Z. The above conditions mean exactly that the mean of $\begin{pmatrix} u \\ v \end{pmatrix}$ on $\Delta(Z)$ is $(Z, P^*(Z))$:

$$E\left[\begin{pmatrix} u \\ v \end{pmatrix} | Z \right] = \begin{pmatrix} Z \\ P^*(Z) \end{pmatrix}.$$

This is illustrated graphically in Figure 2.

Armed with these 3 lemmas we can now establish Proposition 6:

By Lemma 1, Π^* is admissible. By Lemma 3, $(\Pi^* + uv)$ is linear on all the sets $\Delta(Z)$ and :

$$E\left[\Pi^* + uv | Z \right] = \Pi^*(Z, \ P^*(Z)) + ZP^*(Z) = ZP^*(Z).$$

If Π is admissible, then $\Pi + uv$ is convex. Jensen's lemma implies :

$$E\left[\Pi + uv | Z \right] \geq \Pi(Z, P^*(Z)) + ZP^*(Z).$$

Since Π is non–negative, this implies that :

$$E[\Pi^*] \leq E[\Pi],$$

and therefore that Π^* solves $\mathcal{P}_1$.

■

As we already remarked, Proposition 6 implies the *uniqueness* of equilibrium, since any equilibrium is a solution of $\mathcal{P}_1$. In fact, the converse of Proposition 6 is also true : any solution of $\mathcal{P}_1$ is an equilibrium. Since the set of solutions of $\mathcal{P}_1$ is not empty, this implies the *existence* of equilibrium. This is what we now establish :

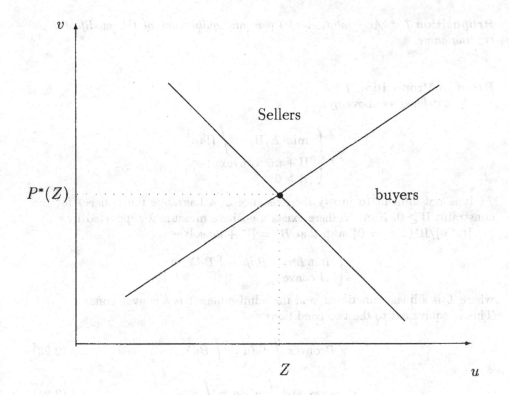

Figure 2 : Graphical illustration of the geometric interpretation of Lemma 3.

Proposition 7 : *Any solution of $\mathcal{P}_1$ is an equilibrium of the modified Kyle trading game.*

Proof of Proposition 7 :
   $\mathcal{P}_1$ is defined as above by :

$$\mathcal{P}_1 \begin{cases} \min \ E(\Pi) = \displaystyle\int \Pi d\mu \\ \Pi + uv \ \text{convex} \\ \Pi \geq 0 \end{cases}$$

It is not difficult to justify the existence of a Lagrange multiplier λ for the constraint $\Pi \geq 0$. Namely, there exists a positive measure λ supported by $\{(u, v)/\Pi^*(u, v) = 0\}$ such that $B^* = \Pi^* + uv$ solves :

$$\begin{cases} \min L = \int B d\mu - \int B d\lambda \\ B \ \text{convex,} \end{cases}$$

where L is a linear functional, and the admissible set is a convex cone. This is equivalent to the two conditions :

$$\forall \ B\text{convex} \int B d\mu \geq \int B d\lambda, \tag{2.23}$$

$$B^*\text{convex and} \int B^* d\mu = \int B^* d\lambda. \tag{2.24}$$

Therefore Π^* is a solution of $\mathcal{P}^1$ if and only if there exists λ (supported by $\{(u, v)/\Pi^*(u, v) = 0\}$ that satisfies (2.23) and such that $B^* = \Pi^* + uv$ satisfies (2.24). We now establish that this implies condition (2.20). We first need a definition :

Definition :

   Let M denote the set of Radon measures on Ω. A measure sweeping on Ω is a mapping T

$$\begin{cases} M & \to & M \\ \lambda & \mapsto & T\lambda \end{cases}$$

defined by a Kernel $K(x, dy)$, i.e., a measurable mapping :

$$\begin{aligned} K : \quad \Omega & \to & P(\Omega) \\ x & \mapsto & K(x, \cdot) \end{aligned}$$

(where $P(\Omega)$ is the set of probability measures on Ω) such that for a.e. x,

$\int yK(x,dy) = x$ (K preserves the expectations).

By definition :

$$T\lambda(dy) = \int K(x,dy)\,d\lambda(x),$$

or for $f \in C(\Omega, \mathbb{R})$:

$$\int f(y)\,d(T\lambda)(y) = \int \int f(y)K(x,dy)d\lambda(x).$$

At this point, it is useful to state Cartier's theorem (see for instance Meyer (1966)) :

Theorem 1 Let μ and λ be two positive measures on Ω. Suppose that for all convex (continuous) B, $\int B d\mu \geq \int B d\lambda$. Then μ is obtained from λ by a sweeping operation.

Remark : The converse result is true, and easy to prove. Indeed, by definition of T:

$$\int B(x)\,d(T\lambda)(x) = \int \int B(y)K(x,dy)d\lambda(x)$$

By Jensen's lemma the convexity of B implies that for a.e. x :

$$B(x) \leq \int B(y)K(x,dy).$$

Since λ is positive this implies the converse to Cartier's theorem :

$$\int B(x)d\lambda(x) \leq \int B(x)d(T\lambda)(x) = \int B(x)d\mu(x).$$

Now, let Π^* solve $\mathcal{P}_1$, $B^* = \Pi^* + uv$ and let Z^* and P^* be defined by (2.21) and (2.20). Considering condition (2.23), we see, by Cartier's theorem, that μ is obtained from λ by a sweeping operation.

We also have to take into account condition (2.24) :

$$\int B^*d\mu = \int B^*d\lambda.$$

Recall that convexity of B^* implies a.e :

$$B^*(u',v') \geq B^*(u,v) + (u'-u)\frac{\partial B^*}{\partial u}(u,v) + (v'-v)\frac{\partial B^*}{\partial v}(u,v). \qquad (2.25)$$

By definition of B^*, this inequality is strict if and only if $Z^*(u',v') \neq Z^*(u,v)$. Now if we integrate both sides of (2.25) with respect to the Kernel $K((u,v),du'dv')$ we find :

$$\int B^*(u', v') K((u, v), du'dv') \geq B^*(u, v), \qquad (2.26)$$

since K preserves expectations. Moreover, by the above remark, the inequality is strict unless the support of $K((u, v), \cdot)$ is included in $\Delta(Z^*(u, v))$. By integration of (2.26) with respect to $d\lambda(u, v)$ we find that $\int B^* d\mu \geq \int B^* d\lambda$ (as already known) with equality only if the support of $K((u, v), \cdot)$ is included in $\Delta(Z^*(u, v))$ for $\lambda-$ a.e. (u, v) a.e. Z. Since the sweeping operation preserves expectations this implies :

$$E_\mu\left[\left(\begin{array}{c} u \\ v \end{array} \right) / Z^*(u, v) = Z \right] = E_\lambda\left[\left(\begin{array}{c} u \\ v \end{array} \right) / Z^*(u, v) = Z \right].$$

Now the support of λ is included in the set $\{(u, v)/\Pi^*(u, v) = 0\}$, which is nothing but the curve $\{u = Z, v = P^*(Z)\}$.
Thus

$$E_\lambda\left[\left(\begin{array}{c} u \\ v \end{array} \right) / Z^*(uv) = Z \right] = \left(\begin{array}{c} Z \\ P^*(Z) \end{array} \right).$$

Combining these two conditions we get in particular condition (2.20) :

$$E_\mu[v/Z^*(u, v) = Z] = P^*(Z).$$

■

3 Price formation and market mechanism with endogenous liquidity trading (Glosten (1989))

We now examine a variant of the Kyle–Rochet–Vila model where we dispense from the noise trading assumption, and endogenize the non–informational (or liquidity) motivation to trade. To do so we rely on the seminal work of Glosten (1989). There are two types of differences between the analysis presented in this subsection and the original work of Glosten (1989). First, as will be outlined in the description of the model below, we posit slightly different (and more general) assumptions on the distribution of the variables on which there is adverse selection (while Glosten (1989) assumes that the private signal and the liquidity shock are normally distributed, we consider general distributions). Second, the mathematical approach we take is different. Relying on techniques drawn from variation calculus and convex analysis we present a simple and powerful solution to this rather difficult problem.

3.1 The model

Assumptions

The main features of the model are the following :

- The structure of the asset pay–off is as presented above in section 2. As in Grossman and Stiglitz (1980) (and unlike in Kyle–Rochet–Vila), the information of the insider is imperfect : after observing s, there is still some residual uncertainty ϵ about the value v. As mentioned in section 2, the two differences between our assumptions and those of the seminal paper of Glosten (1989) are the following : i) we assume $s \perp \epsilon$ while Glosten (1989) assumes $v \perp \epsilon$, and ii) we do not impose any assumptions on the distribution of s while Glosten (1989), as the previous literature, assumes normality. To fix ideas denote the support of s by : $[\underline{s}, \bar{s}]$.

- There are N risk neutral market makers; we will consider in turn the monopolistic case $(N = 1)$ and the "competitive" case where the N market makers are assumed to quote prices equal to their conditional expectation of the value of the asset, as in Kyle (1985).

- As in Kyle–Rochet–Vila or Glosten and in contrast with Grossman and Stiglitz (1980), there is a single strategic insider. In contrast with the Kyle–Rochet–Vila approach, however, the insider is risk averse. As in Grossman and Stiglitz (1980) and Glosten (1989), his utility function is exponential, which combined with the normality of ϵ generates quadratic preferences.

- In contrast with the analyses presented above, there is no noise trading. Instead the motivation for trades is endogenized, in the following way. The insider is endowed with an inventory $-I$ in the risky asset. $-I$ can be positive in which case the insider is "long" the asset, or negative, in which case the insider is "short". Because the insider is risk averse, while the market makers are risk neutral, trading could lead to Pareto improvements through risk sharing. The exact value of the endowment of the insider (I) is not known from the market makers. From their point of view, it is a random variable. Denote its support $[\underline{I}, \bar{I}]$. We do not make assumptions on the distribution of I over this support. This differs from Glosten (1989) whereby I is assumed to be jointly normal with s and ϵ.

The problem from the point of view of the informed agent

Denote by q the trade of the informed agent. If q is positive it refers to a purchase while if it is negative it corresponds to a sale. Denote by $t(q)$ the (algebraic) transfer paid by the insider, corresponding to this trade. If the insider buys this corresponds to the buying price multiplied by q. If the insider sells ($q < 0$), then $t(q)$ will be negative.

The final wealth of the informed agent if she trades q against transfer $t(q)$ is :

$$W_0 + qv - t(q).$$

Her expected utility is :

$$E(-e^{-\rho[W_0+(q-I)v-t(q)]}|s) = -e^{-\rho[W_0+(q-I)s-t(q)-\frac{1}{2}\rho\sigma^2(-I+q)^2]}.$$

The term between brackets (in the right hand side) corresponds to the certainty equivalent of the expected utility of the insider. A crucial parameter is :

$$\theta = s + \rho\sigma^2 \qquad\qquad (3.1)$$

which corresponds to the marginal willingness to pay of the insider for the security. Denote by Θ the set of possible values of θ. This parameter summarizes the mix of informational and risk sharing motivations for trade of the insider. If s is large the insider is inclined to buy since he expects the value of the asset to be high. The insider also desires to buy if I is large, in order to rebalance her portfolio, and share risk. Because the market makers do not know if large buy orders stem from large s or large I, there is adverse selection in the market place.

The expected utility of the insider can be rewritten :

$$-e^{-\rho([\theta q-t(q)-\frac{1}{2}\rho\sigma^2 q^2]+[W_0-Is-\frac{1}{2}\rho\sigma^2 I^2])} \qquad\qquad (3.2)$$

The second term in brackets in the expected utility, (3.2), corresponds to the utility obtained if there was no trade, i.e., the reservation utility. The first term in brackets corresponds to the surplus from trade, or "rent" earned by the insider.

Since the informed agent has always the option not to trade, and since in that case she cannot be compelled to pay anything to the market makers, her rent must always be positive. This is referred to as the "individual rationality condition" or "participation constraint".

The insider chooses q to maximize her expected utility. This is equivalent to maximizing :

$$\theta q - t(q) - \frac{1}{2}\rho\sigma^2 q^2.$$

Since this depends on s and I only through θ, only θ can be revealed in equilibrium. There is "bunching" in equilibrium, i.e., different pairs (s, I) with identical θ give rise to the same allocation $q(\theta)$ and transfer $t(q(\theta))$.

If the different market markers $(i = 1, \ldots, n)$ quote price schedules $t_i(q_i)$, the insider will split her orders $(q = \Sigma_i q_i)$ in such a way that the total cost $\Sigma_i t_i(q_i)$ is minimized. Therefore the only thing that matters is the aggregate schedule :

$$t(q) \stackrel{d}{=} Min_{\Sigma_i q_i=q} t_i(q_i)$$

Given this schedule, and given realization of s and I and hence of θ, the informed agent chooses the trade

$$q \stackrel{d}{=} \Sigma_i q_i$$

maximising her expected utility. Denote by $U(\theta)$ the resulting rent obtained by the informed agent :

$$U(\theta) \overset{d}{=} Max_q[\theta q - \frac{1}{2}\rho\sigma^2 q^2 - t(q)], \tag{3.3}$$

and by $q(\theta)$ the argument maximum, which is a.e. unique.

By condition (3.3) the curve $\theta \to U(\theta)$ is the upper envelope of a family of affine functions $\theta \to \theta q - \frac{1}{2}\rho\sigma^2 q^2 - t(q)$. This is represented in Figure 3. Therefore :

$$U \text{ is convex}, \tag{3.4}$$

$$q(\theta) = \dot{U}(\theta) \text{ for a.e.}, \tag{3.5}$$

and,

$$\theta \to q(\theta) \text{ is non decreasing}. \tag{3.6}$$

Considering the participation constraint $U(\theta) \geq 0$, it is natural to assume $t(0) = 0$ and to define the (unit) price schedule $p(q)$ by :

$$p(q) = \frac{t(q)}{q} \quad (q \neq 0).$$

This definition allows us to offer some comments on the determination of $q(\theta)$, since (3.3) can now be written :

$$U(\theta) = \max_q\{(\theta - p(q))q - \frac{1}{2}\rho\sigma^2 q^2\}. \tag{3.7}$$

- Since $U(\theta) \geq 0$ (the insider can always choose $q = 0$), we see that a purchase $(q(\theta) > 0)$ can occur only when $\theta > p(q(\theta))$, i.e. when the valuation of the trader for the asset is larger than the price.

- If all prices are shifted upward by a constant $m > 0$ (for $q > 0$), equation (3.7) shows that $q(\theta)$ is replaced by $q(\theta - m)$: therefore for all the values of θ such that the insider used to buy $(q(\theta) > 0)$, she now buys less (since q is non decreasing).

- If the risk aversion parameter ρ is allowed to vary, we obtain a more general version of (3.7) :

$$U(\theta, \rho) = \max_q\{(\theta - p(q))q - \frac{1}{2}\rho\sigma^2 q^2\},$$

where the maximum is attained for $q = q(\theta, \rho)$. Applying again the envelope principle we find :

for a.e. (θ, ρ), $\dfrac{\partial U}{\partial \theta}(\theta, \rho) = q(\theta, \rho)$, $\dfrac{\partial U}{\partial \rho}(\theta, \rho) = -\dfrac{1}{2}\sigma^2 q^2(\theta, \rho)$.

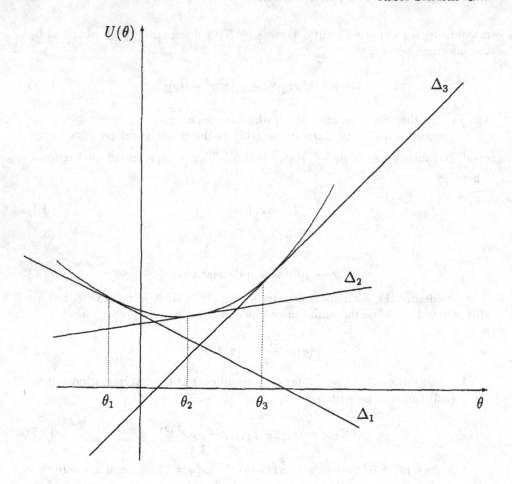

Figure 3 : Expected utility and optimal trade of the informed agent

$$U(\theta) \stackrel{d}{=} \max_{q} \left[\theta q - \frac{1}{2}\rho\sigma^2 q^2 - t(q) \right]$$

$\Delta_1, \Delta_2, \Delta_3$ are the 3 different lines :

$$\theta_i q - \frac{1}{2}\rho\sigma^2 q^2 - t(q), i = 1, 2, 3$$

Therefore the sensitivity of the trade to ρ is given by the cross partial derivative of U :

$$\frac{\partial q}{\partial \rho}(\theta, \rho) = \frac{\partial^2 U}{\partial \rho \partial \theta}(\partial, \rho).$$

Thus at every (θ, ρ) where U is twice differentiable we have :

$$\frac{\partial q}{\partial \rho}(\theta, \rho) = \frac{\partial^2 U}{\partial \theta \partial \rho}(\theta, \rho) = -\sigma^2 q(\theta, \rho) \frac{\partial q}{\partial \theta}(\theta, \rho).$$

This implies that $|q|$ decreases when ρ increases : trade volume decreases when the informed agent becomes more risky averse. By the same argument, $|q|$ decreases when σ^2 increases : trade volume decreases when the asset becomes more risky.

The problem from the point of view of the market makers

The total profit of the market makers is

$$b(\theta) \overset{d}{=} t(q) - qv(\theta) \tag{3.8}$$

where $v(\theta) = E(s|\theta)$. Note that $v(\theta)$ can be viewed as the unit cost of the market maker. To the extent that different qs correspond to different θs, and consequently to different $v(\theta s)$, the unit cost of the market makers varies with the quantity traded.

Note that the market makers are at an informational disadvantage. They only know the distribution of s and I, while the informed agent knows the realization of these variables. In fact the market makers face an adverse selection problem. For example when they observe that q is large they do not know if this stems from an agent desiring to share risk (large realization of I) or an agent wishing to buy to make profits (large realization of s). To cope with this adverse selection problem, however, the market makers can conduct rational inference, based on their observation of $q(\theta)$, to learn about θ. To trace back θ from $q(\theta)$, the market makers can use the above mentioned mathematical properties of the optimal action taken by the informed agent : (3.4), (3.5) and (3.6). For example, (3.6) shows that the market maker can infer from large trades that θ is large.

A benchmark case

Before we turn to the analysis of how the market makers conduct this inference process, consider, as a benchmark, the hypothetical situation where there would be no asymmetric information and no market power. Suppose indeed that a benevolent principal can observe θ and choose q to maximize welfare. The latter is equal to the sum of the rent of the agent and the profit of the market makers :

$$\begin{aligned} SW \overset{d}{=} & \ [\theta q - \frac{1}{2}\rho\sigma^2/q^2 - t(q)] + [t(q) - qv(\theta)] \\ = & \ \theta q - \frac{1}{2}\rho\sigma^2 q^2 - qv(\theta). \end{aligned} \tag{3.9}$$

The program of the social planner is :

$$\max_{q} SW$$

The solution to this problem is

$$\frac{\theta - v(\theta)}{\rho\sigma^2} \stackrel{d}{=} I(\theta). \tag{3.10}$$

If the agent trades $I(\theta)$ at the price $v(\theta)$, the (maximal) welfare is denoted $S(\theta)$ and is equal to :

$$S(\theta) = (\rho\sigma^2/2)I(\theta)^2 \tag{3.11}$$

3.2   Monopolistic market maker

Consider the case where there is a monopolistic market maker, posting the schedule $t(.)$ (associating a transfer $t(q)$ to each trade q with the agent) which maximizes his expected profits. The extensive form of the trading game is as follows :

1. The informed agent privately observes s and I.

2. The monopolistic market maker posts his schedule $t(q)$.

3. The informed agent chooses her trade q.

4. The liquidation value of the asset (v) is realized and consumption takes place.

Note that this differs from the extensive form studied in the previous section in the analysis of the Kyle–Rochet–Vila model, whereby the sequencing of the moves of the insider and the market maker was different (since the insider was assumed to move first). Assuming that the monopolistic market maker moves first contributes to enhancing his market power. Intuitively, by announcing his schedule before the informed agent chooses q, the monopolistic market maker dictates the rules of the game.

We look for a Bayesian Nash equilibrium of this game, defined as follows :

Definition :

A Bayesian Nash equilibrium of the Glosten (1989) monopolistic market maker model is a pair of functions :

$$q(.) : \quad \Theta \to \mathbb{R}$$
$$\theta \mapsto q(\theta)$$

and

$$t(.) : \quad \mathbb{R} \to \mathbb{R}$$
$$q \mapsto t(q)$$

such that,

- *for a given transfer function t(.), and a given observation of θ by the informed agent, q(θ) maximizes her rent*

- *and the market maker chooses t(.) to maximize his expected profit, given his rational expectations about the strategy of the informed agent, q(.).*

A reinterpretation in terms of agency and mechanism design theory

This trading game can be interpreted in terms of agency theory : the insider is the (informed) agent and the market maker is the principal. The latter must propose a menu of contracts to the agent. The informed agent chooses the contract which maximizes her expected utility. In the present case the menu of contracts is the schedule $q \to t(q)$, and each pair $(q, t(q))$ is a single contract. Since the agent holds private information, which leads to adverse selection, the principal must offer incentive compatible contracts, i.e., contracts which elicit truthful revelation from the agent.[10] The incentive compatibility conditions are the optimality conditions for the agent : (3.4), (3.5), and (3.6). Also, the menu of contracts, or schedule, offered by the principal must satisfy the individual rationality condition of the insider (i.e. the condition requiring that the rent be non negative).

For the market maker, the choice of the optimal menu of contracts, among those that elicits truthful revelation, and thus cope with adverse selection, is a mechanism design problem. Thus our analysis in this section is related to that of Rochet and Vila (1994), the difference being that here liquidity trading is endogenous (and results from risk sharing), while there it was exogenous (and resulted from noise trading) in Rochet and Vila (1994).

To the extent that the menu of contracts optimally chosen by the principal defines the set of potential trades and prices, and the information revealed through the trading process, i.e., the market structure, the latter is endogenously analyzed in the present model.

The program of the monopolist

In a Bayesian Nash equilibrium of this model, the monopolist sets the schedule $t(.)$ to maximize his expected profit : under the incentive compatibility condition of the insider (3.4), (3.5), and (3.6), and her individual rationality condition.[11]

---

[10]In agency theory, a contract or mechanism is said to be incentive compatible if it is such that the informed agent finds it optimal to truthfully announce her privately known type.

[11]Any rent obtained by the agent and satisfying these conditions corresponds to a feasible level of expected profits for the market maker, and is said to be "incentive compatible" or "implementable".

Hence to solve for the equilibrium, we need to solve the program of the monopolist :

$$Max_{t(.)}\mathcal{B} \ s.t. \dot{U} = q, U \text{convex}, U \geq 0 \tag{3.12}$$

where :

$$\mathcal{B} \overset{d}{=} E(t(q) - v(\theta)q). \tag{3.13}$$

From equation (3.3), the transfer can be rewritten as :

$$t(q) = \theta q - \frac{1}{2}\rho\sigma^2 q^2 - U(\theta). \tag{3.14}$$

Substituting equation (3.14) and the optimality condition $\dot{U} = q$ in (3.12), the program of the monopolistic market maker can be rewritten as a variation calculus problem in U :

$$Max_{U(.)}E((\theta - v(\theta))\dot{U} - \frac{1}{2}\rho\sigma^2(\dot{U})^2 - U(\theta)), s.t., U\text{convex}. \tag{3.15}$$

Let B denote the maximum profit of the market maker, i.e. the solution of problem (3.15):

$$B = Max_{U\text{convex},\geq 0}E((\theta - v(\theta))\dot{U} - \rho\sigma^2/2(\dot{U})^2 - U(\theta)) \tag{3.16}$$

The characteristics of the solution

We now analyse the properties of the solution of problem (3.12), i.e. the rent $U(\theta)$ left by the market marker to the trader, as a function of his type θ. Feasibility implies that U is a convex non negative function of θ. Moreover, incentive compatibility implies that $q(\theta) = \dot{U}(\theta)$ at every point where U is differentiable.

Let us remark that if we neglect the constraint U convex, problem (3.12) reduces to what is known in physics as an obstacle problem,[12] i.e. a variation calculus problem with an inequality constraint on the unknown function. In our simple framework, the solution to this obstacle problem can be found explicitly, as we will see in a moment. Moreover, a reasonable assumption on the distribution of θ implies that the convexity constraint does not bind. Whenever it does, more complicated tools (derived of the sweeping techniques presented in section 3) have to used (see Rochet-Choné (1996)).

The solution of the relaxed version of (3.12) (obtained by neglecting the constraint that U is convex) can be found by maximizing the Lagrangian :

$$L(U, \dot{U}) = \int_{\underline{\theta}}^{\bar{\theta}} [(\theta - v(\theta))\dot{U} - \frac{1}{2}\rho\sigma^2(\dot{U})^2]dF(\theta) + \int_{\underline{\theta}}^{\bar{\theta}} U[d\Lambda - dF](\theta), \tag{3.17}$$

where Λ is the Lagrange multiplier associated to the constraint $U \geq 0$: it is a positive measure such that $Ud\Lambda = 0$ (i.e. the support of Λ is included in the set $U^{-1}(0)$).

---

[12] There is a large literature on these problems : see for instance Kinderlehrer - Stampacchia (1980).

The first order condition of this problem (Euler equation) states that the derivative of the integrand with respect to $\dot{U}$, i.e.

$$\alpha(\theta) = [\theta - v(\theta) - \rho\sigma^2 q(\theta)]f(\theta), \qquad (3.18)$$

is a primitive of the derivative of the integrand with respect to U, i.e. $(d\Lambda - dF)$. In other words :

$$\alpha(\theta) = \Lambda(\theta) - F(\theta) + \text{ constant.} \qquad (3.19)$$

Since there is no boundary constraint on U, α vanishes both in $\underline{\theta}$ and $\bar{\theta}$, so that the constant in (3.19) is zero and moreover :

$$\Lambda(\bar{\theta}) = F(\bar{\theta}) = 1. \qquad (3.20)$$

This means that the total mass of the measure Λ (the Lagrange multiplier associated to the constraint $U \geq 0$) is one. This implies in particular that $U^{-1}(0)$ is not empty.[13] U being convex and non negative, $U^{-1}(0)$ is an interval $[\theta_1, \theta_2]$, and $U > 0$ outside the interval. Therefore the support of Λ is included in $[\theta_1, \theta_2]$, which means that $\Lambda(\theta) \equiv 0$ on $[\underline{\theta}, \theta_1[$ and $\Lambda(\theta) \equiv 1$ on $]\theta_2, \bar{\theta}]$. Using formulas (3.19) and (3.20), we get the complete expression of $q(\cdot)$, as function of the first best optimal trade $I(\theta) = \dfrac{\theta - v(\theta)}{\rho\sigma^2}$ and the distribution of θ :

$$q(\theta) = I(\theta) + \frac{F(\theta)}{\rho\sigma^2 f(\theta)}, \ \theta < \theta_1, \qquad (3.21)$$

$$q(\theta) = 0, \ \theta_1 \leq \theta \leq \theta_2, \qquad (3.22)$$

$$q(\theta) = I(\theta) - \frac{1 - F(\theta)}{\rho\sigma^2 f(\theta)} \ \theta > \theta_2. \qquad (3.23)$$

Similarly $\Lambda(\theta)$ can be extracted from conditions (3.19) and (3.20) :

$$\left\{ \begin{array}{rcll} \Lambda(\theta) & = & 0 & \theta < \theta_1, \\ & = & [\theta - v(\theta)]f(\theta) + F(\theta) & \theta_1 \leq \theta \leq \theta_2, \\ & = & 1 & \theta > \theta_2. \end{array} \right.$$

Λ being continuous, this implies that :

$$(\theta_1 - v(\theta_1))f(\theta_1) + F(\theta_1) = 0,$$

and :

$$(\theta_2 - v(\theta_2))f(\theta_2) + F(\theta_2) = 1,$$

---

[13] This can also be seen by a direct economic argument : if $U(\theta)$ was positive everywhere, it would have by continuity a positive minimum ϵ on $[\underline{\theta}, \bar{\theta}]$. By raising all prices by ϵ the market marker would increase her profit without violating the participation constraint.

so that q is itself continuous. The last thing to check is that q is indeed non decreasing, so that U is indeed convex. This is satisfied for instance under the following assumption :

Assumption A :

- $\theta \to I(\theta) = \dfrac{\theta - v(\theta)}{\rho\sigma^2}$ is non decreasing.   (A1)

- $\theta \to \log F(\theta)$ and $\theta \to \log(1 - F(\theta))$ are concave [14]   (A2).

Proposition 8 *Under assumption A, the solution of the monopolist's problem is characterized by equations (3.21), (3.22), (3.23), together with the condition $U(\theta_1) = 0$.*

The rent corresponding to this solution is graphically represented in Figure 4.

3.3   The properties of the price schedule

We are now in a position to study the properties of the *monopolistic*[15] price schedule $q \to t(q)$, i.e. the one that implements the mapping $\theta \to q(\theta)$ characterized in Proposition 8. If t is differentiable in $q = q(\theta)$, condition (3.3) implies :

$$t'(q(\theta)) = \theta - \rho\sigma^2 q(\theta).\qquad(3.24)$$

Combined with equations (3.21) and (3.23), this gives :

$$t'(q(\theta)) = v(\theta) - \frac{F(\theta)}{f(\theta)}\qquad \text{if }\ q(\theta) < 0,\qquad(3.25)$$

and

$$t'(q(\theta)) = v(\theta) + \frac{1 - F(\theta)}{f(\theta)}\qquad \text{if }\ q(\theta) > 0.\qquad(3.26)$$

The marginal price paid by the consumer thus equals the sum of $v(\theta)$ (interpreted as the cost of the marginal unit $q(\theta)$ for the market marker) plus a monopolistic mark-up (which is naturally negative when the market maker buys, i.e. when $q(\theta) < 0$, and positive in the other case). The consequence of this mark-up is that the trade volume $|q(\theta)|$ is lower than in the benchmark case (where it is equal to $|I(\theta)|$). This results immediately from Proposition 8 :

---

[14] Assumption (A2) is satisfied for the great majority of common parameterized distributions (uniform, normal, exponential, gamma, chi-squared...). See Bagnoli - Bergstrom (1989) for a proof.

[15] We will contrast it later with the *competitive* price schedule, which emerges when there are many market makers.

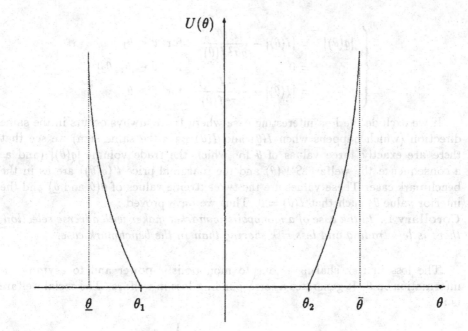

Figure 4 : Optimal rent of the informed agent in the monopolistic market

$$
\begin{cases}
|q(\theta)| &= |I(\theta)| - \dfrac{F(\theta)}{\rho\sigma^2 f(\theta)} & \text{for } \theta < \theta_1 \\[2mm]
&= 0 & \text{for } \theta \in [\theta_1, \theta_2] \\[2mm]
&= |I(\theta)| - \dfrac{1 - F(\theta)}{\rho\sigma^2 f(\theta)} & \text{for } \theta > \theta_2.
\end{cases}
$$

If we exclude the less interesting case where trade always occurs in the same direction (which happens when $I(\underline{\theta})$ and $I(\bar{\theta})$ have the same sign) we see that there are exactly three values of θ for which the trade volume $|q(\theta)|$ (and as a consequence the welfare $SW(\theta)$ and the marginal price $t'(q(\theta))$ are as in the benchmark case. These values are the two extreme values of θ ($\bar{\theta}$ and $\underline{\theta}$) and the interior value θ_0 such that $I(\theta) = 0$. Thus we have proved :

Corollary 1 : *In the case of a monopolistic market maker with adverse selection, there is less trading and less risk sharing than in the benchmark case.*

The loss in risk sharing is due to monopolistic power and to asymmetric information on θ. It generates a corresponding loss in welfare. The social welfare is :

$$
SW(\theta) = (\theta - v(\theta))q(\theta) - \frac{1}{2}\rho\sigma^2 q(\theta)^2 \tag{3.27}
$$

or :

$$
SW(\theta) = S(\theta) - \frac{1}{2}\rho\sigma^2 q(\theta)^2 \tag{3.28}
$$

where $S(\theta)$ is the welfare obtained in the benchmark case, and defined in equation (3.11). Hence we can state the following corollary :

Corollary 2 : *Welfare is lower in the market operated by the monopolistic market marker than in the benchmark case.*

It is interesting to notice that $t(\cdot)$ is never differentiable in 0. Condition (3.24) implies indeed :

$$
t'_+(0) = \lim_{q \to 0+} p(q) = \theta_2, \tag{3.29}
$$

while

$$
t'_-(0) = \lim_{q \to 0-} p(q) = \theta_1, \tag{3.30}
$$

which is strictly lower than θ_2 (since $\theta_2 > \theta_0 > \theta_1$). The difference $t'_+(0) - t'_-(0)$ is naturally to be interpreted as the "bid-ask spread".

Note that this characterization of the bid–ask spread differs from the result obtained in Kyle (1985). In the latter there is a price impact of trades, e.g. large purchases push the (conditional expectation of the value and hence the) price up. Small trades, however, only have a small impact on prices, and in the limit :

$$
\lim_{q \to 0+} p(q) = \lim_{q \to 0-} p(q) = E(v).
$$

In the present approach (where there is a monopolistic market maker) this limiting result does not hold. Rather :

$$\lim_{q\to 0+} p(q) > E(v) > \lim_{q\to 0-} p(q)$$

In the words of Glosten (1989) there is is a "small trade spread."
The trade and welfare functions arising in the case of a monopolistic market maker are illustrated in Figure 5.

3.4 Competitive market makers

Price competition for the order flow

We now turn to the case of competitive market makers. There are $N \geq 2$ risk neutral market makers. The extensive form of the game differs from that assumed in the case of the monopolistic market maker. It is similar to its analogous in the Kyle–Rochet–Vila case :

1. The informed agent privately observes s and I.

2. The informed agent submits a market order : q.

3. The market makers post price schedules.

4. The insider allocates her trade among the market makers (to minimize the cost in the case of a buy order, and maximize the proceeds in the case of a sell order).

5. The liquidation value of the asset (v) is realized and consumption takes place.

Bertrand competition between the rational, homogenously informed and risk neutral market makers, who have identical valuation for the asset, brings prices down to their break even point. If a market marker quoted prices above this break even point, then competitor would find it profitable to undercut, i.e., to post slightly better prices. [16] This leads to the condition :

$$P(q(\theta)) = E(v|q(\theta)). \tag{3.31}$$

In this context we look for a perfect Bayesian equilibrium of this trading game, defined as follows :

Definition :

A perfect Bayesian equilibrium of the Glosten (1989) competitive market makers model is a pair of functions :

---

[16] Biais, Hillion and Spatt (1995) offer empirical illustrations of this undercutting behaviour.

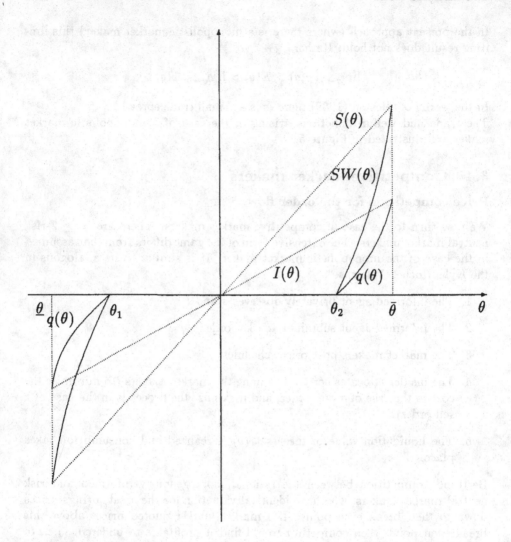

Figure 5 : Trade and welfare in the monopolistic market

$$t(.) \; : \; \mathbb{R} \to \mathbb{R}$$
$$q \mapsto p$$

and

$$q(.) \; : \; \Theta \to \mathbb{R}$$
$$\theta \mapsto q(\theta)$$

such that :

$$t(q) = qE(v|q(\theta)).$$

The interpretation of these conditions is the following :

- the conditional expectation is correctly computed by the market makers, using their rational expectations about the strategy of the insider $q(.)$,

- and the trading strategy is chosen by the insider so as to maximize her rent, taking into account her rational expectation of the response of the market makers : $t(.)$.

It can be shown that many such equilibria exist, among which "semi-separating" equilibria, i.e. equilibria where $q(\cdot)$ is locally constant. Semi-separating equilibria are less interesting from the economic view point and less regular from the mathematical view point (they are necessarily discontinuous). Therefore we focus here on the "separating" equilibria (i.e. equilibria where q is one-to-one) which are more regular. In a separating equilibrium, condition (3.31) becomes very simple :

$$\text{for a.e.} \quad P(q(\theta)) = v(\theta). \tag{3.32}$$

This means that the informed agent captures the entirety of the social welfare :

$$U(\theta) = SW(\theta) = \rho\sigma^2[I(\theta)q(\theta) - \frac{1}{2}q^2(\theta)], \tag{3.33}$$

while the implementability condition holds as before :

$$\dot{U}(\theta) = q(\theta). \tag{3.34}$$

Combining (3.33) and (3.34), we find that separating equilibria are solution of a first order ordinary differential equation of the Hamilton-Jacobi type :

$$U(\theta) = \rho\sigma^2[I(\theta)\dot{U}(\theta) - \frac{1}{2}\dot{U}^2(\theta)]. \tag{3.35}$$

A natural way to solve this equation is to use $\dot{U}(\theta) = q$ as a parameter and look for two functions $\theta(q)$ and $u(q)$ such that $U(\theta(q)) = u(q)$, which implies :

$$u(q) = \rho\sigma^2[I(\theta(q))q - \frac{1}{2}q^2].\tag{3.36}$$

By differentiating (3.36) we find :

$$u'(q) = q\theta'(q) = \rho\sigma^2[I'(\theta(q))q\theta'(q) + I(\theta(q)) - q].$$

Finally $\theta(q)$ has to solve :

$$\theta'(q) = \frac{I(\theta(q)) - q}{q[\frac{1}{\rho\sigma^2} - I'(\theta(q))]},\tag{3.37}$$

and $u(q)$ is then given by (3.36). Since we have assumed that $v(\cdot)$ is increasing (which implies that $I'(\theta(q)) < \frac{1}{\rho\sigma^2}$), we can apply the Cauchy-Lipschitz theorem : for any initial condition $\theta(q_0) = \theta_0$, equation (3.37) has a unique solution defined on a neighborhood of q_0. However we have to remember that the implementability condition implies that $\theta \to q(\theta)$ is non decreasing, which means that $\theta'(q) \geq 0$, so that the solution $(q, \theta(q))$ has to belong to the domain $I(\theta) > q$.

In fact, our competitive trading situation is a particular case of a "signalling game" (like the trading game à la Kyle studied in section 3.2). As already mentioned, such games typically admit many equilibria. More specifically, one can prove the following result :

Proposition 9 *There exists a continuum of separating equilibria, parametrized by the interval $[\theta_-, \theta_+]$ (with $\theta_- < \theta_0 < \theta_+$), on which they are defined. The quantity traded on such an equilibrium, denoted $q_c(\theta)$ is efficient at both extremes and at $\theta_0 : q(\theta) = I(\theta)$ for $\theta = \theta_-, \theta_0, \theta_+$. Anywhere else, we have $|q(\theta)| < I(\theta)$. Finally $\dot{q}(\theta)$ is infinite in θ_- and θ_+, and zero in θ_0. In particular there is no small trade spread.*

The next section is dedicated to a comparison of the monopolistic and competitive solutions in the case where the distribution of I and s is symmetric around the origin.

3.5   Comparison of the competitive and monopolistic solutions when the distribution of I and s is symmetric

When the distribution of (I, s) is symmetric around 0 (as represented in Figure 6), $I(\theta)$ is linear. For instance, suppose that the support of (I, s) is the square $[-\frac{1}{2}, \frac{1}{2}]^2$ and normalize $\rho\sigma^2$ to 1.

Then we have :

$$I(\theta) = v(\theta) = \frac{\theta}{2}; \text{ and } \underline{\theta} = -1, \theta_0 = 0, \bar{\theta} = 1.$$

The monopolistic solution can then be written :

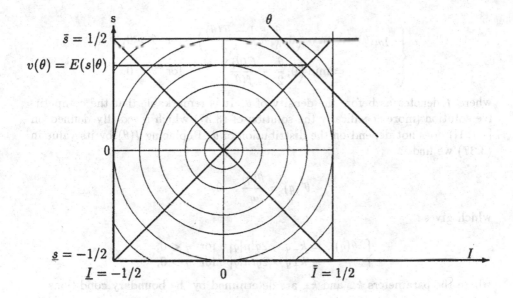

Figure 6 : Distribution of θ on the unit square

$$\begin{cases} q_m(\theta) & = \max\left(0, \dfrac{\theta}{2} - \dfrac{1-F(\theta)}{f(\theta)}\right) \quad \text{for } \theta > 0, \\[2mm] & = \min\left(0, \dfrac{\theta}{2} - \dfrac{F(\theta)}{f(\theta)}\right) \quad \text{for } \theta < 0. \end{cases}$$

where f denotes as before the density of θ. It is remarkable that the competitive solution (more specifically the solution of (3.37) which is exactly defined on $[-1, 1]$) does not depend on the distribution of θ. Replacing $I(\theta)$ by its value in (3.37) we find :

$$\theta'(q) = \frac{\theta(q)}{q} - 2,$$

which gives :

$$\begin{cases} \theta(q) & = k_- q - 2q \ln|q| \quad \text{for } q < 0, \\ & = k_+ q - 2q \ln|q| \quad \text{for } q > 0, \end{cases}$$

where the parameters k_- and k_+ are determined by the boundary conditions :

$$q_c(-1) = I(-1) = -\frac{1}{2}, \quad q_c(1) = I(1) = \frac{1}{2}.$$

This defines implicitly the competitive solution $q_c(\theta)$.
The computation of k_- and k_+ is easy :

$$\begin{cases} -1 = -\dfrac{1}{2}k_- + \ln\dfrac{1}{2}, \\[2mm] 1 = \dfrac{1}{2}k_+ - \ln\dfrac{1}{2}. \end{cases}$$

Finally we obtain :

$$k_- = k_+ = 2(1 - \ln 2), \quad \text{and}$$
$$\theta(q) = 2q - 2q \ln|2q| \quad \text{for } q \neq 0.$$

The graph of this function is given in Figure 7, and its inverse $\theta \to q_c(\theta)$ is represented in Figure 8. The corresponding price function is given in Figure 9.

We now compare the properties of the competitive and monopolistic solution.

First of all, both solutions coincide with the benchmark solution $I(\theta)$ in $\underline{\theta}, \theta_0$, and $\bar{\theta}$. Anywhere else the volume of trade is suboptimal $|q_i(\theta)| < |I(\theta)|$ for $i = c, m$.

Second, $q_c(\theta)$ only vanishes at $\theta = \theta_0$, while $q_m(\theta)$ is null on the whole interval $[\theta_1, \theta_2]$. This corresponds to the existence of a "small trade spread" in the monopolistic case, but not in the competitive case. This implies that the welfare of "small" traders (i.e. those with small values of $|\theta|$) is higher in the competitive case than in the monopolistic situation.

Surprisingly, the converse is true for "large" traders. This comes from the fact that $\dot{q}_c(\theta)$ tends to $+\infty$ when θ tends to $\underline{\theta}$ or $\bar{\theta}$. This is immediately seen

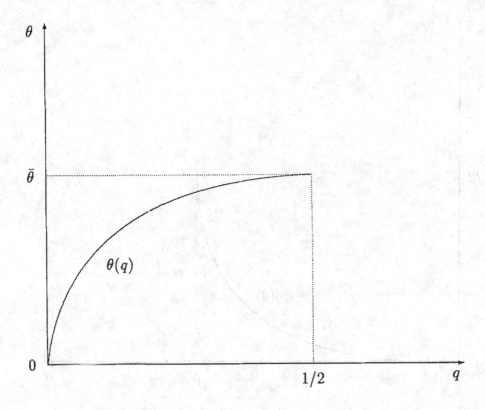

Figure 7 : θ as a function of q when the price is the conditional expectation of v, in the Pareto dominating equilibrium

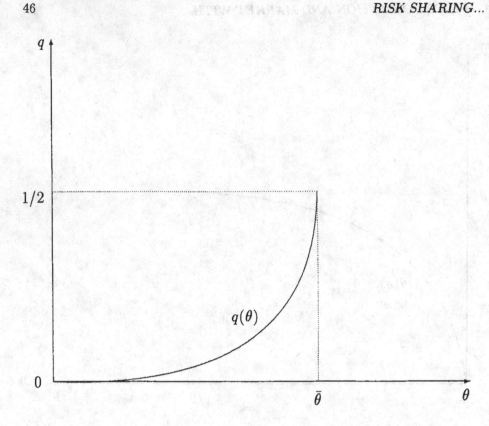

Figure 8 : q **as a function of** θ**, when the price is the conditional expectation of** v**, in the Pareto dominating equilibrium**

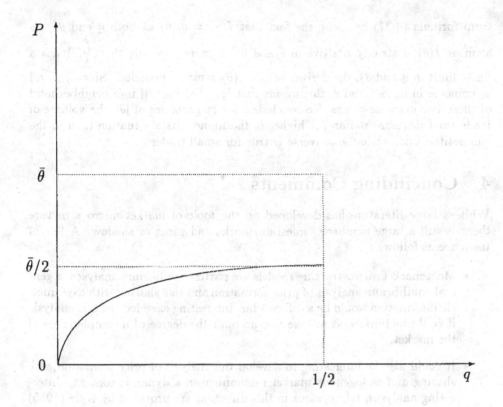

Figure 9 : Prices, as conditional expectations of v, in the Pareto dominating equilibrium

from formula (3.37) by using the fact that $I(\theta) = q_c(\theta)$ at both $\underline{\theta}$ and $\bar{\theta}$. As soon as $f(\theta)$ is strictly positive in $\bar{\theta}$ and $\underline{\theta}$ (or more generally that $\dfrac{\dot{f}}{f}(\theta)$ has a finite limit in $\underline{\theta}$ and $\bar{\theta}$), the derivative of $q_m(\theta)$ remains bounded. Since q_m and q_c coincide in both $\underline{\theta}$ and $\bar{\theta}$, this means that $|q_c(\theta)| < |q_m(\theta)|$ in a neighborhood of these two extreme values. To conclude : for large values of $|\theta|$, the volume of trade (and therefore welfare) is higher in the monopolistic situation than in the competitive one, while the converse is true for small trades.

4   Concluding Comments

While a large literature has developed on the topic of market micro–structure there is still a large number of question–marks and zones of shadow. A few of them are as follows :

- Most market microstructure models are partial equilibrium analyses, a general equilibrium analysis of price formation and risk sharing with asymmetric information would be a difficult but interesting exercice. In this analysis it could be important to take into account the degree of incompleteness of the market.[17]

- It could also be interesting to develop the analysis of price formation, risk sharing and endogenous market mechanism in a dynamic context. Interesting analyses, taking steps in this direction are provided by Kyle (1985) and Back (1992, 1993).[18]

- It would also be interesting to better understand the links and the differences between different types of financial market structures, such as the organization of secondary trading in stocks, Treasury auctions, or initial public offerings.[19]

---

[17] Biais and Hillion (1995) take a preliminary step in that direction by studying how risk–sharing, trading volume and price efficiency are modified when a non redundant option is added.

[18] Back (1993) studies the case where there is an option as well as a stock. Both in Kyle (1985) and in Back (1992, 1993), however, the non informational component of trades is exogenous, i.e., there are exogenous noise traders.

[19] Biais, Bossaerts and Rochet (1996) analyze optimal initial public offerings mechanisms using convex analysis techniques similar to those used in Rochet and Vila (1994).

REFERENCES

Amihud, Y, and H. Mendelson, 1980, Dealership market : market making with inventory, *Journal of Financial Economics*, 8, 31–53.

Aubin, J.P., 1979, *Mathematical Methods of Games and Economic Theory* (Amsterdam : North Holland).

Back, K., 1992, Insider trading in continuous time, *Review of Financial Studies*, 5, 387–410.

Back, K., 1993, Asymmetric information and options, *Review of Financial Studies*, 6, 435–472.

Bagnoli, B. and T. Bergstrom (1989), "Log-Concave probability and its applications", Discussion paper, University of Michigan.

Biais, B., 1993, "Price formation and equilibrium liquidity in fragmented and centralized markets", *Journal of Finance*, 48, 157–184.

Biais, B., P. Bossaerts and J. C. Rochet (1996), "An optimal IPO mechanism, " *Document de Travail, GREMAQ - IDEI.*

Biais, B., and P. Hillion, 1994, "Insider and liquidity trading in stock and options markets," *Review of Financial Studies*, 7, 743–780.

Biais, B., P. Hillion and C. Spatt, 1995, "An empirical analysis of the order book and order flow in the Paris Bourse," *Journal of Finance.*

Biais, B., D. Martimort and J.C. Rochet, 1996 "Competition for the supply of liquidity, risk sharing and asymmetric information" *Document de travail, GREMAQ - IDEI.*

Cho, I.K. and Kreps, D.M., 1987 "Signalling Games and Stable Equilibria", *Quarterly Journal of Economics* 102, 179-221.

Constantinides, G., 1986, Capital market equilibrium with transaction costs, *Journal of Political Economy*, 94, 842–862.

Cvitanic, J., 1996, Optimal trading under constraints, this volume.

Davis, M. and A. Norman, 1990, Portfolio selection with transaction costs, *Mathematics of Operations Research*, 15, 676–713.

Dumas, B., and E. Luciano, 1991, An exact solution to a dynamic portfolio choice problem under transaction costs, *Journal of Finance*, 46, 577–595.

Easley, D., and M. O'Hara, (1987), Trade size and information in securities markets, *Journal of Financial Economics*, 19, 69–90.

El Karoui, N., and M. Quenez, 1996, Non linear pricing theory and backward stochastic differential equations, this volume.

Fudenberg, D. and J. Tirole (1991) *Game Theory*, MIT Press.

Glosten, L., 1989, Insider trading, liquidity and the role of the monopolist specialist, *Journal of Business*, 62, 211–236.

Glosten, L., and P. Milgrom (1985), Bid, Ask and transaction prices in a specialist market with heterogeneously informed traders, *Journal of Financial Economics*, 13, 71–100

Grossman, S. and Stiglitz, J.E. (1980) "On the Impossibility of Informationally Efficient Markets", *American Economic Review* 70, 393-408.

Guesnerie, R., and J.C. Rochet, 1993, "Destabilizing properties of futures markets : an alternative view point" *European Economic Review, 37,5, 1043 - 1064.*

Ho, T., and H. Stoll, 1981, Optimal dealer pricing under transactions and return uncertainty, *Journal of Financial Economics*, 9, 47–73.

Ho, T. and H. Stoll, 1983, The dynamics of dealer markets under competition, *Journal of Finance*, 38, 1053–1074.

Jouini, E., 1996, Market imperfections, equilibrium and arbitrage, this volume.

Kinderlehrer, D. and G. Stampachia, 1980, *An Introduction to Variational Inequalities*, Academic Press, Boston.

Kyle, A.S., 1985, "Continuous Auctions and Insider Trading", *Econometrica*, 53, 1335-1355.

Kyle, A. S., 1989, Informed speculation with imperfect competition, *Review of Economic Studies*, 56, 317–355.

Leland, H., 1985, Option pricing and replication with transaction costs, *Journal of Finance*, 40, 1283–1301.

Magill, M, and G. Constantinides, 1976, Portfolio selection with transaction costs, *Journal of Economic Theory*, 13, 245–263.

Mas Colell, A., 1986, "The Price Equilibrium Existence Problem in Topological Vector lettices", *Econometrica*, 54, 1039-1053.

Meyer, P. A., 1966, *Probability and potentials*, Blaisdell Pub. Co., Waltham Massachussets.

O'Hara, M., 1995, *Market microstructure theory*, Blackwell, Cambridge, Massachussets.

Rochet, J.C. and P. Choné, 1996, "Ironing, Sweeping and Multidimensional Screening" *Document de travail, GREAMQ - IDEI*.

Rochet, J.C. and J.C. Vila, 1994, "Insider Trading without Normality", *Review of Economic Studies*, 61, 131-152.

Rockafellar, T., 1972, Convex analysis, Princeton University Press, Princeton.

Stiglitz, J., and L. Weiss, 1981, Credit rationing in markets with imperfect information, *American Economic Review,* 73 (5), 912–927.

Stoll, H., 1978, "The Supply of dealer services in security markets" *Journal of Finance,* 33, 1133 - 1151.

INTEREST RATE THEORY

Tomas Björk

Department of Finance
Stockholm School of Economics
Box 6501
S-113 83 Stockholm SWEDEN
e-mail: fintb@hhs.se

Contents

1 Preliminaries

1.1 Self financing portfolios

We take as given a filtered probability space $(\Omega, \mathcal{F}, P, \underline{\mathcal{F}})$ carrying a finite number of stochastic processes $S_0, \cdots, S_K$. We assume that all processes are semimartingales, and with very little loss of generality the reader can assume that the S-processes are defined by a system of stochastic differential equations, driven by a finite number of Wiener processes.

The above setup is intended as a model of a financial market with $K+1$ traded assets, where $S_i(t)$ is interpreted as the price of one unit of asset number i at time t. The assets will often be referred to as "stocks", but they could be stocks, bonds, options or any other traded financial securities.

The asset S_0 will play a particular role as "numeraire asset" and we assume henceforth that $S_0(t) > 0$ P-a.s. for all $t > 0$. As a convention we also assume that $S_0(0) = 1$. This numeraire asset can in principle be any asset, like a three year bond, an option, or a share of IBM stock. It is chosen for convenience, and in most cases (but not in all) it is chosen as a so called "risk free" asset, with P-dynamics of the form

$$\begin{cases} dS_0(t) &= r(t)S_0(t)dt, \\ S_0(0) &= 1, \end{cases} \tag{1.1}$$

where r is a given adapted stochastic process. One interpretation is that this describes a bank with the stochastic **short rate of interest** r. The asset S_0 in (1.1) is termed "locally risk free" since its infinitesimal rate of return $r(t)$ is known at time t. When S_0 is the risk free asset we often denote it by B instead of S_0.

Example 1.1 The simplest nontrivial example of a market is probably given by the **Black-Scholes model**. In this model we have a deterministic short rate of interest, and one stochastic asset S_1. Denoting, as is common, S_0 by B and S_1 by S, the model is defined as follows.

$$\begin{cases} dB(t) &= rB(t)dt, \\ dS(t) &= \alpha S(t)dt + \sigma S(t)d\tilde{W}(t), \end{cases} \tag{1.2}$$

where $\tilde{W}$ is a P-Wiener process. The filtration is defined as the internal filtration generated by $\tilde{W}$. The short rate of interest r, the local mean rate of return α, and the local volatility σ, are assumed to be known deterministic constants. The local rate of return on the stock is (formally) given by $\alpha + \sigma\dot{W}$, and we note that (as opposed to the risk free asset) it is stochastic.

We will study portfolios (trading strategies) based on the assets above, and to this end we need some definitions.

Definition 1.1

1. A **portfolio** *is any locally bounded predictable vector process*

$$h(t) = \left[h^0(t), \cdots, h^K(t)\right].$$

2. *The **value process** corresponding to a portfolio h is defined by*

$$V(t; h) = \sum_{i=0}^{K} h^i(t)S_i(t). \tag{1.3}$$

3. *A portfolio h is said to be **self financing** if the following hold*

$$dV(t; h) = \sum_{i=0}^{K} h^i(t)dS_i(t). \tag{1.4}$$

The interpretation of the first two items are fairly straightforward: $h^i(t)$ is the number of shares of type i held in the portfolio at time t, and $V(t; h)$ is the market value of the portfolio h at time t. The self financing condition is intended to formalize the intuitive idea of a trading strategy with no exogenous infusion or withdrawal of money, i.e. a strategy where the purchase of a new asset is financed solely by the sale of assets already in the portfolio. It is not at all self evident that the condition (1.4) really is an appropriate formalization of the intuitive notion of a self financing portfolio but, after having done a discrete time analysis, the definition turns out to be the natural one.

All prices above are interpreted as being given in terms of some a priori given **numeraire**, and typically this numeraire is a local currency like US\$. We may of course equally well express all prices in terms of some other numeraire, and we will in fact choose S_0.

Definition 1.2

1. *The **discounted price process** vector $Z(t) = [Z_0(t), \cdots, Z_K(t)]$ is defined by*

$$Z(t) = \frac{S(t)}{S_0(t)}. \tag{1.5}$$

2. *The **discounted value process** corresponding to a portfolio h is defined as*

$$V^Z(t; h) = \frac{V(t; h)}{S_0(t)}. \tag{1.6}$$

As is to be expected, the property of being self financing does not depend upon the particular choice of numeraire.

Lemma 1.1 *A portfolio h is self financing if and only if the following relation hold.*

$$dV^Z(t;h) = \sum_1^K h_i(t)dZ_i(t). \qquad (1.7)$$

Proof. Itô's formula. ∎

For some applications it is convenient to express a portfolio strategy in terms of the "relative weights" $[u^0, \cdots, u^K]$, where $u(t)$ is defined as the proportion of the portfolio value which at time t is invested in asset i.

Definition 1.3 *Consider a given portfolio h. The corresponding* **relative portfolio** *u is defined by*

$$u^i(t) = \frac{h^i(t)S_i(t-)}{V(t-;h)}.$$

From this definition we at once have the dynamics of a self financing portfolio expressed in terms of the relative weights.

Lemma 1.2 *If h is self financing then*

$$dV(t) = V(t-) \sum_{i=0}^K u^i(t) \frac{dS_i(t)}{S_i(t-)},$$

where u is the corresponding relative portfolio.

1.2 Contingent claims, arbitrage and martingale measures

The basic objects of study in arbitrage theory are the **financial derivatives**. These are assets which in some way are defined in terms of the **underlying** assets $S_0, \cdots, S_K$. Typical examples are options, forwards, futures, bonds, interest rate swaps, caps, and floors. The mathematical formalization of a derivative is straightforward.

Definition 1.4 *A contingent T-claim is any random variable $X \in L^0(\mathcal{F}_T, P)$ (i.e. an arbitrary $\mathcal{F}_T$-measurable random variable). We use the notation $L^0_+(\mathcal{F}_T, P)$ to denote the set of non-negative elements of $L^0(\mathcal{F}_T, P)$, and $L^0_{++}(\mathcal{F}_T, P)$ denotes the set of elements X of $L^0_+(\mathcal{F}_T, P)$ with $P(X > 0) > 0$.*

The interpretation of this definition is that the contingent claim is a contract which specifies that the stochastic amount, X, of money is to be payed out to the holder of the contract at time T. In the sequel we will often have to impose some further integrability restrictions upon the set of contingent claims under consideration.

Example 1.2 For the Black-Sholes model the most popular (and historically the oldest) example of a contingent claim is that of a **European call option** on S with **strike price** K and **date of expiration** T. This T-contract is defined by

$$X = \max[S(T) - K, \, 0],$$

and gives the owner the right, but not the obligation, to by one share of the stock at the prespecified price K, at time T.

The two main problems to be treated in this text are as follows:

- What is a "fair" price for a given contingent claim X?

- Suppose that you have sold the claim X. Then you have promised to pay the stochastic amount X at time T. How do you "hedge" yourself against this financial risk?

We will attack these problems by studying certain self financing portfolios, and in particular we will study "arbitrage portfolios" and "hedging portfolios".

Definition 1.5 *An* **arbitrage portfolio** *is a self financing portfolio h such that the corresponding value process has the properties*

(a) $V(0) = 0$,

(b) $V(T) \in L^0_{++}(\mathcal{F}_T, P)$.

If no arbitrage portfolios exist for any $T \in \mathbf{R}_+$ we say that the model is "free of arbitrage" or "arbitrage-free".

As can be seen from the definition, an arbitrage portfolio is a deterministic money-making machine, and we typically assume that there are no arbitrage possibilities in our market (this is in fact more or less the definition of an efficient market).

A typical example of a potential arbitrage opportunity ocurs if, apart from the usual risk free asset B above, we also have another riskfree asset (or portfolio) Y, with dynamics given by

$$dY(t) = k(t)Y(t), \tag{1.8}$$

where k is some adapted process. We then have the following simple observation, which will be used later on. The intuitive proof given below can easily be formalized.

Lemma 1.3 *Assume that the market is free of arbitrage. Then, with probability one it must hold (for almost all t) that*

$$k(t) = r(t).$$

Proof. If, on some interval, we have $k > r$, then we may obtain an arbitrage by borrowing in the bank and investing in Y. If $k < r$ then we sell Y short and invest in the bank. ∎

Our first project is to investigate when a given model is free of arbitrage, and the main tool is the concept of a martingale measure.

Definition 1.6 *We say that a probability measure is a* **martingale measure** *if*

1. $Q \sim P$

2. The discounted price process Z is a Q-local martingale.

The set of martingale measures is denoted by $\mathcal{P}$. If the discounted price process Z is a Q-martingale we say that Q is a **strong martingale measure.**

We will sometimes have use for the following simple characterization of a martingale measure, in the case when we use the money account B as the numeraire. The proof is an application of the Itô formula.

Lemma 1.4 *Assume that the numeraire asset is chosen as the money account B. Then a measure $Q \sim P$ is a martingale measure if and only if every asset price process S_i has price dynamics under Q of the form*

$$dS_i(t) = r(t)S_i(t)dt + dM_i(t),$$

where M_i is a Q-local martingale.

Example 1.3 In the Black-Scholes model it is easy to determine the class of martingale measures. Denoting $S(t)/B(t)$ by $Z(t)$ and applying the Itô formula we get

$$dZ = Z[\alpha - r]dt + Z\sigma d\tilde{W}.$$

We know that, because the filtration is the internal one, all absolutely continuous measure transformations must be of the Girsanov type, so we define the prospective martingale measure Q on $\mathcal{F}_T$ by $dQ = L(T)dP$, where

$$\begin{cases} dL(t) &= L(t)\varphi(t)d\tilde{W}(t), \\ L(0) &= 1. \end{cases} \tag{1.9}$$

Applying the Girsanov Theorem we have

$$d\tilde{W}(t) = \varphi(t)dt + dW(t),$$

where W is a Q-Wiener process. Thus the Q-dynamics for Z are given by

$$dZ = Z[\alpha - r + \sigma\varphi]dt + Z\sigma dW.$$

We see that for the Black-Scholes model there is in fact a unique martingale measure Q, with Girsanov kernel given by

$$\varphi = \frac{r - \alpha}{\sigma},$$

We note for further use that the Q-dynamics of S are given by

$$dS = rSdt + \sigma SdW. \tag{1.10}$$

We now go back to the general case, and we want to tie absence of arbitrage to the existence of a martingale measure. However, in order to do this we have to slightly modify our set of admissible portfolios.

Definition 1.7 *Consider a given martingale measure Q, and a self financing portfolio h. Then h is called Q -admissible if $V^Z(t; h)$ is a Q-martingale.*

Since by definition Z is a Q-martingale, and since the V^Z-process is the stochastic integral of h with respect to Z, we see that every sufficiently integrable self financing portfolio is in fact admissible. It is of course annoying that the definition of admissibility is dependent upon the particular choice of martingale measure, but the need of the admissibility condition can be seen in the proof of the following proposition, which is one of the basic results in the theory.

Proposition 1.1 *Assume that there exists a martingale measure Q. Then the model is free of arbitrage in the sense that there exists no Q-admissible arbitrage portfolio.*

Proof. Assume that h is an arbitrage portfolio with $V(T) \in L_{++}(\mathcal{F}_T, P)$. Then we also have $V^Z(T) \in L_{++}(\mathcal{F}_T, Q)$ and consequently we have

$$V(0) = V^Z(0) = E^Q \left[V^Z(T) \right] > 0$$

which contradicts the arbitrage condition $V(0) = 0$. ∎

We have seen that the existence of a martingale measure implies absence of arbitrage, and a natural question is whether there is a converse to this statement, i.e. if absence of arbitrage implies the existence of a martingale measure. For models in discrete time with a finite sample space Ω there is indeed such a result (see Harrison & Pliska [24]), which is based on the separation theorem for convex sets. For infinite models the situation is more complicated, basically due to the fact that the definition of admissibility depends on the choice of martingale measure.

This state of things is of course irritating, and a lot of work has been done in order to give an alternative to the standard definition of absence of arbitrage. This alternative should firstly be meaningful from an economic point of view, and secondly it should be equivalent to the existence of a martingale measure. A deep study of these problems can be found in Delbaen-Schachermayer [13]), where the notion of "no arbitrage" is replaced by the notion of "no free lunch with vanishing risk". The Delbaen-Schachermayer paper also contains an extensive bibliography.

Summing up we see that the theory in a sense is somewhat unsatisfactory, but the general consensus seems to be that the existence of a martingale measure is (informally) considered to be more or less equivalent to the absence of arbitrage. In this overview we will take this informal view, so when we speak of absence of arbitrage below we really mean that there exists a martingale measure.

As an immediate corollary of the general result above we see that, since there exists a martingale measure, the Black-Scholes model is free of arbitrage.

1.3 Hedging

In this subsection we assume absence of arbitrage, or to be more precise we assume that there exists a martingale measure. We now turn to the possibility of replicating a given contingent claim in terms of a portfolio based on the underlying assets.

Definition 1.8

1. *A fixed T-claim X is said to be* **attainable** *if there exists a self financing portfolio h such that the corresponding value process have the property that*

$$V(T; h) = X, \quad P - a.s. \tag{1.11}$$

2. *The market is said to be* **complete** *if every claim is attainable.*

Remark 1.1 As a matter of fact we will never, in this text, encounter a truly complete market. What we typically will be able to prove is that, for a given market, some reasonably large subset of of the set of contingent claims can be hedged. Thus the word "complete" will actually be used in an informal manner, and when it occurs as a part of a mathematical result it will be properly specified in each concrete situation.

The portfolio h above is called a **replicating** or **hedging** portfolio for the claim and we see that, from a purely financial point of view, the replicating portfolio is indistinguishable from the claim X.

The main problem is to determine which claims that can be hedged, and this is most conveniently carried out in terms of discounted prices. We have the following useful lemma, which shows that hedging is equivalent to the existence of a stochastic integral representation of the discounted claim.

Lemma 1.5 *Fix a martingale measure Q and assume that the discounted claim $X/S_0(T)$ is integrable. If the Q-martingale M, defined by*

$$M(t) = E^Q \left[\frac{X}{S_0(T)} \middle| \mathcal{F}_t \right], \tag{1.12}$$

admits an integral representation of the form

$$M(t) = x + \sum_{i=1}^{K} \int_0^t h^i(s) dZ_i(s), \tag{1.13}$$

where $h^1, \cdots, h^K$ are locally bounded and predictable, then X is attainable.

Proof. In terms of discounted prices, and using Lemma 1.1, we are looking for a process $h = (h^0, h^1, \cdots, h^K)$ such that

$$V^Z(T; h) = \frac{X}{S_0(T)}, \quad P - a.s. \tag{1.14}$$

$$dV^Z = \sum_{i=1}^{K} h^i dZ_i, \qquad (1.15)$$

where the discounted value process is given by

$$V^Z(t;h) = h^0(t) \cdot 1 + \sum_{i=1}^{K} h^i(t) Z_i(t). \qquad (1.16)$$

A reasonable guess is that $M = V^Z$, so we let M be defined by (1.12). Furthermore we define $(h^1, \cdots, h^K)$ by (1.13), and we define h^0 by

$$h^0(t) = M(t) - \sum_{i=1}^{K} h^i(t) Z_i(t). \qquad (1.17)$$

Now, from (1.16) we obviously have $M = V^Z$, and from (1.13) we get

$$dV^Z = dM = \sum_{i=1}^{K} h^i dZ_i,$$

which shows that the portfolio is self financing. Furthermore we have

$$V^Z(T;h) = M(T) = E^Q \left[\frac{X}{S_0(T)} \bigg| \mathcal{F}_T \right] = \frac{X}{S_0(T)},$$

which shows that X is replicated by h. ∎

We thus see that, modulo some integrability considerations, completeness is equivalent to the existence of a martingale representation theorem for the discounted price process. Thus we may draw on the results (see Jacod [28]) from semimartingale theory which connect martingale representation properties for Z with the set of extremal points of $\mathcal{P}$. We quote the following result from Harrison & Pliska [24].

Theorem 1.1 *If the martingale measure Q is unique, then the market is complete in the restricted sense that every claim X satisfying*

$$\frac{X}{S_0(T)} \in L^1(Q, \mathcal{F}_T)$$

is attainable.

Example 1.4 We consider again the Black-Scholes model, and for illustrative purposes we consider a T-claim $X \in L^2(\mathcal{F}_T, Q)$. Defining, as above, the martingale M by

$$M(t) = E^Q \left[e^{-rT} X \big| \mathcal{F}_t \right],$$

we know from the martingale representation theorem for Wiener processes that there exist an optional process g such that, under Q we have

$$M(t) = M(0) + \int_0^t g(s) dW(s).$$

On the other hand, the Q-dynamics of Z are given by

$$dZ = Z\sigma dW,$$

so in fact we have

$$M(t) = M(0) + \int_0^t h^1(s)dZ(s),$$

with $h^1 = \frac{g}{\sigma Z}$. We may now apply Lemma 1.5 to conclude that X is attainable using the portfolio defined by

$$
\begin{aligned}
h^1(t) &= g(t)/\sigma Z(t), \\
h^0(t) &= M(t) - h^1(t)Z(t).
\end{aligned}
$$

In the case of a so called **simple** contingent claim, i.e. an X of the form $X = \Phi(S(T))$, where Φ is some (sufficiently integrable) given real valued function, we can obtain even more explicit results. In this case we have

$$M(t) = E^Q\left[e^{-rT}\Phi(S(T))\bigg| \mathcal{F}_t\right],$$

and from the Kolmogorov backward equation (or from a Feynman-Kač representation) we have $M(t) = f(t, S(t))$ where f solves the boundary value problem

$$
\begin{cases}
\frac{\partial f}{\partial t}(t, s) + rs\frac{\partial f}{\partial s}(t, s) + \frac{1}{2}\sigma^2 s^2 \frac{\partial^2 f}{\partial s^2}(t, s) &= 0, \\
f(T, s) &= e^{-rT}\Phi(s).
\end{cases}
$$

Itô's formula now gives us

$$dM = \sigma S(t)\frac{\partial f}{\partial s}(t, S(t)),$$

so in terms of the notation above we have

$$g(t) = \sigma S(t) \cdot \frac{\partial f}{\partial s}(t, S(t)),$$

which gives us the replicating portfolio h as

$$
\begin{aligned}
h^0(t) &= f(t, S(t)) - S(t)\frac{\partial f}{\partial s}(t, S(t)), \\
h^1(t) &= B(t)\frac{\partial f}{\partial s}(t, S(t)).
\end{aligned}
$$

We have the interpretation $f(t, S(t)) = V^Z(t)$, but it is natural to express everything in terms of the undiscounted value process $V(t)$ rather than in terms of V^Z. Therefore we define $F(t, s)$ by $F(t, s) = e^{rt} f(t, s)$ which gives us the following result.

Proposition 1.2 *Consider the Black-Scholes model and a T-claim of the form $X = \Phi(S(T))$. Then X can be replicated by the portfolio*

$$
\begin{cases}
h^0(t) &= \frac{F(t, S(t)) - S(t)\frac{\partial F}{\partial s}(t, S(t))}{B(t)}, \\
h^1(t) &= \frac{\partial F}{\partial s}(t, S(t)),
\end{cases}
\tag{1.18}
$$

*where F solves the **Black-Scholes equation***

$$\begin{cases} \frac{\partial F}{\partial t}(t,s) + rs\frac{\partial F}{\partial s}(t,s) + \frac{1}{2}\sigma^2 s^2 \frac{\partial^2 F}{\partial s^2}(t,s) - rF(t,s) &= 0, \\ F(T,s) &= \Phi(s). \end{cases} \qquad (1.19)$$

Furthermore the value process for the replicating portfolio is given by

$$V(t;h) = F(t, S(t)).$$

1.4 Completeness - Absence of Arbitrage

From the previous results we know that the model is free of arbitrage if there exists a martingale measure, and that the model is complete if the martingale measure is unique. In this subsection we will give some general rules of thumb for quickly determining whether a certain model is complete and/or free of arbitrage. The arguments will be purely heuristic.

Let us assume that the price processes of the underlying assets are driven by R "random sources". We can not give a precise definition of what constitutes a "random source" here, but the typical example is a driving Wiener process. If we e.g. have 5 independent Wiener processes driving our prices, then $R = 5$. Another example of a random source would be a counting process such as e.g. a Poisson process. In this context it is important to note that if the prices are driven by a point process with different jump sizes, then the appropriate number of random sources equals the number of different jump sizes.

When discussing completeness and absence of arbitrage it is important to realize that these concepts work in "opposite directions". Let e.g. the number of random sources R be fixed. Then every new underlying asset added to the model (without increasing R) will of course give us a potential opportunity of creating an arbitrage portfolio so in order to have an arbitrage free market the number K of underlying assets (apart from the numeraire asset) must be small in comparison with the number of random sources R.

On the other hand we see that every new underlying asset added to the model gives us new possibilities of replicating a given contingent claim, so completeness requires M to be large in comparison with R.

We cannot formulate and prove a precise result here, but the following rule of thumb, or "Meta-Theorem" is nevertheless extremely useful. In concrete cases it can in fact be given a precise formulation and a precise proof. We will later use the meta-theorem when dealing with problems connected with non-traded underlying assets in general and interest rate theory in particular.

Meta-Theorem 1.1 *Let K denote the number of underlying assets in the model excluding the numeraire asset, and let R denote the number of random sources. Generically we then have the following relations.*

1. The model is arbitrage free if and only if $K \le R$.

2. The model is complete if and only if $K \ge R$.

9. The model is complete and arbitrage free if and only if $K = R$.

As an example we take the Black-Scholes model, where we have one underlying asset S plus the risk free asset so $K = 1$. We have one driving Wiener process, giving us $R = 1$ so in fact $K = R$. Using the Meta-Theorem above we thus expect the Black-Scholes model to be arbitrage free as well as complete and this is indeed the case.

1.5   Pricing

We now turn to the problem of determining a "reasonable" price process $\Pi(t, X)$ for a fixed contingent T-claim X. We assume that $\mathcal{P} \neq \emptyset$, i.e that there exists a martingale measure. There are two main approaches:

- The derivative should be priced in a way that is **consistent** with the prices of the underlying assets. More precisely we should demand that the extended market $\Pi(\ ,X), S_0, S_1, \cdots, S_K$ is free of arbitrage possibilities.

- If the claim is **attainable**, with hedging portfolio h, then the only reasonable price is given by $\Pi(t, X) = V(t; h)$.

To keep things simple let us, in the first approach above, more specifically demand that there should exist a strong martingale measure for the extended market $[\Pi(X), S_0, S_1, \cdots, S_K]$. Letting Q denote such a measure and applying the definition of a strong martingale measure we obtain

$$\frac{\Pi(t, X)}{S_0(t)} = E^Q \left[\frac{\Pi(T, X)}{S_0(T)} \middle| \mathcal{F}_t \right] = E^Q \left[\frac{X}{S_0(T)} \middle| \mathcal{F}_t \right] \tag{1.20}$$

We thus have the pricing formula

$$\Pi(t, X) = S_0(t) E^Q \left[\frac{X}{S_0(T)} \middle| \mathcal{F}_t \right], \tag{1.21}$$

where Q is a martingale measure for the a priori given market $S_0, S_1, \cdots, S_K$. Note that different choices of Q will generically give rise to different price processes.

For the second approach to pricing let us assume that X can be replicated by h. Since the holding of the derivative contract and the holding of the replicating portfolio are equivalent from a financial point of view, we see that price of the derivative must be given by the formula

$$\Pi(t, X) = V(t; h). \tag{1.22}$$

One problem here is what will happen in a case when X can be replicated by two different portfolios, and one would also like to know how this formula is connected to (1.21).

Defining $\Pi(t, X)$ by (1.22) we see that $\Pi(t, X) / S_0(t) = V^Z(t)$ and since , assuming enough integrability, V^Z is a Q-martingale, we see that also $\Pi(t, X) / S_0(t)$

is a Q-martingale. Thus we again obtain the formula (1.21) and for a attainable claim we have in particular the formula

$$V(t; h) = S_0(t) E^Q \left[\left. \frac{X}{S_0(T)} \right| \mathcal{F}_t \right],$$
(1.23)

which will hold for any replicating portfolio and for any martingale measure Q. Thus we have the following result.

Proposition 1.3

1. *If we assume the existence of a strong martingale measure for the market* $[\Pi(X), S_0, S_1, \cdots, S_K]$ *then* X *must be priced according to the formula*

$$\Pi(t, X) = S_0(t) E^Q \left[\left. \frac{X}{S_0(T)} \right| \mathcal{F}_t \right],$$
(1.24)

where Q *is a strong martingale measure for* $[S_0, S_1, \cdots, S_K]$.

2. *Different choices of* Q *will generically give rise to different price processes, but if* X *is attainable then all choices of* Q *will produce the same price process.*

3. *For an attainable claim the price process will also be given by the formula*

$$\Pi(t, X) = V(t; h),$$
(1.25)

where h *is the hedging portfolio. Different choices of hedging portfolios (if such exist) will produce the same price process.*

Summing up we see that in a complete market the price of any derivative will be **uniquely** determined by the requirement of absence of arbitrage. The price is unique precisely because the derivative is in a sense superfluous - it can equally well be replaced by its replicating portfolio. In particular we see that the price does not depend on any assumptions made about the risk-preferences of the agents in the market. The agents can have any attitude towards risk, as long as they prefer more (deterministic) money to less.

In an incomplete market the requirement of no arbitrage is no longer sufficient to determine a unique price for the derivative. We have several martingale measures, all of which can be used to price derivatives in a way consistent with no arbitrage. The question which martingale measure one should use for pricing has a very simple answer: The martingale measure is chosen by the market.

Schematically speaking the price of a derivative is thus determined by two major factors.

1. We require that the derivative should be priced in such a way as to not introduce arbitrage possibilities into the market. This requirement is reflected by the fact that all derivatives must be priced by formula (1.24) where the same Q is used for all derivatives.

2. In an incomplete market the price is also partly determined by aggregate supply and demand on the market. Supply and demand for a specific derivative are in turn determined by the aggregate risk aversion on the market, as well as by liquidity considerations and other factors. All these aspects are aggregated into the particular martingale measure used by the market.

Example 1.5 For the Black-Scholes model and a simple claim, the pricing problem is in fact already solved since we have earlier determined the value process for the replicating portfolio. Using Proposition 1.2 we have the following classic result.

Proposition 1.4 *Consider the Black-Scholes model and a square integrable T-claim X. Then the price process* $\Pi\,(t, X)$ *is given by*

$$\Pi\,(t, X) = e^{-r(T-t)} E^Q\,[X|\,\mathcal{F}_t]\,, \tag{1.26}$$

where the Q-dynamics of S are given by

$$dS = rSdt + \sigma SdW. \tag{1.27}$$

If furthermore X is of the form $X = \Phi(S(T))$, *then the price process is is also given by* $\Pi\,(t, X) = F(t, S(t))$ *where F solves the Black-Scholes equation*

$$\begin{cases} \frac{\partial F}{\partial t}(t, s) + rs\frac{\partial F}{\partial s}(t, s) + \frac{1}{2}\sigma^2 s^2 \frac{\partial^2 F}{\partial s^2}(t, s) - rF(t, s) &= 0, \\ F(T, s) &= \Phi(s). \end{cases} \tag{1.28}$$

In particular we may use any of the results above in order to abtain the price of a European call. The easiest way is in fact to compute the expected value in (1.26), rather than to solve the PDE (1.28).

Proposition 1.5 *The price of a European call option with strike price K and time of maturity T is given by the formula* $\Pi\,(t) = F(t, S(t))$, *where*

$$F(t, s) = sN\,[d_1(t, s)] - e^{-r(T-t)} KN\,[d_2(t, s)]\,. \tag{1.29}$$

Here N is the cumulative distribution function for the $N\,[0, 1]$-*distribution and*

$$d_1(t, s) = \frac{1}{\sigma\sqrt{T-t}} \left\{ \ln\left(\frac{s}{K}\right) + \left(r + \frac{1}{2}\sigma^2\right)(T-t) \right\}, \tag{1.30}$$

$$d_2(t, s) = d_1(t, s) - \sigma\sqrt{T-t}. \tag{1.31}$$

2 The bond market

2.1 Generalities

We again take as given a filtered probability space $(\Omega, \mathcal{F}, P, \underline{\mathcal{F}})$, and the market we will study is mainly the market of zero coupon bonds.

Definition 2.1 *A zero coupon bond with maturity date T, also called a T-bond, is a contract which guarantees the holder 1 \$ to be paid on the date T. The price at time t of a bond with maturity date T is denoted by $p(t, T)$.*

We now make an assumption to guarantee the existence of a sufficiently rich bond market.

Assumption 2.1 *We assume that*

1. *There exists a (frictionless) market for T-bonds for every $T > 0$.*

2. *For every fixed T, the process $\{p(t, T); \ 0 \leq t \leq T\}$ is an optional stochastic process with $p(t, t) = 1$ for all t.*

3. *For every fixed t, $p(t, T)$ is P-a.s. continuously differentiable in the T-variable. This partial derivative is often denoted by*

$$p_T(t, T) = \frac{\partial p(t, T)}{\partial T}.$$

Given the bond market above, we may now define a number of interest rates, and the basic construction is as follows. Suppose that we are standing at time t, and let us fix two other points in time S and T with $t < S < T$. The immediate project is to write a contract at time t which allows us to make an investment of 1\$ at time S, and to have a **deterministic** rate of return, determined at the contract time t, over the interval $[S, T]$. This can easily be achieved as follows.

1. At time t we sell one S-bond. This will give us $p(t, S)$\$.

2. For this money we can buy exactly $p(t, S)/p(t, T)$ T-bonds. Thus our net investment at time t equals zero.

3. At time S the S-bond matures, so we are obliged to pay out 1\$.

4. At time T the T-bonds mature at 1\$ a piece, so we will receive the amount $p(t, S)/p(t, T) \cdot 1\$$.

5. The net effect of all this is that, based on a contract at t, an investment of 1\$ at time S has yielded $p(t, S)/p(t, T)$\$ at time T. We now determine the equivalent constant short rate of interest over this period as the solution R to the equation

$$e^{R \cdot (T-S)} \cdot 1 = \frac{p(t, S)}{p(t, T)}$$

Based on this argument we proceed to the formal definitions.

Definition 2.2

1. *The **forward rate** for $[S, T]$ contracted at t is defined as*

$$R(t; S, T) = -\frac{\log p(t, T) - \log p(t, S)}{T - S}.$$

2. The **spot rate**, $R(S, T)$, for the period $[S, T]$ is defined as

$$R(S, T) = R(S; S, T).$$

3. The **instantaneous forward rate with maturity** T, **contracted at** t is defined by

$$f(t, T) = -\frac{\partial \log p(t, T)}{\partial T}.$$

4. The **instantaneous short rate at time** t is defined by

$$r(t) = f(t, t).$$

We now go on to define the money account process B.

Definition 2.3 The **money account** process is defined by

$$B_t = \exp\left\{ \int_0^t r(s)ds \right\},$$

i.e.

$$\begin{cases} dB(t) &= r(t)B(t)dt, \\ B(0) &= 1. \end{cases}$$

The interpetation of the money account is the same as before, i.e you may think of it as describing a bank with a stochastic short rate of interest. It can also be shown (see a discussion the end of subsection 2.2), that investing in the money account is equivalent to investing in a self financing "rolling over " trading strategy, which at each time t consists entirely of 'just maturing" bonds, i.e. bonds which will mature at $t + dt$.

As an immediate consequence of the definitions we have the following useful formulas.

Lemma 2.1 For $t \le s \le T$ we have

$$p(t, T) = p(t, s) \cdot \exp\left\{ -\int_s^T f(t, u)du \right\},$$

and in particular

$$p(t, T) = \exp\left\{ -\int_t^T f(t, s)ds \right\}.$$

2.2 Relations between df, dp, and dr

In this subsection we shall specialize to the case where all randomness is generated by a (vector) Wiener process W, and we will present a small "toolbox" of results which will come in handy below. We follow [3], [4] and [25].

We will, either by implication or by assumption, consider dynamics of the following type.

Short rate dynamics

$$dr(t) = a(t)dt + b(t)dW(t), \qquad (2.1)$$

Bond price dynamics

$$dp(t,T) = p(t,T)m(t,T)dt + p(t,T)v(t,T)dW(t), \qquad (2.2)$$

Forward rate dynamics

$$df(t,T) = \alpha(t,T)dt + \sigma(t,T)dW_t. \qquad (2.3)$$

In the above formulas the coefficients are assumed to meet standard conditions required to guarantee that the various processes are well defined.

We will now study the formal relations which must hold between bond prices and interest rates. These relations hold regardless of the measure under consideration, and in particular we do *not* assume that markets are free of arbitrage. We shall, however, need a number of technical assumptions which we collect below in an "operational" manner.

Assumption 2.2

1. *For each fixed ω, t all the objects $m(t,T)$, $v(t,T)$, $\alpha(t,T)$, and $\sigma(t,T)$ are assumed to be continuously differentiable in the T-variable. This partial T-derivative sometimes is denoted by $m_T(t,T)$ etc.*

2. *All processes are assumed to be regular enough to allow us to differentiate under the integral sign as well as to interchange the order of integration.*

3. *For any t the price curves $p(\omega, t, \cdot)$ are bounded for almost all ω.*

This assumption is rather *ad hoc* and one would, of course, like to give conditions which *imply* the desired properties above. This can be done but at a fairly high price as to technical complexity. For the stochastic Fubini theorem for the interchange of integration with respect to dW and dt see Protter [44] or Heath–Jarrow–Morton [25].

Proposition 2.1

1. *If $p(t,T)$ satisfies (2.2), then for the forward rate dynamics we have*

$$df(t,T) = \alpha(t,T)dt + \sigma(t,T)dW(t),$$

where α, and σ are given by

$$\begin{cases} \alpha(t,T) &= v_T(t,T) \cdot v(t,T) - m_T(t,T), \\ \sigma(t,T) &= -v_T(t,T). \end{cases} \qquad (2.4)$$

2. *If $f(t,T)$ satisfies (2.3) then the short rate satisfies*

$$dr(t) = a(t)dt + b(t)dW(t),$$

where

$$\begin{cases} a(t) &= f_T(t,t) + \alpha(t,t), \\ b(t) &= \sigma(t,t). \end{cases} \tag{2.5}$$

3. *If $f(t,T)$ satisfies (2.3) then $p(t,T)$ satisfies*

$$dp(t,T) = p(t,T)\left\{ r(t) + A(t,T) + \frac{1}{2}\|S(t,T)\|^2 \right\}dt + p(t,T)S(t,T)dW(t),$$

where

$$\begin{cases} A(t,T) &= -\int_t^T \alpha(t,s)ds, \\ \\ S(t,T) &= -\int_t^T \sigma(t,s)ds. \end{cases} \tag{2.6}$$

Proof. The first part of the Proposition follows immediately if we apply the Itô formula to the process $\log p(t,T)$, write this in integrated form and differentiate with respect to T.

For the second part we integrate the forward rate dynamics to get

$$r(t) = f(0,t) + \int_0^t \alpha(s,t)ds + \int_0^t \sigma(s,t)dW(s). \tag{2.7}$$

Now we can write

$$\alpha(s,t) = \alpha(s,s) + \int_s^t \alpha_T(s,u)du,$$

$$\sigma(s,t) = \sigma(s,s) + \int_s^t \sigma_T(s,u)du,$$

and, inserting this into (2.7) we have

$$r(t) = f(0,t) + \int_0^t \alpha(s,s)ds + \int_0^t \int_s^t \alpha_T(s,u)duds$$

$$+ \int_0^t \sigma(s,s)dW_s + \int_0^t \int_s^t \sigma_T(s,u)dudW_s.$$

Changing the order of integration and identifying terms gives us the result.

For the third part we adapt a technique from Heath–Jarrow–Morton [25]. Using the definition of the forward rates we may write

$$p(t,T) = \exp\{Y(t,T)\} \tag{2.8}$$

where Y is given by

$$Y(t,T) = -\int_t^T f(t,s)ds. \tag{2.9}$$

Writing (2.3) in integrated form, we obtain

$$f(t,s) = f(0,s) + \int_0^t \alpha(u,s)du + \int_0^t \sigma(u,s)dW(u).$$

Inserting this expression into (2.9), splitting the integrals and changing the order of integration gives us

$$
\begin{aligned}
Y(t,T) &= -\int_t^T f(0,s)ds - \int_0^t \int_t^T \alpha(u,s)dsdu - \int_0^t \int_t^T \sigma(u,s)dsdW(u) \\
&= -\int_0^T f(0,s)ds - \int_0^t \int_u^T \alpha(u,s)dsdu - \int_0^t \int_u^T \sigma(u,s)dsdW(u) \\
&\quad + \int_0^t f(0,s)ds + \int_0^t \int_u^t \alpha(u,s)dsdu + \int_0^t \int_u^t \sigma(u,s)dsdW(u) \\
&= Y(0,T) - \int_0^t \int_u^T \alpha(u,s)dsdu - \int_0^t \int_u^T \sigma(u,s)dsdW(u) \\
&\quad + \int_0^t f(0,s)ds + \int_0^t \int_0^s \alpha(u,s)duds + \int_0^t \int_0^s \sigma(u,s)dW(u)ds
\end{aligned}
$$

Now we can use the fact that $r(s) = f(s,s)$ and, integrating the forward rate dynamics (2.3) over the interval $[0, s]$, we see that the last two lines above equal $\int_0^t r(s)ds$ so we finally obtain

$$Y(t,T) = Y(0,T) + \int_0^t r(s)ds - \int_0^t \int_u^T \alpha(u,s)dsdu - \int_0^t \int_u^T \sigma(u,s)dsdW(u). \tag{2.10}$$

Thus, with A and S as in the statement of the proposition, the stochastic differential of Z is given by

$$dY(t,T) = \{r(t) + A(t,T)\}\, dt + S(t,T)dW(t),$$

and an application of the Itô formula to the process $p(t,T) = \exp\{Y(t,T)\}$ completes the proof. ∎

We end the subsection by giving a heuristic proof of the claim made earlier, that the existence of the locally risk free asset B is implied by the possibility of rolling over just-maturing bonds. Such a strategy is defined by the fact that, at every instant of time t, the entire value of the portfolio is invested in just-maturing bonds (i.e. bonds maturing at $t + dt$). Denoting the portfolio weights, at time t, for bonds with maturity T, by $u(t,T)$ we thus have $u(t,T) = \delta(T-t)$ (with the Kronecker delta), and the dynamics of the corresponding value process are given by Lemma 1.2 as

$$dV(t) = V(t) \cdot u(t,t) \cdot \frac{dp(t,t)}{p(t,t)}. \tag{2.11}$$

From Proposition 2.1 we have

$$dp(t,t) = p(t,t)\left\{r(t) + A(t,t) + \frac{1}{2}\|S(t,t)\|^2\right\}dt + p(t,t)S(t,t)dW(t).$$

By definition $p(t,t) = 1$ and from (2.6) we see that $S(t,t) = A(t,t) = 0$, so in fact $dp(t,t) = r(t)dt$. We thus obtain

$$dV(t) = V(t)r(t)dt, \tag{2.12}$$

which we recognize as the dynamics of the money account.

It is worth noticing that this argument is not quite precise, and the reason is that the rolling over portfolio strategy is not really included in the general framwork of portfolios outlined in Section 1. In the general theory we assumed that there were only a finite number of traded assets in our portfolios, but this is no longer the case with the rolling over strategy. At each point of time it contains only one asset, namely the just-maturing bond, but *over an arbitrarily short time interval the strategy uses an infinite number of assets.*

This type of portfolio is thus formally not covered by our earlier theory, and we see that for the theory of bonds there is a clear need of extending the definition of a portfolio. A trading strategy on the bond market would then be naturally defined as a pair of processes $\{g(t), h(t, dT)\}$, where $g(t)$ denotes the number of units of the risk free asset held in the portfolio at time t, whereas $g(t, dT)$ denotes the number of bonds with maturities in the interval $[T, T + dT]$ which are being held in the portfolio at time t. This is the approach taken in [4]-[5], and we will come back to these issues below.

2.3   Bond pricing and martingale measures

We shall now introduce martingale measures into the bond market model above, and in order to do this we have to specify our choice of numeraire. The obvious and most common choice is to use the risk free asset B as numeraire. This is of course not the only possible choice, and later on we will in fact also use the price process for a T-bond (for a fixed T) as numeraire. In order to keep track of the choice of numeraire we add a prefix to the notion of a martingale measue.

Definition 2.4 *Take the filtered space $(\Omega, \mathcal{F}, P, \underline{\mathcal{F}})$ and the bond market as given. A martingale measure Q is called a* **risk neutral** *martingale measure if it uses the risk free asset as numeraire, i.e. if for every fixed T, the process*

$$Z(t,T) = \frac{p(t,T)}{B(t)}, \quad 0 \le t \le T, \tag{2.13}$$

is a Q-martingale.

We immediately have the following basic pricing formula.

Proposition 2.2 *Consider a fixed T-claim X, and assume that Q is a risk neutral martingale measure. Then the price process $\Pi(t, X)$ is given by*

$$\Pi(t, x) = E^Q \left[X \cdot e^{-\int_t^T r(s)ds} \,\middle|\, \mathcal{F}_t \right]. \tag{2.14}$$

In particular the price process for a T-bond is given by

$$p(t, T) = E^Q \left[e^{-\int_t^T r(s)ds} \,\middle|\, \mathcal{F}_t \right]. \tag{2.15}$$

Proof. Using B as S_0 the result follows at once from Proposition 1.3. ∎

There is an important moral to be learned from these formulas.

- The dynamics of prices and interest rates under the objective probability measure P are, to a high degree, irrelevant. The important objects to study are the dynamics of prices and interest rates *under the relevant martingale measure Q*.

When building a model it is thus natural, and in most cases time saving, to model all objects directly under a martingale measure Q. This approach is called "martingale modelling", and as the reader will see in the forthcoming sections, it will be the main approach of the present study. The drawback of martingale modelling is mainly one of parameter estimation, which we will return to below.

Assuming that all randomness is generated by a Wiener process we may also derive the structural form (2.2) for the bond dynamics.

Proposition 2.3 *Assume that the filtration is generated by a P-Wiener process $\tilde{W}$, and that Q is a risk neutral martingale measure. Then there exist optional processes $m(t, T)$ and $v(t, T)$ such that the P dynamics for $p(t, T)$ are given by*

$$dp(t, T) = m(t, T)p(t, T)dt + p(t, T)v(t, T)d\tilde{W}(t). \tag{2.16}$$

Furthermore, Q-dynamics are given by

$$dp(t, T) = r(t)p(t, T)dt + p(t, T)v(t, T)dW(t). \tag{2.17}$$

Proof. We know by definition that, for any fixed T,

$$Z(t, T) = \frac{p(t, T)}{B(t)},$$

is a Q-martingale. Since the filtration is generated by a Wiener process we also know that Q is obtained from P by a Girsanov transformation with a likelihood process of the form

$$\begin{cases} dL(t) & = & L(t)g(t)dW(t), \\ L(0) & = & 1. \end{cases}$$

Thus W is a Q-Wiener process , where

$$d\tilde{W}(t) = g(t)dt + dW(t),$$

and we may use a martingale representation theorem in order to deduce the existence of an optional process $v(t,T)$ such that

$$dZ(t,T) = Z(t,T)v(t,T)dW(t).$$

From Itô's formula and the relation $p(t,T) = B(t)Z(t,T)$ we now obtain (2.17), and from this result and the Girsanov Theorem we obtain the P dynamics as

$$dp(t,T) = [r(t) - g(t)v(t,T)]\, p(t,T)dt + p(t,T)v(t,T)d\tilde{W}(t),$$

which proves (2.16). ∎

2.4   Swaps

We end this section by analyzing one of the most used interest rate derivatives - the interest rate swap. The content of such a contract is basically that of exchanging a floating interest rate for a fixed interest rate. Swaps come in many forms, but we will confine ourselves to **Forward swap settled in arrears**. We follow [41], which contains an encyclopedical treatment of interest rate theory, and where also a number of other swap contracts are discussed.

Consider a fixed time t, which is the time at which the contract is made. Furthermore we fix a sequence of equally distanced points $T_0 < T_1 < \cdots T_n$, where we use the notation $\delta = T_{i+1} - T_i$, $i = 0, \cdots, n-1$. We denote the prespecified **fixed rate of interest** by R, and the **notional amount** by K.

A swap contract with K and R fixed for the period $T_0, \cdots, T_n$ is now defined as a sequence of payments, where the amount of money paid out at time $T_{i+1}, i = 0, \cdots, n-1$ is defined by

$$X_{i+1} = K\delta\left(L(T_i) - R\right).$$

Here the floating rate $L(T_i)$ which is working over the period $[T_i, T_{i+1}]$ is quoted as a **simple rate**, and defined by

$$p(T_i, T_{i+1}) = \frac{1}{1 + \delta L(T_i)}.$$

Without loss of generality we may confine ourselves to the case where the nominal amount $K = 1$. Using the above definitions, the price, at t, of the entire swap contract is then given by

$$\Pi(t) = \sum_{i=1}^{n} E^Q\left[e^{-\int_t^{T_i} r(s)ds}\delta\left(L(T_i) - R\right)\bigg|\mathcal{F}_t\right]$$

$$= \sum_{i=1}^{n} E^{Q} \left[e^{-\int_{t}^{T_i} r(s)ds} \left(\frac{1}{p(T_{i-1}, T_i)} - (1 + \delta R) \right) \middle| \mathcal{F}_t \right]$$

$$= \sum_{i=1}^{n} E^{Q} \left[e^{-\int_{t}^{T_{i-1}} r(s)ds} E^{Q} \left[e^{-\int_{T_{i-1}}^{T_i} r(s)ds} \middle| \mathcal{F}_{T_{i-1}} \right] \left(\frac{1}{p(T_{i-1}, T_i)} - (1 + \delta R) \right) \middle| \mathcal{F}_t \right]$$

$$= \sum_{i=1}^{n} [p(t, T_{i-1}) - (1 + \delta R)p(t, T_i)].$$

Thus we have the price of the swap as

$$\Pi(t) = p(t, T_0) - \sum_{i=1}^{n} c_i p(t, T_i),$$

where $c_i = R\delta$ for $i = 1, \cdots, n-1$ and $c_n = 1 + \delta R$.

We notice that the entire swap contract can be priced with only the knowledge of the bond prices at the time t, and it is easily seen that the swap contract can be replicated by a portfolio of bonds.

Above we took the fixed rate of interest R as given. We may now go on to define the **swap rate** for the period above, contracted at t as the value of R which, at the time of contract t, gives the price zero to the swap contract. Using the formulas above the swap rate is easily determinded as

$$\hat{R} = \frac{p(t, T_0) - p(t, T_n)}{\delta \sum_{i=1}^{n} p(t, T_i)}.$$

3   Short rate models

3.1   Generalities

This section is entirely devoted to models of the bond market where the only explanatory variable is the short rate of interest r. This is of course a very limited subclass of the family of bond market models, but historically it is the oldest approach, and it also has some interesting computational features. In particular it turns out that a considerable amount of pricing and hedging theory can be carried out within a PDE framework.

In order to get a feeling of things, let us assume that the short rate is modelled, under the objective probability measure P, as

$$dr(t) = \mu(t, r(t)) \, dt + \sigma(t, r(t)) \, d\tilde{W}(t), \tag{3.1}$$

where $\mu(t, r)$ and $\sigma(t, r)$ are given real valued functions, which are assumed to be regular enough to allow the SDE (3.1) to have a unique strong solution. As usual $\tilde{W}$ is a P-Wiener process. The filtration is by definition the internal one generated by $\tilde{W}$. In order to avoid confusion we have to be rather clear about which objects are given a priori, as opposed to the objects that are derived endogenously.

Assumption 3.1 *We consider as given the short rate of interest r with dynamics given by (3.1). The only a priori given price process on the market is the locally risk free asset B, with*

$$
\begin{cases}
dB(t) &= r(t)B(t)dt, \\
B(0) &= 1.
\end{cases}
$$

It is natural to view bonds as derivatives, with the short rate of interest as the underlying object, in the same way that options are derivatives of the stock price process in the Black-Scholes model. The first natural question to ask is then whether bond prices are uniquely determined by a specification of the P-dynamics of r as in (3.1), plus a requirement that the bond market should be free of arbitrage.

The answer to this question is emphatically no, and the reason is that our a priori specified market is not complete. The only exogenously given asset is the risk free one, so the only portfolios we are allowed to form consist in putting all our money into the bank, and then to sit back and wait passively. In particular there is no possiblity of replicating any interesting contingent claim, not even the simple claim associated with a zero coupon bond.

This lack of completeness can also be seen from the fact that the martingale measure for the a priori given market is not unique. On the contrary, since in this case the discounted price process is the trivial process

$$
Z_0(t) = \frac{B(t)}{B(t)} \equiv 1,
$$

we have the following immediate observation.

- For the market specified above, **every** measure Q with $Q \sim P$ is a martingale measure.

- More precisely, every $Q \sim P$ will generate an arbitrage free bond market, with bond prices defined by

$$
p(t,T) = E^Q \left[e^{-\int_t^T r(s)ds} \middle| \mathcal{F}_t \right].
$$

We may also compare with the Meta-Theorem 1.1. In the present case we have $K = 0$ whereas $R = 1$, so we should expect the market to be free of arbitrage but incomplete. Comparing with the Black-Scholes model we see that in both models we have one source of randomness and one state process. The important difference between the models is the fact that, as opposed to S, *the short rate r is **not** the price of a traded asset.*

Summing up:

- The price of a particular bond is *not* uniquely determined by a specification of the P-dynamics of the short rate of interest, plus the requirement of an arbitrage free bond market.

- This is due to the fact that arbitrage pricing is a matter of pricing a derivative *in terms of* a given underlying price process. In the a priori given market above we do not have a sufficiently rich family of underlying assets.

Thus we can say very little about the price of a particular bond. The situation is however not hopeless. We see (at least informally) from the Meta-Theorem that:

- Bonds of *different* maturities must satisfy certain *internal concistency* relations in order to avoid arbitrage possibilities.

- If we include one single bond in the exogenoulsy given market then we ought to be able to price all other bonds in terms of this "benchmark bond".

3.2   Locally risk free portfolios

This subsection constitutes a digression. The object is to use the informal ideas from the previous subsection in order to derive bond pricing results, by constructing so called locally risk free portfolios. This approach is historically the oldest one, and it gives some valuable insights, but from a logical point of view it can be skipped.

We keep all assumptions from the previous subsection and we add one new assumption.

Assumption 3.2 *We assume that there is an arbitrage free market for T-bonds for all possible values of T and the short rate of interest is assumed to have the P dynamics given by*

$$dr(t) = \mu\left(t, r(t)\right) dt + \sigma\left(t, r(t)\right) d\tilde{W}(t),$$

The price process of a T-bond is furthermore assumed to be of the form

$$p(t, T) = F\left(t, r(t); T\right), \tag{3.2}$$

where F is a smooth function of three variables.

We will often write $F^T(t, r)$ instead of $F(t, r; T)$ and our project is to find out what the relations must be between pricing functions for different maturities. To this end we will carry out the following program.

1. Fix two maturities S and T, and then form a portfolio consisting of T-bonds and S-bonds.

2. Since by the assumption above, the bond prices will be perfectly correlated, we may by choosing the relative weights appropriately be able to form a portfolio with no driving Wiener noise. Thus the value dynamics of the profolio will be of the form

$$dV(t) = k(t)V(t)dt.$$

3. SHORT RATE MODELS

3. This means that we have created a "synthetic bank" with $k(t)$ as its short rate of interest. In order to avoid arbitrage (se Lemma 1.3) we must now have the following relation P-a.s.

$$k(t) = r(t), \quad \text{for all } t,$$

and this condition will in fact turn out to have the form of a PDE.

We now carry out this program, and using Itô's formula we obtain the T-bond dynamics as

$$dF^T = F^T \alpha_T dt + F^T \sigma_T d\tilde{W}, \tag{3.3}$$

where

$$\begin{cases} \alpha_T = \frac{F_t^T + \mu F_r^T + \frac{1}{2}\sigma^2 F_{rr}^T}{F^T}, \\ \sigma_T = \frac{\sigma F_r^T}{F^T}. \end{cases} \tag{3.4}$$

Denoting the relative portfolio by (u^S, u^T) we have the following value dynamics for our portfolio.

$$dV = V \left\{ u^T \frac{dF^T}{F^T} + u^S \frac{dF^S}{F^S} \right\}, \tag{3.5}$$

and plugging in the differentials from (3.3) gives us, after some reshuffling of terms,

$$dV = V \cdot \left\{ u^T \alpha_T + u^S \alpha_S \right\} dt + V \cdot \left\{ u^T \sigma_T + u^S \sigma_S \right\} dW. \tag{3.6}$$

Now we choose the portfolio weights as the solution (for each t) to the system

$$\begin{cases} u^T + u^S = 1, \\ u^T \sigma_T + u^S \sigma_S = 0. \end{cases} \tag{3.7}$$

Using this portfolio, equation (3.6) turns into

$$dV = V \cdot \left\{ u^T \alpha_T + u^S \alpha_S \right\} dt. \tag{3.8}$$

The solution of (3.7) is given by

$$\begin{cases} u^T = -\frac{\sigma_S}{\sigma_T - \sigma_S}, \\ u^S = \frac{\sigma_T}{\sigma_T - \sigma_S}, \end{cases} \tag{3.9}$$

and we end up with the following V-dynamics.

$$dV = V \cdot \left\{ \frac{\alpha_S \sigma_T - \alpha_T \sigma_S}{\sigma_T - \sigma_S} \right\} dt. \tag{3.10}$$

In order to avoid arbirage possibilites the process within brackets must equal the short rate of interest. Thus we have, after some manipulations,

$$\frac{\alpha_S(t) - r(t)}{\sigma_S(t)} = \frac{\alpha_T(t) - r(t)}{\sigma_T(t)}, \tag{3.11}$$

The left hand side of this equation does not depend upon the choice of T, whereas the right hand side does not depend on S. Thus we have the following result.

Proposition 3.1 *Assume that the bond market is free of arbitrage. Then there exists a process λ such that the relation*

$$\frac{\alpha_T(t) - r(t)}{\sigma_T(t)} = \lambda(t), \tag{3.12}$$

holds for all t and for every choice of maturity time T.

There is a natural interpretation of this result. In the numerator $\alpha_T(t) - r(t)$ we recognize the "risk premium" for the T-bond, i.e. the the rate of return over the risk free rate commanded by a T-bond. In the denominator we have the volatility for the T-bond, and thus the quotient can be interpreted as "the risk premium per unit of volatility" for the T-bond. This quotient is often termed "the market price of risk" and we may express the content of Proposition 3.1 as saying that "on an arbitrage free market all bonds will have the same market price of risk".

Inserting (3.4) into (3.12), and noticing that we must have

$$F(T, r; T) = 1$$

for all values of r we at last end up with the following result.

Proposition 3.2 (Term structure equation) *If the bond market is free of arbitrage then F^T will satisfy the equation*

$$F_t^T + \{\mu - \lambda\sigma\} F_r^T + \frac{1}{2}\sigma^2 F_{rr}^T - rF^T = 0, \tag{3.13}$$

$$F^T(T, r) = 1. \tag{3.14}$$

This equation is closely related to the Black-Scholes equation, but it is more complicated due to the appearence of the market price of risk term λ. The important fact to notice is that λ is not determined within the model, so in order to solve the equation we have to specify λ a priori.

We can easily obtain Feynman-Kač representation of the solution to the term structure equation.

Proposition 3.3 *The pricing function $F(t, r; T)$ can be expressed as*

$$F(t, r; T) = E_{t,r}^Q \left[\exp\left\{ -\int_t^T r(s)ds \right\} \right], \tag{3.15}$$

where the Q-dynamics dynamics for the short rate are given by

$$dr(s) = \{\mu - \lambda\sigma\} ds + \sigma dW(s), \tag{3.16}$$

$$r(t) = r \tag{3.17}$$

We see that the Q-dynamics of r are obtained from the P-dynamics by means of a Girsanov transformation, and Q is of course a martingale measure for the bond market. Specifying the market price of risk is thus equivalent to specifying the martingale measure, and we again stress the fact that the martingale measure is *not* determined by the model. It is the agents on the concrete market who (implicitly) determine Q and λ.

3.3 Martingale modelling

We now go back to the martingale based theory, and following the general philosophy of Subsection 2.3 we model the short rate directly under a fixed martingale measure Q.

Assumption 3.3 *We assume that Q is a fixed martingale measure for the bond market, and that the short rate have the following Q-dynamics*

$$dr(t) = \mu\left(t, r(t)\right) dt + \sigma\left(t, r(t)\right) dW(t), \tag{3.18}$$

where W is a Q-Wiener process.

By definition we see that the short rate of interest is a Markov process, and this implies that, at least for simple contingent claims, the problem of pricing can be transformed into a parabolic boundary value problem.

Proposition 3.4 (Term Structure Equation) *Consider a fixed T-claim of the form $X = \Phi\left(r(T)\right)$. Then the arbitrage free price process is given by $\Pi\left(t, X\right) = F\left(t, r(t)\right)$, where F is a solution to the boundary value problem*

$$\begin{cases} \frac{\partial F}{\partial t}(t, r) + \mu(t, r)\frac{\partial F}{\partial r}(t, r) + \frac{1}{2}\sigma^2(t, r)\frac{\partial^2 F}{\partial r^2}(t, r) - rF(t, r) & = & 0, \\ F(T, r) & = & \Phi(r). \end{cases} \tag{3.19}$$

In particular, bond prices are given by $p(t, T) = F^T\left(t, r(t)\right)$, where

$$\begin{cases} \frac{\partial F^T}{\partial t} + \mu\frac{\partial F^T}{\partial r} + \frac{1}{2}\sigma^2\frac{\partial^2 F^T}{\partial r^2} - rF^T & = & 0, \\ F^T(T, r) & = & 1. \end{cases} \tag{3.20}$$

Proof. The result follows at once from (2.14) and from the Kolmogorov backward equation. ∎

Contrary to the stock-price case, where the Black-Scholes model (despite its shortcomings) is the canonical model, there are a large number of different models for the short rate process. A far from complete list is given below, with the convention that if a parameter is time dependent this is written explicitly. Otherwise all parameters are constant.

1. Vasiček

$$dr = (b - ar) dt + \sigma dW, \tag{3.21}$$

2. Cox-Ingersoll-Ross

$$dr = (b - ar) dt + \sigma\sqrt{r}dW, \tag{3.22}$$

3. Dothan

$$dr = ar\,dt + \sigma r\,dW, \tag{3.23}$$

4. Black-Derman-Toy

$$dr = a(t)rdt + \sigma(t)rdW, \tag{3.24}$$

5. Ho-Lee

$$dr = \Phi(t)dt + \sigma dW, \tag{3.25}$$

6. Hull-White (extended Vasiček)

$$dr = (\Phi(t) - a(t)r)\, dt + \sigma(t)dW, \tag{3.26}$$

7. Hull-White (extended CIR)

$$dr = (\Phi(t) - a(t)r)\, dt + \sigma(t)\sqrt{r}dW, \tag{3.27}$$

The problem of parameter estimation for these models will be treated below.

3.4   Affine term structures

Suppose that we have chosen a particular short rate model model of the form (3.18) above, and suppose that we want to compute the price of a particular contingent claim - say a European option on an underlying T-bond, with delivery date S and strike price K (K, S and T are of course fixed). This means that we have to price the S-contract

$$X = \max[p(S,T) - K,\ 0],$$

and schematically we proceed as follows.

1. Solve, for the fixed T above, the term structure equation

$$\begin{cases} F_t^T(t,r) + \mu(t,r)F_r^T(t,r) + \tfrac{1}{2}\sigma^2(t,r)F_{rr}^T(t,r) - rF^T(t,r) &=& 0, \\ F^T(T,r) &=& 1, \end{cases}$$

where subscripts denote partial derivatives.

2. For the given T above solve the boundary value problem

$$\begin{cases} G_t(t,r) &+& \mu(t,r)G_r(t,r) + \tfrac{1}{2}\sigma^2(t,r)G_{rr}(t,r) - rG(t,r) = 0, \\ G(T,r) &=& \max[F^T(S,r) - K,\ 0], \end{cases}$$

3. The price of the option is now given by $\Pi(t,X) = G(t,r(t))$.

If this program is to be carried out for large number of options within reasonable time it is of course vital that the PDE:s are easy to solve. This leads to the question when a given interest rate model have nice computational properties, and the main result in this direction concerns the existence of a so called Affine Term Structure. For references see Brown-Schaefer (1994) [10], Duffie (1992) [16], and Duffie-Kan (1993) [17].

Definition 3.1 *If bond prices are given by $p(t,T) = F^T(t,r(t))$, where F^T has the form*

$$F^T(t,r) = e^{A(t,T) - B(t,T)r}, \tag{3.28}$$

*and where $A(t,T)$ and $B(t,T)$ are deterministic functions, then the model is said to possess an **Affine Term Structure** (ATS).*

It turns out to be extremely easy to give sufficient conditions for the existence of an ATS. Assume therefore that the short rate model is given by (3.18), and assume that we in fact have an ATS. Using equation (3.28) we may easily compute the various partial derivatives of F^T, and plugging these into the term structure equation (3.20) leaves us with the equation

$$A_t(t,T) - \{1 + B_t(t,T)\}\, r - \mu(t,r)B(t,T) + \frac{1}{2}\sigma^2(t,r)B^2(t,T) = 0. \quad (3.29)$$

The boundary value $F^T(r;T) \equiv 1$ implies

$$A(T,T) = 0, \quad (3.30)$$
$$B(T;T) = 0. \quad (3.31)$$

We see immediately that if μ and σ^2 both are affine in r, then equation (3.29) becomes separable. We thus make the additional assumption that μ and σ have the form

$$\mu(t,r) = \alpha(t)r + \beta(t), \quad (3.32)$$
$$\sigma(t,r) = \sqrt{\gamma(t)r + \delta(t)}. \quad (3.33)$$

Plugging this into (3.29) and collecting terms gives us the equation

$$A_t(t,T) - \beta(t)B(t,T) + \frac{1}{2}\delta(t)B^2(t,T) -$$
$$- \left\{1 + B_t(t,T) + \alpha(t)B(t,T) - \frac{1}{2}\gamma(t)B^2(t,T)\right\} r = 0. \quad (3.34)$$

If this equation holds for all t, T and r then it is easy to see that we must have

$$B_t(t,T) = -\alpha(t)B(t,T) + \frac{1}{2}\gamma(t)B^2(t,T) - 1, \quad (3.35)$$

$$A_t(t,T) = \beta(t)B(t,T) - \frac{1}{2}\delta(t)B^2(t,T). \quad (3.36)$$

We can easily turn this informal reasoning around to obtain the following result.

Proposition 3.5 *Assume that μ and σ are of the form*

$$\mu(t,r) = \alpha(t)r + \beta(t), \quad (3.37)$$
$$\sigma(t,r) = \sqrt{\gamma(t)r + \delta(t)}. \quad (3.38)$$

Then the model has an affine term structure of the form 3.28, where A and B satisfy the system

$$B_t(t,T) + \alpha(t)B(t,T) - \frac{1}{2}\gamma(t)B^2(t,T) = -1, \quad (3.39)$$
$$B(T;T) = 0. \quad (3.40)$$

$$A_t(t,T) = \beta(t)B(t,T) - \frac{1}{2}\delta(t)B^2(t,T), \quad (3.41)$$
$$A(T;T) = 0. \quad (3.42)$$

We note that (3.39) is a Riccati equation for B (for each fixed T). Having solved (3.39) we can then easily obtain A by simply integrating (3.41)-(3.42).

We have thus shown that an affine structure for μ and σ^2 is a sufficient condition for the existence of an ATS. It can be shown fairly easily (see Duffie [16]) that if μ and σ are time independent, then this condition is also necessary. For a considerable extension of the ideas above see Duffie-Kan (1993) [17].

Going back to the models in the previous subsection we see that all of them, with the exception of the Dothan model and the Black-Derman-Toy model, have an affine term structure.

There are, in fact, good probabilistic reasons why the linear models above give rise to an ATS. Taking the Vasiček model as an example, and looking at $p(0,T)$ we can easily solve the SDE to obtain

$$r(s) = e^{-as}r(0) + b\int_0^s e^{-a(s-u)}du + \sigma\int_0^s e^{-a(s-u)}dW(u).$$

This gives us

$$\int_0^T r(s)ds = \int_0^T\int_0^s e^{-a(s-u)}duds$$
$$+ \int_0^T \sigma\int_0^s e^{-a(s-u)}dW(u)ds + \int_0^T e^{-as}r(0)ds.$$

Using the formula

$$p(0,T) = E^Q\left[e^{-\int_0^T r(s)ds}\right]$$

the bond price is given by

$$p(0,T) = A_0(T)\cdot e^{-B(T)r(0)},$$

where

$$A_0(T) = \exp\left\{\int_0^T\int_0^s e^{-a(s-u)}duds\right\}\cdot E^Q\left[\int_0^T \sigma\int_0^s e^{-a(s-u)}dW(u)ds\right]$$

$$B(T) = \int_0^T e^{-as}r(0)ds.$$

Thus we have an Affine Term Structure $p(0,T) = \exp\{A(T) - B(T)r(0)\}$ with $A = \log A_0$, and the same procedure will provide us with bond prices at an arbitrily running time t. Notice that the expected value above is easy to compute, since we have an integtral of a Gaussian process.

For the Cox-ingersoll-Ross model the probabilistic situation is a bit more complicated. The CIR short rate model is a Bessel process, i.e. the short rate is essentially the square of a Wiener process with drift. It is therefore not surprising that the analytical formulas for bond prices and bond option prices given in Cox-Ingersoll-Ross (1985) [11] are expressed in terms of the distribution function of a non central chi-square distribution.

If we try to imitate the calculations above for the geometric models (Dothan and Black-Derman-Toy), we will again have to compute the distribution of $\int_0^T r(s)ds$. This time, however, the short rate is lognormally distributed, and since sums of lognormal variables are notoriously diffcult to handle we find ourselves in computational trouble. See [15] and [6].

It should also be noted that the geometric models above have the unpleasant property that $E^Q[B(t)] = \infty$.

Example 3.1 *We end this subsection by computing the term structure for the Vasiček model. The equations (3.39)-(3.42) becomes*

$$B_t(t,T) - aB(t,T) = -1 \tag{3.43}$$
$$B(T,T) = 0. \tag{3.44}$$

$$A_t(t,T) = bB(t,T) - \frac{1}{2}\sigma^2 B^2(t,T), \tag{3.45}$$
$$A(T,T) = 0. \tag{3.46}$$

Since (3.43)-(3.44) is a simple linear ODE (for fixed T) we immediately obtain

$$B(t,T) = \frac{1}{a}\left\{1 - e^{-a(T-t)}\right\} \tag{3.47}$$

Plugging this into (3.45) and integrating we obtain

$$A(t,T) = \frac{\sigma^2}{2}\int_t^T B^2(s,T)ds - b\int_t^T B(s,T)ds. \tag{3.48}$$

The computations of the integrals above are left to the reader.

3.5 Inverting the yield curve

In this subsection we will discuss some questions concerning parameter estimation problems in connection with short rate models. Suppose, to be specific, that we have chosen to model the short rate by the Vasiček model. In order to implement this model in real life we have to estimate the parameters a, b and σ, and to this end it would seem natural to use historical time series data for the short rate process. Such a procedure would however be nonsensical, and the reason is as follows.

Using the martingale modelling paradigm, we have chosen to specify our model under a fixed martingale measure Q, and in particular this means that all parameters are the Q-parameters. Our observations, on the other hand, are of course not made under the martingale measure Q, but under the objective probability measure P. Thus we see that parameter estimation in a martingale model is *not* a standard problem in statistical estimation theory.

To avoid this problem we could of course specify our model under the objective measure P, which would leave us free to use the standard machinery of statistical estimation theory for diffusion processes in order to obtain point estimates for all parameters. Then, on the other hand, we can not use the model in order to compute bond prices, since these are computed under Q. We thus run into the problem of estimating, in some sense, the Girsanov kernel which takes us from P to Q, but this problem is more or less equivalent to that of estimating the Q-parameters. (Note, however, that the volatility process is the same under P as under Q.)

All these problems have to do with the fact that, if the only object specified a priori is the short rate of interest, then the bond market is incomplete. This, as we recall, implies that there are several martingale measures, so we have to estimate which Q we are dealing with.

Since the martingale measure is *not* specified within the model, but rather chosen by the market, we must in some way obtain price information from the market in order to determine Q. More specifically this is done by "calibrating the model to data" or "fitting the yield curve", and schematically it is done as follows.

- Fix a concrete model involving a parameter vector denoted by α. To emphasize parameter dependence we write the r-dynamics as

$$dr(t) = \mu(t, r(t); \alpha)dt + \sigma(t, r(t); \alpha)dW(t), \tag{3.49}$$

- Solve, for all fixed times of maturity T, the term structure equation

$$F_t^T + \mu F_r^T + \frac{1}{2}\sigma^2 F_{rr}^T - rF^T = 0, \tag{3.50}$$

$$F^T(T, r) = 1. \tag{3.51}$$

This will provide us with theoretical bond prices as

$$p(t, T; \alpha) = F^T(t, r; \alpha).$$

- We now go to the bond market in order to get price data. In particular we may today (i.e. at $t = 0$) observe $p(0, T)$ for all values of T. Denote this *empirical term structure* by $\{p^\star(0, T); \ T \geq 0\}$

- We now choose the parameter vector α in such a way that the theoretical curve $\{p(0, T; \alpha); \ T \geq 0\}$ fits the empirical curve $\{p^\star(0, T); \ T \geq 0\}$ as close as possible (according to some objective function). This gives us our estimated parameter vector $\alpha^\star$.

- Using the procedure above we have now pinned down our martingale meaure Q, by the requirement that the Q-dynamics for r are given by

$$dr(t) = \mu(t, r(t); \alpha^\star)dt + \sigma(t, r(t); \alpha^\star)dW(t), \tag{3.52}$$

Ideally we would of course like to find an $\alpha^\star$ such that

$$p(0, T; \alpha^\star) = p^\star(0, T) \tag{3.53}$$

for all $T \geq 0$. We observe, however, that (3.53) is an infinite dimensional system of equations (one equation for each T), so if we are working in a model containing only a finite dimensional parameter vector (like the Vasiček model) there is no hope of obtaining a complete fit between observed and theoretical bond prices. This is a big problem, since one of our main goals is to compute prices for interest rate derivatives like bond-options, and it is well known that option prices often are very sensitive with respect to changes in the underlying price process. It would thus be somewhat disturbing to have a model for derivative pricing which is not even able to price the underlying bonds correctly.

Thus it is natural to construct models having an infinite parameter vector, and one way of doing this is to let one or several of the parameters in a finite dimensional model be a deterministic function of time. This is exactly the why we have the time dependent parameter in the Ho-Lee model, and it is also the approach used by Hull-White in their extensions of the models by Vasiček and Cox-Ingersoll-Ross. Whether it really is possible to obtain a perfect fit for a particular model is not at all clear a priori, and we will now study this problem for the Hull-White extension of the Vasiček model.

We study a slightly simplified version of the Hull-White model. The Q dynamics are given by

$$dr = (\Phi(t) - ar)\, dt + \sigma dW, \tag{3.54}$$

where a and σ are deterministic constants, which furthermore are assumed to be known. We are also given an observed term structure $\{p^*(0, T);\ T \geq 0\}$ and the problem is to choose Φ in order to fit theoreteical bond prices to the observed term structure.

We have an affine structure so by Proposition 3.5 bond prices are given by

$$p(t, T) = e^{A(t,T) - B(t,T)r(t)}, \tag{3.55}$$

where A and B solve the system

$$B_t(t, T) = aB(t, T) - 1, \tag{3.56}$$

$$B(T; T) = 0. \tag{3.57}$$

$$A_t(t, T) = \Phi(t)B(t, T) - \frac{1}{2}\sigma^2 B^2(t, T), \tag{3.58}$$

$$A(T; T) = 0. \tag{3.59}$$

The solution to this system is given by

$$B(t, T) = \frac{1}{a}\left\{1 - e^{-a(T-t)}\right\}, \tag{3.60}$$

$$A(t, T) = \int_t^T \left\{\frac{1}{2}\sigma^2 B^2(s, T) - \Phi(s)B(s, T)\right\} ds. \tag{3.61}$$

Now we want to fit the theoretical prices above to the observed prices and it is convenient to do this using forward rates rather than bond prices. Since we have an affine term structure, forward rates are given by

$$f(0,T) = B_T(0,T)r(0) - A_T(0,T), \tag{3.62}$$

and, after inserting (3.60)-(3.61) we have

$$f(0,T) = e^{-aT}r(0) + \int_0^T e^{-a(T-s)}\Phi(s)ds - \frac{\sigma^2}{2a^2}\left(1 - e^{-aT}\right)^2. \tag{3.63}$$

Given our observed forward rate curve $\{f^*(0,T); \; T \geq 0\}$, defined by

$$f^*(0,T) = -\frac{\partial \log p^*(0,T)}{\partial T},$$

we now look for a function Φ solving, for each T the equation

$$f^*(0,T) = e^{-aT}r(0) + \int_0^T e^{-a(T-s)}\Phi(s)ds - \frac{\sigma^2}{2a^2}\left(1 - e^{-aT}\right)^2. \tag{3.64}$$

We may now write (3.64) as

$$f^*(0,T) = x(T) + g(T), \tag{3.65}$$

where x and g are defined by

$$\dot{x} = -ax(t) + \Phi(t), \tag{3.66}$$
$$x(0) = r(0), \tag{3.67}$$

$$g(t) = \frac{\sigma^2}{2a^2}\left(1 - e^{-aT}\right)^2 = \frac{\sigma^2}{2a^2}B^2(0,t). \tag{3.68}$$

We now have

$$\begin{aligned}
\Phi(T) &= \dot{x}(T) + ax(T) = f_T^*(0,T) - \dot{g}(T) + ax(T) = \\
&= f_T^*(0,T) + \dot{g}(T) + a\left\{f^*(0,T) + g(T)\right\},
\end{aligned} \tag{3.69}$$

Thus we have the following result.

Lemma 3.1 *Fix an arbitrary bond curve $\{p^*(0,T); \; T > 0\}$, such that $p^*(0,T)$ is twice differentiable w.r.t T. Assume that a and σ are fixed and choose Φ as in (3.69). Then $p(0,T) = p^*(0,T)$ for all $T > 0$.*

Defining Φ as above for a fixed choice of a and σ will completely determine our martingale measure Q. In order to obtain the theoretical bond prices under Q we now have to plug Φ into (3.60) and then plug A and B, given by (3.60)-(3.61) into formula (3.55). The result is as follows.

Proposition 3.6 *Consider the Hull-White model (3.54) for a fixed choice of a and σ, and assume that the observed term structure is twice differentiable. After* *fitting the model to the observed term structure, the theoretical bond prices are given by*

$$p(t, T) = \frac{p^\star(0, T)}{p^\star(0, t)} \exp \left\{ B(t, T) f^\star(0, t) - \frac{\sigma^2}{4a} B^2(t, T) \left(1 - e^{-2at} \right) - B(t, T) r(t) \right\},$$

(3.70)

where B is given by (3.60).

It is easy to show that the Ho-Lee model also can be made to fit any smooth observed term structure. We give the result.

Proposition 3.7 *Consider the Ho Lee model (3.25) for a fixed choice of σ, and assume that the observed bond prices are twice differentiable. Then the following hold.*

1. *If Φ is defined by*

$$\Phi(t) = f_T^\star(0, t) + \sigma^2 t,$$

   then $p(0, T) = p^\star(0, T)$ for all $T \geq 0$.

2. *Choosing Φ as above will give us bond prices as*

$$p(t, T) = \frac{p^\star(0, T)}{p^\star(0, t)} \exp \left\{ (T - t) f^\star(0, t) - \frac{\sigma^2}{2} t(T - t)^2 - (T - t) r(t) \right\}.$$

Suppose now that we use the Hull-White model above, and that we want to compute the price of a bond option as described in the beginning of Subsection 3.4. Then the price of the option is given by $\Pi(t) = G(t, r(t))$, where G solves

$$\begin{cases} G_t(t, r) + \mu(t, r) G_r(t, r) + \frac{1}{2} \sigma^2(t, r) G_{rr}(t, r) - r G(t, r) = 0, \\ G(T, r) = \max \left[F^T(S, r) - K, 0 \right], \end{cases}$$

with

$$F^T(S, r) = \frac{p^\star(0, T)}{p^\star(0, S)} \exp \left\{ (T - S) f^\star(0, S) - \frac{\sigma^2}{2} S(T - S)^2 - (T - S) r \right\}.$$

The important methodological point to notice here is that we are not computing bond prices per se. Instead we "only" use the theoretical bond prices in order to fit our interest rate model to observed data (i.e. to determine Q). We implicitly assume that the bond market in fact is efficient, and we then compute prices for interest rate derivatives *in terms of* the bond prices given by the market. This is again quite in accordance with our general philosophy that derivative pricing concerns the pricing of derivatives in *terms of* a set of given underlying prices.

The prospect of actually having to solve the PDE above is not all that appealing and we will in fact not attempt to do it. Instead, in Section 5 below, we will develop a a theory of so called "forward measures" which radically will facilitate the computation of interest rate derivatives.

4   Heath-Jarrow-Morton

4.1   Existence of martingale measures

Up to this point we have studied interest models where the short rate r is the only explanatory variable. The main advantages with such models are:

- Specifying r as the solution of an SDE allows us to use Markov process theory, so we may work within a PDE framework.

- In particular it is often possible to obtain analytical formulas for bond prices and derivatives.

The main drawbacks of short rate models are as follows.

- From an economic point of view it seems unreasonable to assume that the entire money market is governed by only one explanatory variable.

- It is hard to obtain a realistic volatility structure for the forward rates without introducing a very complicated short rate model.

- As the short rate model becomes more realistic, the inversion of the yield curve described above becoms increasingly more difficult.

These, and other considerations has led various authors to propose models which use more than one state variable. One obvious idea would e.g. be to present an a priori model for the short rate as well as for some long rate, and one could of course also model one or several intermediary interest rates. The method proposed by Heath-Jarrow-Morton in [25] is at the far end of this spectrum - they choose the entire forward rate curve as their (infinite dimensional) state variable.

Assumption 4.1 *Consider $(\Omega, \mathcal{F}, P, \underline{F})$ to be given, and let $\tilde{W}$ be a finite dimensional standard Wiener process on this space. The filtration is by definition the internal one, generated by $\tilde{W}$. We assume that, for every fixed T, the forward rate $f(t, T)$ has a stochastic differential which, under the objective measure P is given by*

$$df(t, T) = \alpha(t, T)dt + \sigma(t, T)d\tilde{W}_t, \qquad (4.1)$$

where, for each T, $\alpha(t, T)$ and $\sigma(t, T)$ are adapted processes. We also assume the existence of the usual locally risk free asset with price process B.

We see that (4.1) is an infinite dimensional stochastic system (one equation for each fixed T). As boundary value at $t = 0$ we use the observed forward curve, i.e.

$$f(0, T) = f^\star(0, T), \quad \forall T \geq 0, \qquad (4.2)$$

so by construction we will automatically obtain a perfect fit to the observed term structure, and the problem of inverting the yield curve is thus completely avoided.

Since there is a one to one correspondence between forward rates and bond prices, given by

$$p(t,T) = \exp\left\{-\int_t^T f(t,s)ds\right\},\qquad(4.3)$$

we see that an exogenous specification of the family of forward rates $\{f(t,T);\ T>0\}$, is equivalent to a specification of the entire family of bond prices $\{p(t,T);\ T>0\}$. Our first task is to investigate when the bond market induced by the forward rate model (4.1) above is free of arbitrage, in the sense that there exists an equivalent martingale measure. The main result is as follows.

Proposition 4.1 *Assume that the family of forward rates are given by (4.1). Then there exists an equivalent martingale measure if and only if there exists an adapted process $g(t)$, with the properties that*

1. *The Doleans exponential $\mathcal{E}(g \star \tilde{W})$ is a P-martingale.*

2. *For all $T \geq 0$ and for all $t \leq T$, we have*

$$\alpha(t,T) = \sigma(t,T)\int_t^T \sigma(t,s)ds - \sigma(t,T)g(t).\qquad(4.4)$$

Proof. From Proposition 2.1 we know that the induced bond price system is of the form

$$dp(t,T) = p(t,T)\left\{r(t) + A(t,T) + \frac{1}{2}\|S(t,T)\|^2\right\}dt + p(t,T)S(t,T)d\tilde{W}(t),$$
$$(4.5)$$

where

$$\begin{cases} A(t,T) &= -\int_t^T \alpha(t,s)ds, \\ S(t,T) &= -\int_t^T \sigma(t,s)ds. \end{cases}\qquad(4.6)$$

Working with the internal filtration we know that every measue Q, equivalent to P is obtained by a Girsanov transformation, and, denoting the Girsanov kernel by g, we see that the Q-dynamics of the bonds are given by

$$dp(t,T) = p(t,T)\left\{r(t) + A(t,T) + \frac{1}{2}\|S(t,T)\|^2 + S(t,T)g(t)\right\}dt$$
$$+ p(t,T)S(t,T)dW(t),\qquad(4.7)$$

where W is a Q-Wiener process. Furthermore Q is a martingale measure if and only if the local rate of return of all bonds equal the short rate of interest, i.e. if and only if we have

$$A(t,T) + \frac{1}{2}\|S(t,T)\|^2 + S(t,T)g(t) = 0.\qquad(4.8)$$

for all $T > 0$ and all $t \leq T$. Since the equation holds identically for all T we may take the T-derivative, which gives us the desired relation. The martingale condition for the Doleans exponential guarantees that Q is indeed as probability measure. ∎

Looking more closely at equation (4.4) we have

$$\alpha(t,T) = \sum_{i=1}^{d} \sigma_i(t,T) \int_0^T \sigma_i(t,s)ds - \sum_{i=1}^{d} \sigma_i(t,T)g_i(t). \qquad (4.9)$$

Taking α and σ as given, this is, for each t, an infinite dimensional linear system of equations for the determination of the d-dimensional vector $g(t)$. The system is highly overdetermined, which means that we can not freely specify all drift terms $\alpha(\cdot,T)$ and volatilities $\sigma(\cdot,T)$ if we wish to have an arbitrage free bond market. Roughly (and somewhat loosely) speaking, the situation is as follows.

- We may generically specify arbitrary volatilities $\sigma(t,T)$ for all times of maturities.

- Given the volatility structure above, fix d "benchmark" maturities $T_1, \cdots, T_d$. For these maturities we may also exogenously specify the drift terms $\alpha(t,T_1), \cdots \alpha(t,T_1)$.

- For each fixed t the matrix $\{\sigma_i(t,T_j)\}_{i,j}$ will be invertible (in the generic case), so the Girsanov kernels $[g_1, \cdots, g_d]$ are uniquely determined (for each fixed t) as the solution to the linear system of equations

$$\alpha(t,T_j) = \sum_{i=1}^{d} \sigma_i(t,T_j) \int_0^T \sigma_i(t,s)ds - \sum_{i=1}^{d} \sigma_i(t,T_j)g_i(t), \quad j=1,\cdots d.$$
$$(4.10)$$

Thus the martingale measure is uniquely determined.

- Now all drift terms with maturity times different from $T_1, \cdots, T_d$ will be uniquely defined by the relation (4.4).

In terms of bond prices this means that we are allowed to specify the bond price volatilities for exactly d benchmark bonds. All other bond prices will then be determined by the price structure for the benchmarks bonds and by the requirement of an arbitrage free market. This is also quite in accordance with the Meta-Theorem 1.1, since in this case we have d independent sources of randomness, so we may expect the market to be complete if we specify d bond prices.

A natural question is now to investigate in detail under which conditions the matrix $\{\sigma_i(t,T_j)\}_{i,j}$ is invertible for each t. We will return to this question in Subsections 4.3 and 4.4.

4.2   Martingale modelling

We now turn to the task of modelling the forward rate family directly under a martingale measure Q. We saw earlier that for a short rate model, we may specify the short rate dynamics arbitrarily under Q, and that an arbitrage free bond market will then be generated by the formula

$$p(t,T) = E^Q \left[\int_t^T r(s)ds \middle| \mathcal{F}_t \right].$$

In the case of modelling the forward rates the situation is different, since a specification of the forward rates is equivalent to a specification of all bond prices. Thus, under a martingale measure Q there must be some relations between the various infinitesimal characteristics for the forward rates, and the main result in this direction, known as "the Heath-Jarrow-Morton drift condition", follows in fact directly from the results of the preceding subsection.

Proposition 4.2 *Assume that Q is a martingale measure for the bond market, and assume that the forward rate dynamics under Q are given by*

$$df(t,T) = \alpha(t,T)dt + \sigma(t,T)dW(t), \tag{4.11}$$

where W is a Q-Wiener process. Then we have :

1. For all $T > 0$ and all $t \leq T$ the following holds Q-a.s.

$$\alpha(t,T) = \sigma(t,T) \int_0^T \sigma(t,s)ds. \tag{4.12}$$

2. Bond price dynamics under Q are given by

$$dp(t,T) = p(t,T)r(t)dt + p(t,T)S(t,T)dW(t), \tag{4.13}$$

where S is given by (4.6).

Proof. The fact that we are working directly under a martingale measure simply means that, in terms of the preceeding subsection, $P = Q$. Thus the Girsanov kernel g is zero, and the result now follows immediately from Proposition 4.1. ∎

The main point of Proposition 4.2 is that when we specify the forward rate dynamics (under Q) we may freely specify the volatility structure. The drift parameters are then uniquely determined. Notice also that, since P and Q are equivalent, the volatility process is the same under P as under Q.

For practical purposes the use of a HJM model can schematically be written as follows.

1. Specify (this is the modelling part) the volatilities $\sigma(t,T)$.

2. The drift parameters $\alpha(t,T)$ are then uniquely determined as

$$\alpha(t,T) = \sigma(t,T) \int_t^T \sigma(t,s)ds. \tag{4.14}$$

3. Go to the market and observe today's forward rate structure

$$\{f^*(0,T); \ T \geq 0\}$$

4. Integrate in order to get the forward rates as

$$f(t,T) = f^*(0,T) + \int_0^t \alpha(s,T)ds + \int_0^t \sigma(s,T)dV(s). \qquad (4.15)$$

5. Bond prices can now be computed using the formula

$$p(t,T) = \exp\left\{ -\int_t^T f(t,s)ds \right\}. \qquad (4.16)$$

6. We may now (in principle) go on to price other derivatives.

To see how this works in a concrete case we consider an example.

Example 4.1 The simplest case possible is when the volatility process $\sigma(t,T)$ is a constant, which, with some misuse of the language, we shall call σ. Thus, using Proposition 4.2, the drift term is given by

$$\alpha(t,T) = \sigma \int_t^T \sigma ds = \sigma^2(T-t),$$

so the forward dynamics under the measure Q are given by

$$\begin{aligned} df(t,T) &= \sigma^2(T-t)dt + \sigma dW(t), & (4.17) \\ f(0,T) &= f^*(0,T). & (4.18) \end{aligned}$$

This can be integated directly (for each T) to

$$f(t,T) = f^*(0,T) + \int_0^t \sigma^2(T-s)ds + \int_0^t \sigma dW(s),$$

i.e. we have

$$f(t,T) = f^*(0,T) + \sigma^2 t\left(T - \frac{t}{2}\right) + \sigma W(t). \qquad (4.19)$$

In particular we see that the short rate of interest $r(t) = f(t,t)$ is given by

$$r(t) = f^*(0,t) + \sigma^2\frac{t^2}{2} + \sigma W(t), \qquad (4.20)$$

so the stochastic differential of r is given by

$$dr(t) = \left\{ f_T^*(0,t) + \sigma^2 t \right\} dt + \sigma dW(t).$$

We thus see that what we have here is in fact the Ho-Lee model

$$dr(t) = \Phi(t)dt + \sigma dW(t),$$

with Φ chosen so as to fit the model perfectly to initial data. Notice how the HJM formulation of the model relieves us from the task of actually inverting the yield curve.

Using (4.16) and (4.19) we now go on to compute bond prices. An easy calculation gives us

$$\int_t^T f(t,s)ds = \int_t^T f^*(0,s)ds + \frac{\sigma^2}{2}tT(T-t) + \sigma(T-t)W(t),$$

so

$$p(t,T) = \frac{p^*(0,T)}{p^*(0,t)}e^{-\frac{1}{2}\sigma^2 tT(T-t)-\sigma(T-t)W(t)}. \tag{4.21}$$

This expression gives us the distribution of the bond prices, but we may perhaps want to express the bond price more explicitly in terms of the short rate of interest. This can easily be achieved by using (4.20) to write

$$\sigma W(t) = r(t) - f^*(0,t) - \sigma^2\frac{t^2}{2}. \tag{4.22}$$

Substituting this into (4.21) gives us the formula

$$p(t,T) = \frac{p^*(0,T)}{p^*(0,t)}\exp\left\{(T-t)f^*(0,t) - \frac{\sigma^2}{2}t(T-t)^2 - (T-t)r(t)\right\}, \tag{4.23}$$

which we recognize from Proposition 3.7.

4.3   Uniqueness of Q

Assume that we have specified the forward rate dynamics under a fixed martingale measure Q as

$$df(t,T) = \alpha(t,T)dt + \sigma(t,T)dW(t), \tag{4.24}$$

where as before W is a d-dimensional Q-Wiener process, and where we assume that the HJM drift condition

$$\alpha(t,T) = \sigma(t,T)\int_0^T \sigma'(t,s)ds, \tag{4.25}$$

is satisfied.

From Harrison-Pliska [24] we may now expect that the bond market is complete if and only if Q is the unique martingale measure. This is however not entirely obvious, since [24] only consider a market with finitely many assets, whereas we are considering the bond market, which contains a continuum of assets (one for each time of maturity T).

The abstract uniqueness result is as follows.

Proposition 4.3 *The following conditions are equivalent*

1. *The martingale measure Q is unique.*

2. *For each fixed t, there exist maturities $T_1, \cdots, T_d$ (which may depend on t) such that the matrix $D(t; T_1, \cdots, T_d)_{i,j} = \{\sigma_i(t,T_j)\}$ is nonsingular.*

3. *For each fixed t, there exist maturities $T_1, \cdots, T_d$ (which may depend on t) such that the matrix $H(t; T_1, \cdots, T_d)_{i,j} = \{S_i(t, T_j)\}$ is nonsingular.*

Proof. Follows immediately from the statement of Proposition 4.1 and from formula (4.8). ∎

The natural problem is now to give conditions on the volatility structure which guarantee that D above is invertible. We have the following result, which is a special case of a more general result in [4].

Proposition 4.4 *Assume that*

1. *For each t, ω the functions $\sigma_1(t, T), \cdots, \sigma_d(t, T)$ are real analytic in the T-variable.*

2. *For each t, ω the functions $\sigma_1(t, T), \cdots, \sigma_d(t, T)$ are linearly independent (as functions of T).*

Then, for each fixed t, it is possible to choose volatilities $T_1, \cdots, T_d$ such that the matrix $\{S_i(t, T_j)\}_{i,j}$ is nonsingular. Apart from a finite set of forbidden points these volatilities can be chosen freely as long as they are distinct.

Proof. See [4] for a full proof. The idea is to use induction on d, to consider the corresponding determinants, and to use the fact that an analytic function can only have a finite number of zeroes on a compact. ∎

To illustrate we take an example from from Heath-Jarrow-Morton [25]. This model has two driving Wiener processes, and a corresponding volatility structure of the form

$$\begin{aligned}
\sigma_1(t, T) &= \sigma_1 > 0, \\
\sigma_2(t, T) &= \sigma_2 e^{-\lambda(T-t)}.
\end{aligned}$$

Using Proposition 4.4 we see at once that for this model Q is indeed unique.

4.4 Completeness

As in the previous subsection we consider, as given, the model (4.24)-(4.25) under a fixed martingale measure Q. The object under study is the completeness of the bond market, and since in this case we have d sources of randomness, we may expect to be able to hedge any (sufficiently integrable) contingent claim with a portfolio consisting of bonds with d different maturities.

Consider a fixed T-claim X, and for simplicity assume that $X/B(T) \in L^2$. We would now like to hedge against X and we look for a hedging portfolio consisting, at each time t, of bonds with (no more than) d different times of maturity $T_1, \cdots, T_d$. Intuitively this seems to be a reasonable requirement, but it may of course very well happen that this set of "basic maturities" vary with running time. This would be unpleasant from a practical point of view, since in reality we only have a finite number of traded bonds. It would also be unpleasant from a mathematical point of view, since it could then happen that the portfolio would use a continuum of bonds over a finite time interval, and such a situation is not covered by standard stochastic integration theory. A natural way out of these problems is to introduce measure valued portfolios, and a stochastic integration theory allowing for integrators which are Banach space valued semimartingales, see [4], [5].

From several points of view we would thus like to choose the "basic maturities" that stay fixed as running time t varies. Let us consequently fix d maturities $T_1, \cdots, T_d$. We may now use Lemma 1.12 to infer that we can hedge against X with bonds of these prespecified maturities if and only if we can find processes $h^1, \cdots, h^d$ such that

$$dM(t) = \sum_{i=1}^{d} h^i(t) dZ(t, T_i), \qquad (4.26)$$

where the martingale M is defined by

$$M(t) = E^Q \left[\frac{X}{B(T)} \middle| \mathcal{F}_t \right],$$

and the discounted bond prices are defined by

$$Z(t, T) = \frac{p(t, T)}{B(t)}.$$

Since we are working with the internal filtration, the standard martingale representation theorem tells us that there exist processes $\gamma_1, \cdots, \gamma_d$ such that

$$dM(t) = \sum_{j=1}^{d} \gamma_j(t) dW_j(t). \qquad (4.27)$$

Furthermore, from (4.13) we obtain the Z dynamics as

$$dZ(t, T_i) = Z(t, T_i) \sum_{j=1}^{d} S_j(t, T_i) dW_j(t). \qquad (4.28)$$

Plugging (4.27) and (4.28) into (4.26) and equating coefficients, we see that we may, at a fixed t, solve for $h^1, \cdots, h^d$ if and only the "hedging equation"

$$\sum_{i=1}^{d} h^i(t) Z(t, T_i) S_j(t, T_i) = \gamma_j(t), \quad j = 1, \cdots, d. \qquad (4.29)$$

admits a solution for each fixed t, i.e. if and only if the matrix

$$H(t, T_1, \cdots, T_d)_{i,j} = S_i(t, T_j)$$

is invertible.

From Proposition 4.3 we now see that the martingale measure is unique if and only if the model is complete in the sense that the hedging equation can, for each t, be solved for **some** choice of maturities $T_1, \cdots, T_d$ (which unfortunately may depend on t). Looking closer we see that in fact uniqueness of Q is equivalent to the operator $H^*(t; T_1, \cdots, T_d)$ being injective for each t, whereas completeness is equivalent to the operator $H(t; T_1, \cdots, T_d)$ being surjective. Since we are in a finite dimensional setting we have the trivial result $\{\text{Ker}A\}^\perp = \text{Im}A^*$, so the equivalence between uniqueness of Q and completeness is nothing more than this simple duality for $A = H^*$.

Now we want explictit conditions on the volatility structure which guarantees market completeness, and to this end we may of course use Proposition 4.4. Notice however, that if we insist on having a fixed set of maturities in our portfolio strategy, then Proposition 4.4 is not good enough. From [4] we cite the following stronger result, which guarantees that we can indeed work with a fixed system of basic bonds.

Proposition 4.5 *Suppose that the assumptions of Proposition 4.4 are in force. Assume furthermore that the volatility functions $\sigma_1(t, T), \cdots, \sigma_d(t, T)$ are deterministic and real analytic in the t-variable. Then the maturities in Proposition 4.4 can be chosen to be the same for every t.*

Thus, under the conditions of Proposition 4.5, we may hedge any claim using bonds with a prespecified set of d maturities. As an example we can again look at the Heath-Jarrow-Morton example at the end of the previous subsection. We see at once that the conditions above are satisfied, so for this example we may hedge any claim with a portfolio consisting of bonds with two prespecified maturities (plus of course the risk free asset).

4.5 The Musiela parametrization

Up to this point all bond prices and forward rates have been parameterized using the time of maturity, T, as the fundamental parameter. For practical purposes it is often much more natural to express all objects by using instead the time to maturity, x, as the parameter. This motivates the following notation.

Definition 4.1 *For all $x \geq 0$ the forward rates $r(t, x)$ are defined by the relation*

$$r(t, x) = f(t, t + x). \tag{4.30}$$

Suppose now that we have the following model for the forward rates under a martingale meaure Q

$$df(t, T) = \alpha(t, T)dt + \sigma(t, T)dW(t). \tag{4.31}$$

The question is to find the Q-dynamics for $r(t, x)$, and from the definition above we immediately have

$$dr(t, x) = df(t, t + x) + \frac{\partial f}{\partial T}(t, t + x) \tag{4.32}$$

$$= r_x(t,x)dt + \alpha(t,t+x)dt + \sigma(t,t+x)dW(t) \qquad (4.33)$$

Using this expression and the HJM drift condition (Proposition4.2) we thus get the following result, henceforth referred to as the Musiela equation.

Proposition 4.6 (The Musiela Equation) *Assume that the forward rate dynamics under Q are given by (4.31). Then*

$$dr(t,x) = \{Ar(t,x) + D(t,x)\}\, dt + \sigma_0(t,x)dW(t), \qquad (4.34)$$

where

$$A = \frac{\partial}{\partial x},$$

$$D(t,x) = \sigma_0(t,x) \int_0^x \sigma_0(t,y)dy,$$

$$\sigma_0(t,x) = \sigma(t,t+x).$$

This approach was first taken in a more systematic way by Musiela [40] and Brace & Musiela [8]. One of its main points is that we see much more clearly the nature of the forward rate equation as an infinite dimensional stochastic equation evolving over time. In particular we see that if the volatility $\sigma_0(t,x)$ is deterministic, then the Musiela equation is indeeed an infinite dimensional **linear** equation.

5 Change of numeraire

5.1 Generalities

Consider as given a financial market (not necessarily a bond market) with the usual locally risk free asset B, and a risk neutral maringale measure Q. As noted in Section 1 a measure is a martingale measure only relative to some chosen numeraire asset, and we recall that the risk neutral martingale measure, with the money account B as numeraire, has the property of martingalizing all processes of the form $S(t)/B(t)$ where S is the arbitrage free price process of any traded asset.

In many concrete situations the computational work needed for the determination of arbitrage free prices can be drastically reduced by a clever change of numeraire, and the purpose of the present section is to analyze such changes.

The first to use a numeraire different from the risk free asset B were, independently, Geman [21] and (in a Gaussian framework) Jamshidian [30], who both used a bond maturing at a fixed time T as numeraire. A systematic study has been carried out by Geman and El Karoui in a series of papers, and here we follow basically Geman-El Karoui-Rochet [22].

Assumption 5.1 *Assume that Q is a fixed risk neutral martingale measure, and $S_0(t)$ is a strictly positive process with the property that the process $S_0(t)/B(t)$ is a Q-martingale.*

The economic interpretation of this assumption is of course that $S_0(t)$ is the arbitrage free price process of a traded asset. We are now searching for a measure Q^* with the property that, for every arbitrage free price process $\Pi(t)$, the process $\Pi(t)/S_0(t)$ is a Q^*-martingale.

In order to get an idea of what Q^* must look like, let us consider a fixed time T and a T-contract X. Assuming enough integrability we then know that the arbitrage free price of X at time $t = 0$ is given by

$$\Pi(0, X) = E^Q \left[\frac{X}{B(T)} \right]. \tag{5.1}$$

Assume, on the other hand, that the measure Q^* actually exists, with a Radon-Nikodym derivative process

$$L(t) = \frac{dQ^*}{dQ}, \quad \text{on } \mathcal{F}_t.$$

Then we know that, because of the assumed Q^*-martingale property of the process process $\Pi(t, X)/S_0(t)$, we have

$$\frac{\Pi(0, X)}{S_0(0)} = E^* \left[\frac{\Pi(T, X)}{S_0(T)} \right] = E^* \left[\frac{X}{S_0(T)} \right] = E^Q \left[L(T) \frac{X}{S_0(T)} \right]$$

Thus we have

$$\Pi(0, X) = E^Q \left[L(T) \frac{X \cdot S_0(0)}{S_0(T)} \right], \tag{5.2}$$

and, comparing (5.1) with (5.2), we see that a natural candidate as likelihood process for the intended change of measure is given by $L(t) = S_0(t)/S_0(0) \cdot B(t)$.

We now go on to the formal definitions and results.

Definition 5.1 *Under Assumption 5.1 define, for any fixed t, the measure Q^* on $\mathcal{F}_t$ by*

$$\frac{dQ^*}{dQ} = L(t), \tag{5.3}$$

where the likelihood process L is defined by

$$L(t) = \frac{S_0(t)}{S_0(0) \cdot B(t)}. \tag{5.4}$$

We note at once that L is a positive Q-martingale with $L(0) = 1$, so the measure Q^* is indeed a probability measure. We now want to prove that Q^* martingalizes every process of the form $\Pi(t)/S_0(t)$, where $\Pi(t)$ is any arbitrage free price proces. The formalization of this idea is the following result.

Proposition 5.1 *Define Q^* as above. Assume that $\Pi(t)$ is a process such that $\Pi(t)/B(t)$ is a Q-martingale. Then the process $\Pi(t)/S_0(t)$ is a Q^*-martingale.*

Proof. Denoting integration with respect to $Q^\star$ by $E^\star$, and using the abstract Bayes's formula, we obtain

$$E^\star \left[\frac{\Pi(t)}{S_0(t)} \Big| \mathcal{F}_s \right] = \frac{E^Q \left[L(t) \frac{\Pi(t)}{S_0(t)} \Big| \mathcal{F}_s \right]}{L(s)} = \frac{E^Q \left[\frac{\Pi(t)}{B(t)S_0(0)} \Big| \mathcal{F}_s \right]}{L(s)}$$

$$= \frac{\Pi(s)}{B(s)S_0(0)L(s)} = \frac{\Pi(s)}{S_0(s)}. \quad \blacksquare$$

As an immediate corollary we have the following.

Proposition 5.2 *Define $Q^\star$ as above and consider a T-claim X such that $X/B(T) \in L^1(Q)$. Then the price process, $\Pi(t, X)$ is given by*

$$\Pi(t, X) = S_0(t) E^\star \left[\frac{X}{S_0(T)} \Big| \mathcal{F}_t \right]. \tag{5.5}$$

This formula is particularly useful when X is of the form $X = S_0(T) \cdot Y$, since then we obtain the simple expression

$$\Pi(t, X) = S_0(t) E^\star [Y | \mathcal{F}_t].$$

A typical example when this situation occurs is when dealing with derivatives defined in terms of several underlying assets. Assume for example that we are given two asset prices S_0 and S_1, and that the contract X to be priced is of the form $X = \Phi(S_0(T), S_1(T))$, where Φ is a given **linearly homogenous** function. Using the standard machinery we would have to compute the price as

$$\Pi(t, X) = E^\star \left[e^{-\int_t^T r(s) ds} \Phi(S_0(T), S_1(T)) \Big| \mathcal{F}_t \right],$$

which essentially amounts to the calculation of a triple integral. If we instead use S_0 as numeraire we have

$$\Pi(t, X) = S_0(t) E^\star [\varphi(Z(T)) | \mathcal{F}_t], \tag{5.6}$$

where $\varphi(z) = \Phi(1, z)$ and $Z(T) = S_1(T)/S_0(T)$. Note that the factor $S_0(t)$ is the price of the traded asset S_0 at time t, so this quantity does not have to be computed - it can be directly observed on the market. Thus the computational work is reduced to computing a single integral.

Remark 5.1 Note that it is easy to find the Girsanov transformation which carries Q into $Q^\star$. Since $Q^\star$ martingalizes the process $S_0(t)/B(t)$, the $Q^\star$-dynamics of S_0 must be of the form

$$dS_0(t) = r(t)S_0(t)dt + S_0(t)v(t)dM(t) \tag{5.7}$$

where M is the driving Q-martingale of S_0 (typically M is a Wiener process), and v is the volatility for S_0. From (5.7) and (5.4) it now follows that the likelihood process L has the Q-dynamics

$$dL(t) = L(t)v(t)dM(t), \tag{5.8}$$

so we can easily read off the relevant Girsanov kernel directly from the volatility of the S_0-process.

Example 5.1 Assume that we have two stocks, S_0 and S_1, with price processes of the following form under the objective probability P.

$$
\begin{aligned}
dS_0 &= \alpha S_0 dt + \sigma S_0 d\tilde{W}_0, & (5.9)\\
dS_1 &= \beta S_1 dt + \delta S_1 d\tilde{W}_1. & (5.10)
\end{aligned}
$$

Here $\tilde{W}_0$ and $\tilde{W}_1$ are assumed to be independent P-Wiener processes, but it would also be easy to treat the case when there is a coupling between the two assets.

Under Q the price dynamics will be given by

$$
\begin{aligned}
dS_0 &= r S_0 dt + \sigma S_0 dW_0, & (5.11)\\
dS_1 &= r S_1 dt + \delta S_1 dW_1, & (5.12)
\end{aligned}
$$

and from the remark above it follows that the Girsanov transformation from Q to Q^* is given explicitly by

$$
\begin{aligned}
dL(t) &= L(t)\sigma dW_1,\\
L(0) &= 1.
\end{aligned}
$$

The T-claim to be priced is an **exchange option**, which gives the holder the right, but not the obligation, to exchange one S_0 share for one S_1 share at time T. Formally this means that the claim is given by $X = \max[S_1(T) - S_0(T),\, 0]$, and we note that we have a linearly homogenous contract function. From (5.6) the price process is given by

$$\Pi(t, X) = S_0(t) E^* \left[\max \left[Z(T) - 1,\, 0 \right] \big| \mathcal{F}_t \right],$$

with $Z(t) = S_1(t)/S_0(t)$. We are thus in fact valuing a European call option on $Z(T)$, with strike price $K = 1$. By construction Z will be a Q^*-martingale, and since a Girsanov transformation will not affect the volatility, it follows easily from (5.9)-(5.10) that the Q^*-dynamics of Z are given by

$$dZ = Z\sqrt{\sigma^2 + \delta^2}\, dW^*$$

where W^* is a standard Q^*-Wiener process. The price is thus given by the formula

$$\Pi(t, X) = S_0(t) \cdot c(t, Z(t)).$$

Here $c(t, z)$ is given directly by the Black-Scholes formula as the price of a European call option, valued at T, with time of maturity T, strike price $K = 1$, short rate of interest $r = 0$, on a stock with volatility $\sqrt{\sigma^2 + \delta^2}$ and price z.

5.2 Forward measures

In this subsection we specialize the theory developed in the previous subsection to the case when the new numeraire chosen is a bond maturing at time T. As can be expected this choice of numeraire is particularly useful when dealing with interest rate derivatives.

Suppose therefore that we are given a specified bond market model with a fixed martingale measure Q. For a fixed time of maturity T we now choose the process $p(t, T)$ as our new numeraire.

Definition 5.2 *The T-forward measure Q^T is defined by*

$$dQ^T = L^T(t)dQ$$

on $\mathcal{F}_t$ for $0 \leq t \leq T$ where

$$L^T(t) = \frac{p(t, T)}{B(t)p(0, T)}. \tag{5.13}$$

Observing that $P(T, T) = 1$ we have the following useful pricing formula as an immediate corollary of Proposition 5.2.

Proposition 5.3 *Assume that the T-claim X has the property that $X/B(T) \in L^1(Q)$. Then*

$$\Pi(t, X) = p(t, T)E^T[X|\mathcal{F}_t], \tag{5.14}$$

where E^T denotes integration w.r.t. Q^T.

Note again that the price $p(t, T)$ does not have to be computed. It can be observed directly on the market at time t.

A natural question to ask is when Q and Q^T coincide. This occurs if and only if Q-a.s. we have $L^T(T) = 1$, i.e. when

$$1 = \frac{p(T, T)}{B(T)p(0, T)} = \frac{e^{-\int_0^T r(s)ds}}{E^Q\left[e^{-\int_0^T r(s)ds}\right]}$$

i.e. if and only if r is deterministic.

In this subsection we shall demonstrate the usefulness of the pricing equation (5.14) by studying the pricing of (bond-) options, but first we note the following martingale property of the forward rate process $f(t, T)$.

Lemma 5.1 *Assume that, for all $T > 0$ we have $r(T)/B(T) \in L^1(Q)$. Then, for every fixed T, the process $f(t, T)$ is a Q^T-martingale for $0 \leq t \leq T$, and in particular we have*

$$f(t, T) = E^T[r(T)|\mathcal{F}_t]. \tag{5.15}$$

Proof. Using Proposition 5.14 with $X = r(T)$ we have

$$\Pi(t, X) = E^Q\left[r(T)e^{-\int_t^T r(s)ds}\bigg|\mathcal{F}_t\right] = p(t, T)E^T\left[r(T)|\mathcal{F}_t\right].$$

This gives us

$$E^T\left[r(T)|\mathcal{F}_t\right] = \frac{1}{p(t, T)}E^Q\left[r(T)e^{-\int_t^T r(s)ds}\bigg|\mathcal{F}_t\right]$$

$$= -\frac{1}{p(t, T)}E^Q\left[\frac{\partial}{\partial T}e^{-\int_t^T r(s)ds}\bigg|\mathcal{F}_t\right] = -\frac{1}{p(t, T)}\frac{\partial}{\partial T}E^Q\left[e^{-\int_t^T r(s)ds}\bigg|\mathcal{F}_t\right]$$

$$= -\frac{p_T(t, T)}{p(t, T)} = f(t, T).\blacksquare$$

Remark: It is sometimes claimed that "the forward rate is an unbiased estimate of the future spot rate". In view of the lemma above we see that in general this conjecture is false, not only under the objective measure P (which is not surprising), but also under the risk neutral measure Q.

5.3   Option pricing

We will now apply the theory developed above to give a fairly general formula for the pricing of European call options. Assume therefore that we are given a financial market with a (possibly stochastic) short rate of interest r, and a strictly postitive asset price process $S(t)$. We also assume the existence of a risk neutral martingale mesure Q.

Consider now a fixed time T, and a European call on S with date of maturity T and strike price K. We are thus considering the T-claim

$$X = \max[S(T) - K,\, 0],\tag{5.16}$$

and to simplify notation we restrict ourselves to computing the price $\Pi(t, X)$ at time $t = 0$. The main trick when dealing with options is to write X as

$$X = [S(T) - K]\cdot I\{S(T) \geq K\}.$$

We obtain

$$\begin{aligned}\Pi(0, X) &= E^Q\left[B^{-1}(T)[S(T) - K]I\{S(T) \geq K\}\right]\\ &= E^Q\left[B^{-1}(T)S(T)\cdot I\{S(T) \geq K\}\right]\\ &\quad- KE^Q\left[B^{-1}(T)\cdot I\{S(T) \geq K\}\right].\end{aligned}$$

For the first term we change to the measure Q^S having S as numeraire, and for the second term we use the T-forward measure. Using Propositions 5.2 and 5.3 we obtain the following basic option pricing formula, where we recognize the structure of the standard Black-Scholes formula.

Proposition 5.4 *Given the assumtions above, the option price is given by*

$$\Pi\left(0, X\right) = S(0)Q^S\left(S(T) \geq K\right) - Kp(0,T)Q^T\left(S(T) \geq K\right). \qquad (5.17)$$

In order to get more concrete results we make an additional assumption.

Assumption 5.2 *Assume that*

1. *The filtration is generated by a d-dimensional Q-Wiener process W.*

2. *The process $Z_{S,T}$ defined by*

$$Z_{S,T}(t) = \frac{S(t)}{p(t,T)}, \qquad (5.18)$$

has a stochastic differential of the form

$$dZ_{S,T}(t) = Z_{S,T}(t)m_T^S(t)dt + Z_{S,T}(t)\sigma_{S,T}(t)dW, \qquad (5.19)$$

*where the volatility process $\sigma_{S,T}(t)$ is **deterministic**.*

The crucial point here is of course the assumption that the d-dimensional row vector process $\sigma_{S,T}$ is deterministic. Also note that the volatility process is unaffected by a continuous change of measure.

In order to analyze the option formula (5.17) we start with the second term which we write as

$$Q^T\left(S(T) \geq K\right) = Q^T\left(\frac{S(T)}{p(T,T)} \geq K\right) = Q^T\left(Z_{S,T}(T) \geq K\right). \qquad (5.20)$$

By construction we know that $Z_{S,T}$ is a martingale under Q^T, so its Q^T-dynamics are given by

$$dZ_{S,T}(t) = Z_{S,T}(t)\sigma_{S,T}(t)dW^T, \qquad (5.21)$$

with the solution

$$Z_{S,T}(T) = \frac{S(0)}{p(0,T)}\exp\left\{-\frac{1}{2}\int_0^T \sigma_{S,T}^2(t)dt + \int_0^T \sigma_{S,T}(t)dW^T\right\} \qquad (5.22)$$

The stochastic integral in the exponent is Gaussian with zero mean and variance

$$\Sigma_{S,T}^2(T) = \int_0^T \|\sigma_{S,T}(t)\|^2 dt. \qquad (5.23)$$

We thus have, for the second term in (5.17),

$$Q^T\left(S(T) \geq K\right) = N[d_2],$$

where

$$d_2 = \frac{\ln\left(\frac{S(0)}{Kp(0,T)}\right) - \frac{1}{2}\Sigma_{S,T}^2(T)}{\sqrt{\Sigma_{S,T}^2(T)}} \qquad (5.24)$$

For the first term in (5.17) we write

$$Q^S\left(S(T) \geq K\right) = Q^S\left(\frac{p(T,T)}{S(T)} \leq \frac{1}{K}\right) = Q^S\left(Y_{S,T}(T) \leq \frac{1}{K}\right), \qquad (5.25)$$

Where the process $Y_{S,T}$ is defined by

$$Y_{S,T}(t) = \frac{p(t,T)}{S(t)} = \frac{1}{Z_{S,T}(t)}.$$

Under the measure Q^S the process $Y_{S,T}$ is a martingale, so its Q^S-dynamics are of the form

$$dY_{S,T}(t) = Y_{S,T}(t)\delta_{S,T}(t)dW^S.$$

Since $Y_{S,T} = Z_{S,T}^{-1}$ it is easily seen that in fact $\delta_{S,T}(t) = -\sigma_{S,T}(t)$. Thus we have

$$Y_{S,T}(T) = \frac{p(0,T)}{S(0)} \exp\left\{-\frac{1}{2}\int_0^T \sigma_{S,T}^2(t)dt - \int_0^T \sigma_{S,T}(t)dW^S\right\},$$

and with exactly the same reasoning as above we have, after some simplifications,

$$Q^S\left(S(T) \geq K\right) = N[d_1],$$

where

$$d_1 = d_2 + \sqrt{\Sigma_{S,T}^2(T)} \qquad (5.26)$$

We have thus proved the following result.

Proposition 5.5 *Under the conditions given in Assumption 5.2, the price of the call option defined in (5.16) is given by the formula*

$$\Pi(0,X) = S(0)N[d_2] - K \cdot p(0,T)N[d_1], \qquad (5.27)$$

where d_2 and d_1 are given in (5.24) and (5.26) respectively, whereas $\Sigma_{S,T}^2(T)$ is given by (5.23).

5.4   The Hull-White model

To illustrate the technique further we will now discuss pricing of interest rate derivatives in the Hull-White model, and we recall that in this model the Q-dynamics of r are given by

$$dr = \{\Phi(t) - ar\}\,dt + \sigma dW. \qquad (5.28)$$

For the Hull-White model we have an affine term structure (see Subsection 3.4), and bond prices are given by

$$p(t,T) = e^{A(t,T)-B(t,T)r(t)}, \qquad (5.29)$$

where A and B are deterministic functions, and where B is given by (see 3.60)

$$B(t,T) = \frac{1}{a}\left\{1 - e^{-a(T-t)}\right\}. \tag{5.30}$$

We start by pricing a European call option with date of maturity T_1 and strike price K, on an underlying bond with date of maturity T_2, where $T_1 < T_2$. In terms of the preceeding subsection this means that $T = T_1$ and that $S(t) = p(t, T_2)$, and first we have to check Assumption 5.2, i.e. if the volatility, σ_z, of the process

$$Z(t) = \frac{p(t, T_2)}{p(t, T_1)}, \tag{5.31}$$

is deterministic. (in terms of the preceeding subsection Z corresponds to $S_{S,T}$ and σ_z corresponds to $\sigma_{S,T}$).

From (5.29) we have

$$Z(t) = \exp\left\{A(t, T_2) - A(t, T_1) - [B(t, T_2) - B(t, T_1)]\, r(t)\right\},$$

so under Q we have

$$dZ(t) = Z(t)\left\{\cdots\right\}dt + Z(t) \cdot \sigma_z(t)dW, \tag{5.32}$$

where

$$\sigma_z(t) = -\sigma\left[B(t, T_2) - B(t, T_1)\right] = \frac{\sigma}{a}e^{at}\left[e^{-aT_1} - e^{-aT_2}\right] \tag{5.33}$$

Thus we see that σ_z is in fact deterministic, and we may now use Proposition 5.5 to have the following result.

Proposition 5.6 *Consider the Hull-White model (5.28). The price, at $t = 0$, of a European call with strike price K, and time of maturity T_1, on a bond maturing at T_2 is given by the formula*

$$\Pi\left(0, X\right) = S(0)N[d_2] - K \cdot p(0, T)N[d_1], \tag{5.34}$$

where

$$d_2 = \frac{\ln\left(\frac{p(0, T_2)}{Kp(0, T_1)}\right) - \frac{1}{2}\Sigma^2}{\sqrt{\Sigma^2}}, \tag{5.35}$$

$$d_1 = d_2 + \sqrt{\Sigma^2}, \tag{5.36}$$

$$\Sigma^2 = \frac{\sigma^2}{2a^3}\left\{1 - e^{-2aT_1}\right\}\left\{1 - e^{-a(T_2 - T_1)}\right\}^2. \tag{5.37}$$

We end this subsection by analyzing a T-claim of the form $X = H\left(r(T)\right)$, where H is some given (sufficiently integrable) function. From proposition 5.3 we have the pricing formula

$$\Pi\left(t, X\right) = p(t, T)E^T\left[H\left(r(T)\right)|\mathcal{F}_t\right], \tag{5.38}$$

so we must find the distribution of $r(T)$ under Q^T. This means that we have to identify the Girsanov transormation from Q to Q^T and to this end we will use Remark 5.1.

From the affine term structure formula (5.29) the Q-dynamics of bond prices are given by

$$dp(t,T) = r(t)p(t,T)dt + v(t,T)p(t,T)dW, \qquad (5.39)$$

where W is a Q-Wiener process and the volatility $v(t,T)$ is given by

$$v(t,T) = -\sigma B(t,T). \qquad (5.40)$$

Using Remark 5.1 the Girsanov transformation from Q to Q^T is obtained by using a likelihood process L^T defined by

$$
\begin{aligned}
dL^T(t) &= v(t,T)dW(t), \\
L^T(0) &= 1.
\end{aligned}
$$

Thus the Q^T-dynamics of the short rate are given by

$$dr = \left[\Phi(t) - ar - \sigma^2 v(t,T)\right]dt + \sigma dW^T, \qquad (5.41)$$

where W^T is a Q^T-Wiener process.

Since $v(t,T)$ and Φ are deterministic, r is a Gaussian process, so the distribution of $r(T)$ is determined by its mean and variance under Q^T. The SDE (5.41) is easily solved as

$$
\begin{aligned}
r(T) &= e^{-a(T-t)} + \int_t^T e^{-a(T-s)}\left[\Phi(s) - \sigma^2 v(s,T)\right]ds \qquad (5.42) \\
&\quad + \sigma \int_t^T e^{-a(T-s)}dW^T(s).
\end{aligned}
$$

From this equation the conditional Q^T-variance of $r(T)$, $\sigma_r^2(t,T)$, is given by the expression

$$\sigma_r^2(t,T) = \sigma^2 \int_t^T e^{-2a(T-s)}ds = \frac{\sigma^2}{2a}\left\{1 - e^{-2a(T-t)}\right\}. \qquad (5.43)$$

The Q^T-mean of $r(T)$, $m_r(t,T)$, is obtained directly from (5.15) as

$$m_r(t,T) = f(t,T),$$

which can be observed directly from market data.

Summing up we see that under Q^T the conditional distribution of $r(T)$ has the normal distribution $\mathcal{N}[f(t,T), \sigma_r(t,T)]$, and performing the integration in (5.38) we have the final result.

Proposition 5.7 *Given the assumptions above, the price of the claim $X = H(r(T))$ is given by*

$$\Pi(t,X) = p(t,T)\frac{1}{\sqrt{2\pi\sigma_r^2(t,T)}}\int_{-\infty}^{\infty} h(z)\exp\left\{-\frac{[z - f(t,T)]^2}{2\sigma_r^2(t,T)}\right\}dz, \qquad (5.44)$$

where $\sigma_r^2(t,T)$ is given by (5.43).

5.5 Caps and floors

The object of this subsection is to present the most important interest rate derivative - the cap, and to show how it can be priced within the framework above.

An interest rate **cap** is a contract where the seller of the contract promises to pay a certain amount of cash to the holder of the contract if the interest rate exceeds a certain predetermined level (the "cap rate") at some future date (or dates). If you take a loan, at a floating rate of interst, you may e.g. buy a cap from the bank in order to ensure that you will never have to pay more than, say 6%. In the same way, the seller of a **floor** contract promises to pay cash if some future interest rate falls below a certain level.

Technically a cap contract is a sum of elementary contracts called *caplets*, and we will now give a precise description of a caplet.

We denote the time at which the contract is written by t, and the time period for which the caplet is to be in force is denoted by $[T_0, T_1]$, with length $\delta = T_1 - T_0$. In the cap contract, some underlying **nominal amount** of money, is always specified, and we will denote this amount by K. The **cap rate** is denoted by R.

The interest rate which (in real life) determines the payments of the cap is not the instantaneous rate r, but some market rate like $LIBOR$ (London Interbank Offer Rate). Such an interest rate is not quoted as a continuously compounded rate of interest but rather as a **simple rate** over the period $[T_0, T_1]$. This simple rate, which we denote by $L(T_0, T_1)$ (sometimes omitting the arguments), is determined at T_0 and defined by the relation

$$p(T_0, T_1) = \frac{1}{1 + \delta L} \qquad (5.45)$$

This simply means that if you borrow K dollars over the interval $[T_0, T_1]$ at the $LIBOR$ rate above, then you have to pay back $(1 + \delta L)K$ at the end of the period.

Finally we define the caplet as a contingent T_1-claim, which at time T_1 will pay the amount

$$X = K\delta (L - R)^+ = K \cdot \delta \max (L - R, \, 0)$$

to the holder of the contract. The problem is now how to price the caplet, and without loss of generality we may of course assume that $K = 1$.

From (5.45) we see that, using the notation $p = p(T_0, T_1)$, $R^* = 1 + \delta R$, we have

$$L = \frac{1 - p}{p\delta},$$

which gives us the caplet as

$$X = \delta (L - R)^+ = \delta (L - R)^+ = \delta \left(\frac{1 - p}{p\delta} - R \right)^+$$

$$= \left(\frac{1}{p} - (1 + \delta R)\right)^+ = \left(\frac{1}{p} - R^\star\right)^+.$$

The price of the caplet at time t can now be computed as

$$
\begin{aligned}
\Pi(t, X) &= E^Q\left[e^{-\int_t^{T_1} r(s)ds} \left(\frac{1}{p} - R^\star\right)^+ \bigg| \mathcal{F}_t\right] \\
&= E^Q\left[e^{-\int_t^{T_0} r(s)ds} \left(\frac{1}{p(T_0, T_1)} - R^\star\right)^+ e^{-\int_{T_0}^{T_1} r(s)ds} \bigg| \mathcal{F}_t\right] \\
&= E^Q\left[e^{-\int_t^{T_0} r(s)ds} \left(\frac{1}{p(T_0, T_1)} - R^\star\right)^+ E^Q\left[e^{-\int_{T_0}^{T_1} r(s)ds} \bigg| \mathcal{F}_{T_0}\right] \bigg| \mathcal{F}_t\right] \\
&= E^Q\left[e^{-\int_t^{T_0} r(s)ds} \left(\frac{1}{p(T_0, T_1)} - R^\star\right)^+ p(T_0, T_1) \bigg| \mathcal{F}_t\right] \\
&= E^Q\left[e^{-\int_t^{T_0} r(s)ds} [1 - P(T_0, T_1)R^\star]^+ \bigg| \mathcal{F}_t\right] \\
&= R^\star E^Q\left[e^{-\int_t^{T_0} r(s)ds} \left[\frac{1}{R^\star} - P(T_0, T_1)\right]^+ \bigg| \mathcal{F}_t\right]
\end{aligned}
$$

Thus we see a caplet is equivalent to $R^\star$ put options on a T_1-bond, with delivery date T_0 and strike price $1/R^\star$. Thus we may use the results earlier in this section in order to obtain analytical formulas for caps and floors.

6   Some new directions

In this section we shall, very briefly, give an introduction to some of the recent papers in the area of interest rates. The choice of subjects is of course highly subjective, and the list below is far from complete.

6.1   Lognormal models

In many popular interest rate models, either the short rate or the forward rates are modelled as Gaussian processes. This is the case with such models as Ho-Lee, Vasiček, and and the Hull-White extension of the Vasiček model. On the forward rate side, any model within the Heath-Jarrow-Morton framework possessing a deterministic volatility, will give rise to a Gaussian forward rate curve.

The reason for the popularity of the Gausian models lies of course in their analytical tractability, but there are also drawbacks, the main one being the fact that in a Gaussian model we will have negative interest rates with positive probability. From a philosophical point of view most of us do not want negative interest rates, and a strong argument against Gaussian models can be found in Rogers [46]. In fact, if we assume the existence of cash, then negative interest rates would lead to theoretical arbitrage possibilities.

This has led a number of authors to propose models for which the interest rates are positive, and among those models we find the Cox-ingersoll-Ross (for certain choices of parameters), Dothan, and Black-Derman-Toy.

A natural idea is now to model the interest rates lognormally, thus avoiding negative rates, and this is the way taken by Dothan [16] and Black-Derman-Toy [6]. Lognormal models are nice in the sense that the interest rates stay positive, but the main drawback with a lognormal short rate is that the money account will have infinte expected value, i.e.

$$E[B(T)] = \infty, \quad \forall T.$$

This is a very unpleasant propoerty, and it will lead to the nonsensical result that the price of a Eurodollar future will be minus infinity.

One may also be tempted to model the forward rates in a "lognormal" fashion, by specifying the volatily structure as

$$df(t, T) = \alpha(t, T)dt + \sigma f(t, T)dW. \tag{6.1}$$

It can however be shown (see [25] for references) that in this case the forward rates will explode to plus infinity. This will force bond prices to zero and thus introduce arbitrage possibilities.

The possiblities of modelling *instantaneous* rates in lognormal fashion is thus at a dead end, but lately a lot of work has been done on the modelling of *finite* forward rates within a lognormal framework. See Miltersen-Sandmann-Sondermann 1995 [39] and Brace-Gatarek-Musiela (1995) [7]. Here we follow [39].

The main object of study in [39] is the α-**compounding forward rate** $f(t, T, \alpha)$. This is a forward rate, contracted at t, for the interval $[T, t + \alpha]$, and it is defined by the relation

$$P(t, T + \alpha) = p(t, T) \cdot \frac{1}{1 + \alpha f(t, T, \alpha)}. \tag{6.2}$$

The basic idea in [39] is now to model, for a fixed compounding period α, the α-compounding rates lognormally as

$$df(t, T, \alpha) = \mu(t, T)f(t, T, \alpha)dt + \gamma(t, T)f(t, T, \alpha)dW,$$

where μ and γ are **deterministic**.

Constructing explicit hedging strategies, based on forward contracts on bonds maturing at $T + \alpha$ the authors are then able to produce analytical formulas for the valuation of bond options, caps, and floors. Furthermore, for caps and floors the formulas obtained are similar to the Black formula used in market practice, but without having to make the unrealistic assumptions underlying the Black formula.

6.2   Modelling state price densities

Take as given an arbitrage free model with a stochastic short rate of interest r, and a fixed martingale measure Q. We recall that for a T-claim Y the arbitrage free price process is given by

$$\Pi\left(t, X\right) = E^Q \left[e^{-\int_t^T r(s)ds} \cdot Y \middle| \mathcal{F}_t \right],$$

and, in particccular, we have the price at $t = 0$ as

$$\Pi\left(0, X\right) = E^Q \left[e^{-\int_0^T r(s)ds} \cdot Y \right]. \tag{6.3}$$

We denote the likelihood process for the transition from the objective measure P to the martingale measure Q by L, i.e.

$$L(t) = \frac{dQ_t}{dP_t}, \tag{6.4}$$

where subindex t denotes the restriction to $\mathcal{F}_t$. We may of course also write the price in (6.3) as an expected value under P:

$$E^P \left[e^{-\int_0^T r(s)ds} \cdot L(T) \cdot Y \right] = E^P \left[Z(T) \cdot Y \right], \tag{6.5}$$

where the **state price density process** Z is defined by

$$Z(t) = e^{-\int_0^t r(s)ds} \cdot L(t). \tag{6.6}$$

The idea in Rogers (1996) [45] is to model the state price density process directly under the objective measure P (or in fact under any measure equivalent to Q). This approach has of course also been used in economic theory, but the point of Rogers' paper is that he is able to produce an extremely flexible scheme for producing a great variety of analytically tractable interest rate models. We now go on to give an idea of the structure of Rogers' scheme.

First we note that the state price density has a stochastic differential given by

$$dZ = Z \left[-rdt + L^{-1}dL \right], \tag{6.7}$$

so we can always determine the short rate from a description of the process Z.

The task is now to produce a nice class of state price densities, and to this end we note that L is a positive P-martingale. Furthermore, assuming that r is postive, we see from (6.6) that Z is a positive supermartingale. If, in addition, we have a term structure with the property that $p(0, T) \to 0$ as $T \to \infty$ then Z will in fact be a potential. This is good news, since we can produce a great variety of potentials using the theory of Markov processes and resolvents.

Definition 6.1 *Let X be a Markov process with generator G. Then, for any real α the resolvent R_α is an operator defined by*

$$R_\alpha g(x) = E_x^P \left[\int_0^\infty e^{-\alpha t} g(X_t) dt \right],$$

for any sufficiently nice function g.

It is easy to see that, for any α and any nonnegative g, the process

$$e^{-\alpha t} \frac{R_\alpha g(X_t)}{R_\alpha g(X_0)}, \tag{6.8}$$

is a nonnegative supermartingale, so the basic scheme works as follows.

First we fix a Markov process X, a nonnegative function g, and a number α. Then we define the state price density Z, by

$$Z(t) = e^{-\alpha t} \frac{R_\alpha g(X_t)}{R_\alpha g(X_0)}$$

With this definition Z will indeed be a nonnegative supermartingale, and the short rate can be recaptured by

$$r(t) = \frac{g(X_t)}{R_\alpha g(X_t)}. \tag{6.9}$$

The problem with this scheme is that, for a concrete case, it may be very hard to compute the quotient in (6.9). Using the identity $R_\alpha = (\alpha - G)^{-1}$ we see however that with $f = R_\alpha g$ we have

$$\frac{g(X_t)}{R_\alpha g(X_t)} = \frac{(\alpha - G)f(X_t)}{f(X_t)},$$

where it usually is a trivial task to comput the last quotient.

Thus Rogers instead uses the following scheme

1. Fix a Markov process X, number α and a nonnegative function f.

2. Define g by
$$g = (\alpha - G)f$$

3. Choose α (and perhaps the parameters of f) such that g is nonnegative.

4. Now we have $f = R_\alpha g$, and the short rate can be recaptured by

$$r(t) = \frac{(\alpha - G)f(X_t)}{f(X_t)}.$$

In this way Rogers produces a surprising variety of concrete analytically tractable nonnegative interest rate models. Exchange rates are also considered within the same framework.

6.3   Point process models

Previously in the text all models have been driven by one or several Wiener processes. It is also natural to investigate the consequences of introducing one or several jump processes into the dynamics of the various interest trates.

The introduction of jumps can be thought of as modelling such real world phenomena as interventions by central banks and natural disasters (e.g. a major earthquake in Tokyo). Another reason for introducing jumps is the empirical fact that in many cases the interest trajectories do not at all look like diffusion trajectories, but rather like pure jump trajectories. See e.g. the discussion of repo rates in [36].

Very little seems to have been written about interest rate models driven by point processes. Shirakawa [48], Björk [3], and Jarrow–Madan [32] all consider interest rate models driven by a finite number of counting processes. (Jarrow–Madan also consider the interplay between the stock- and the bond market). See also Babbs-Webber [2] and Lindberg-Orzag-Perraudin [36].

The case of an infinite mark space is treated in [4] and, for Levy processes, in [5] where also the appropriate stochastic integration theory for Banach space valued integrators is developed. See also [33] where the authors consider a fairly general model for asset prices driven by semimartingales.

We follow [4], where the setup is basically as follows. The forward rates are modelled, usually under Q, as

$$df(t,T) = \alpha(t,T)dt + \sigma(t,T)dW_t + \int_E \delta(t,x,T)\mu(dt,dx), \qquad (6.10)$$

where μ is a multivariate point process. This covers the case of a finite number of driving counting processes as well as the case of an infinite mark space E.

The introduction of an infinite mark space into the model will introduce an infinite number of random sources, so in order to have any chance of hedging we are forced to work with measure-valued portfolios which at each point in time contain bonds with a continuum of maturities. Denoting by $h(t,dT)$ the number of bonds held at time t with maturities in the interval $[T, T + dT]$, and by $g(t)$ the number of units of the risk free asset B, the natural definition of the value process is

$$V(t) = g(t)B(t) + \int_t^\infty p(t,T)h(t,dT),$$

The formal generalization of the standard self-financing condition is given by

$$dV(t) = g(t)dB(t) + \int_t^\infty h(t,dT)dp(t,T)$$

and one of the problems is now to give a precise mathematical meaning to the self financing condition above. For driving point processes this can be done in a standard manner ([4]) but for the case of a driving Levy process we need to develop a new theory of integration for measure-valued processes with respect to jump-diffusion processes with values in some Banach space of continuous functions. See [5].

The main ideas and results in [4] are as follows.

- The standard portfolio concept is extended to include measure valued portfolios.

- The no arbitrage condition on the interest rate dynamics provides an extension of the HJM drift condition.

- As opposed to the standard models (with only a finite set of assets) it turns out that market completeness is no longer equivalent to uniqueness of the martingale measure. Instead it is shown that uniqueness of the martingale measure is equivalent to "approximate completeness" of the market (you can only hedge claims belonging to a dense subspace of the space of all claims).

- The hedging equation is generically ill posed in the sense of Hadamard.

- Conditions are given which guarantees the existence of an affine term structure.

- The computational problems turns out to be quite hard.

6.4 Risky bonds

Up to this point we have silently assumed that at the time of maturity the principal amount is actually payed out. Whereas this assumption may be a good approximation for government bonds, it certainly does not hold for the case of corporate bonds, which may (and not unfrequently do) **default**, i.e. the principal amount is not delivered.

The question now arises how we are to value defaultable bonds, and there are two main lines of research.

In the first approach, bonds are seen as derivatives of the underlying assets of the firm, and default occurs typically when the value of the firm hits a barrier. See e.g. Merton [38], Shimko-Tejima-van Deventer [47], and Longstaff-Schwartz [37].

In another approach the time of default is often modelled by an exogenously specified default process. Typical examples are the papers by Duffie-Singleton [18], Jarrow-Turnbull [34], Jarrow-Lando-Turnbull [31] and Lando [35].

To give a feeling for what is going we will present a simple version of the second approach. We follow Lando [35].

Let us thus consider a fixed nominal contingent T-claim X in a market with the risk free interest rate r. Note that r is the rate of interest for the *riskless* bonds.

Assumption 6.1 *We assume that the time of default, τ, is the first jump time of a Cox process with intensity process λ. The process λ is assumed to be independent of r and X.*

We recall that, by definition, the Cox process has the following property:

$$Q\left(N(t) = k \,\middle|\, \mathcal{F}_t^\lambda\right) = e^{-\int_0^T \lambda(s)ds} \cdot \frac{\left(\int_0^T \lambda(s)ds\right)^k}{k!}, \quad k = 0, 1, 2, \cdots \quad (6.11)$$

The claim X was defined as the *nominal* claim, i.e. the claim which is paid if no default occurs. If default occurs before time T then, by definition, nothing is paid to the holder of the contract. The actual payment at time T, X_R, is thus given by

$$X_R = X \cdot I\left\{\tau > T\right\}.$$

We now have the following basic result.

Proposition 6.1 *Denote the risky price of X at $t = 0$ by $\Pi_R[X]$, and the price of a risky zero coupon bond by $q(0,T)$. Then we have*

$$\Pi_R[X] = E^Q\left[e^{-\int_0^T R(s)ds} \cdot X\right], \quad (6.12)$$

$$q(0,T) = E^Q\left[e^{-\int_0^T R(s)ds} \cdot 1\right], \quad (6.13)$$

where the risk adjusted *interest rate R is defined by*

$$R(s) = r(s) + \lambda(s).$$

Proof. Using the Cox property (6.11) we obtain

$$\Pi_R[X] = E^Q\left[e^{-\int_0^T r(s)ds} \cdot X \cdot I\left\{N_T = 0\right\}\right]$$

$$= E^Q\left[E^Q\left[e^{-\int_0^T r(s)ds} \cdot X \cdot I\left\{N_T = 0\right\}\middle|\mathcal{F}_t^\lambda\right]\right]$$

$$= E^Q\left[e^{-\int_0^T r(s)ds} \cdot X \cdot E^Q\left[I\left\{N_T = 0\right\}\middle|\mathcal{F}_t^\lambda\right]\right]$$

$$= E^Q\left[e^{-\int_0^T r(s)ds} \cdot X \cdot e^{-\int_0^T \lambda(s)ds}\right]$$

$$= E^Q\left[e^{-\int_0^T R(s)ds} \cdot X\right] \blacksquare$$

We see that we have the same structure in the valuation formula as in the standard non defaultable case. The difference is that we have to discount using the risk adjusted rate of interest. These results have also been generalized to more complicated payoff structures, see Lando (1994), [35], and Duffie-Singleton (1995), [18]

6.5   Minimizing arbitrage information

In all models considered so far, we have taken the dynamics of the interest rates as exogenously given. A completely different approach is taken by Platen (1996) [42], where the interest rate is derived as a consequence of an entropy related principle. See also [43].

Platen starts by modelling asset prices $X_1(t), \cdots, X_d(t)$, under the objective measure P as exponential processes.

Price dynamics:

$$X_i(t) = X_i(0)e^{L_i(t)}, \quad i = 1, \cdots, d,$$
$$dL_i = c[l_i - L_i] dt + \sigma_i dW_i.$$

Thus the logarithmic prices are mean reverting processes, and the mean reversion processes are also given dynamics of the form

Mean reversion dynamics:

$$\frac{dl_i}{dt} = I(t) + g_i(t)$$

where I is inflation and g_i is the *net growth rate* of X_i. The price volatility is also dynamical and modelled in the following way.

Volatility dynamics:

$$\frac{d\sigma_i}{dt} = \sigma_i \tau_i,$$

where τ_i is the so called *volatility trend*. At last the inflation rate is assumed to satisfy the following SDE.

Inflation dynamics:

$$dI = q[m - I] dt + p dW$$

All Wiener processes involved are assumed to be independent (this can easily be generalized), and all the constants above are actually allowed to be determinsitic functions of time.

Now consider the following scheme.

1. Specify an interest rate process r, with money account

$$dB(t) = r(t)B(t)dt.$$

2. Change measure **on the price factors** $W_1, \cdots, W_d$ only to get a martingale measure Q. In this way Q will be uniquely determined, and we denote the likelihood process by Φ:

$$\Phi(t) = \frac{dQ_t}{dP_t}$$

3. Compute the Kullback-Leibler information process

$$h(t) = \Phi^{-1}(t) \cdot \log \Phi^{-1}(t)$$

4. Define the information functional H by

$$H(t) = E^Q [h(t)]$$

Platen does not actually specify a short rate of interest as in item 1 above. Instead he proposes the following basic principle.

Principle: *The interest rate chosen by the market is the choice of r which, for each t, minimizes $\frac{dH}{dt}$, the so called "increase of arbitrage information".*

Thus the short rate of interest is determined within the system, and by applying this principle Platen shows that the interest rate dynamics are of the form

$$dr = c\,[\bar{r} - r]\,dt - c\,dM,$$

where M is a Wiener martingale. The "approximate short rate" $\bar{r}$ turns out to be the inflation rate plus a deterministic term that can be computed. The model provide a nice fit to US, Australian and German market data.

Bibliography

[1] Artzner, P. & Delbaen, F. (1989) Term structure of interest rates. *Advances in Applied Mathematics* **10**, 95-129.

[2] Babbs, S. & Webber, N. (1993) A theory of the term structure with an official short rate. Working paper, University of Warwick.

[3] Björk, T. (1995) On the term structure of discontinuous interest rates. *Surveys in Industrial and Applied Mathematics* **2** No.4, 626-657.

[4] Björk, T. & Kabanov, Y. & Runggaldier, W. (1995) Bond market structure in the presence of marked point processes. Submitted to *Mathematical Finance*.

[5] Björk, T. & Di Masi, G. & Kabanov, Y. & Runggaldier, W. (1995) Towards a general theory of bond markets. *Finance and Stochastics*. To appear.

[6] Black, F. & Derman, E & Toy, W. (1990) A one-factor model of interest rates and its application to treasury bond options. *Finan. Analysts J.* jan-feb, 33-39.

[7] Brace, A. & Gatarek, D. & Musiela, M. (1995) The market model of interest rate dynamics. preprint. Dept. of Statistics. University of New South Wales, Australia.

[8] Brace, A. & Musiela M. (1994) A multifactor Gauss Markov implementation of Heath, Jarrow, and Morton. *Mathematical Finance* **4**, 3, 259-283.

[9] Brémaud, P. (1981) *Point Processes and Queues: Martingale Dynamics*. Springer-Verlag, Berlin.

[10] Brown, R.H. & Schaefer, S.M. (1994) Interest rate volatility and the shape of the term structure. *Phil. Trans. R. Soc. Lond. A* **347**, 563-576.

[11] Cox, J. & Ingersoll, J. & Ross, S. (1985) A theory of the term structure of interest rates. *Econometrica* **53**, 385-408.

[12] Dana, R.-A. & Jeanblanc-Picqué M. (1994) *Marchés Financiers en Temps Continu. Valorisation et Equilibre*. Economica, Paris.

[13] Delbaen, F. & Schachermayer, W. (1994) A general version of the fundamental theorem on asset pricing. *Mathematische Annalen* **300**, 463-520.

[14] Dellacherie, C & Meyer, P-A. (1972) *Probabilités et Potentiel.* Hermann,
Paris.

[15] Dothan, M. (1978) On the Term Structure of Interest Rates. *Journal of
Financial Economics* **7**, 229-264.

[16] Duffie, D. (1992) *Dynamic Asset Pricing Theory.* Princeton Univ. Press.

[17] Duffie, D. & Kan, R (1993) A yield factor model of interest rates. Preprint,
GSB Stanford University.

[18] Duffie, D. & Singleton, K. (1995) Modeling of term structures of defaultable
bonds. Working paper, Stanford University.

[19] El Karoui, N. & Myneni, R. & Viswanathan, R. (1992) Arbitrage pricing
and hedging of interest rate claims with state variables. I, II. Preprints,
Laboratoire de Probabilités, Paris 6.

[20] Elliott, R.J. (1982) *Stochastic Calculus and Applications.* Springer-Verlag,
Berlin.

[21] Geman, H. (1989) The importance of the forward neutral probability in a
stochastic approach of interest rates. Working paper, ESSEC.

[22] Geman, H. & El Karoui, N & Rochet, J-C. (1995) Changes of numéraire,
changes of probability measure and option pricing. *Journal of Applied Probability* **32**, 443-458.

[23] Harrison, J.M. & Kreps, D. (1979) Martingales and arbitrage in multiperiod
securities markets. *Journal of Economic Theory* **20**, 381-408.

[24] Harrison, J.M. & Pliska, S. (1981) Martingales and stochastic integrals in
the theory of continuous trading. *Stochastic Processes & Appl.* **11**, 215-260.

[25] Heath, D. & Jarrow, R. & Morton, A. (1992) Bond pricing and the term
structure of interest rates. *Econometrica* **60** No.1, 77-106.

[26] Ho, T. & Lee, S. (1986) Term structure movements and pricing interest rate
contingent claims. *Journal of Finance* **41**, 1011-1029.

[27] Hull, J & White, A. (1990) Pricing interest-rate-derivative securities. *The
Review of Financial Studies* **3**, 573-592.

[28] Jacod, J. (1979) *Calcul Stochastique et Problèmes de Martingales.* LNM
714, Springer-Verlag, Berlin.

[29] Jacod, J. & Shiryaev A.N. (1987) *Limit Theorems for Stochastic Processes.*
Springer-Verlag, Berlin.

[30] Jamshidian, F. (1989) An exact bond option formula. *Journal of Finance* **44**, 205-209.

[31] Jarrow, R. & Lando D. & Turnbull S.M. (1995) A Markov model for the term stucture of credit risk spreads. *Review of Financial Studies.* To appear.

[32] Jarrow, R. & Madan D. (1995) Option pricing using the term structure of interest rates to hedge systematic discontinuities in asset returns. *Mathematical Finance* **5**, 311-336.

[33] Jarrow, R. & Madan D. (1995) Valuing and hedging contingent claims on semimartingales. Preprint.

[34] Jarrow, R. & Turnbull. (1995) Pricing derivatives on financial securities subject to credit risk. *Journal of Finance* **50**, 53-86.

[35] Lando, D. (1994) On Cox processes and credit risky bonds. Institute of Mathematical Statistics, University of Copanehagen.

[36] Lindberg, H. & Orszag, M. & Perraudin, W. (1995) Yield curves with jump short rates. Preprint.

[37] Longstaff, F. & Schwartz, E (1992) Valuing risky debt: A new approach.Working paper, UCLA.

[38] Merton, R. (1974) On the pricing of corporate debt: The risk structure of interest rates. *Journal of Finance* **2**, 449-470.

[39] Miltersen, K. & Sandmann, K. & Sondermann, D. (1995) Closed form solutions for term structure derivatives with log-normal interest rates. Dept. of management, Odense University, Denmark.

[40] Musiela, M. (1993) Stochastic PDEs and term structure models. Preprint.

[41] Musiela, M. & Rutkowski, M. Arbitrage Pricing of Derivative Securities. Theory and Applications. Preprint.

[42] Platen, E. (1996) Explaining interest rate dynamics. Preprint. Centre for Financial mathematics, Australian National University, Canberra.

[43] Platen, E. & Rebolledo, R. (1995) Principles for modelling financial markets. To appear in *Advances in Applied Probability.*

[44] Protter, P. (1990) *Stochastic Integration and Differential Equations.* Springer-Verlag, Berlin.

[45] Rogers, L.C.G. (1995) The potential approach to the term structure of interest rates and foreign exchange rates. Preprint, University of Bath.

[46] Rogers, L.C.G. (1995) Which model for term-structure of interest rates should one use?' *Mathematical Finance, IMA Volume 65,* 93–116, Springer, New York.

[47] Shimko, D. & Tejima, N. & van Deventer, D (1993) The pricing of risky debt when interest rates are stochastic. *Journal of Fixed Income*, September, 58-66.

[48] Shirakawa, H. (1991) Interest rate option pricing with Poisson–Gaussian forward rate curve processes. *Mathematical Finance* 1, 77-94.

[49] Vasiček, O. (1977) An equilibrium characterization of the term structure. *Journal of Financial Economics* 5, 177-188.

OPTIMAL TRADING UNDER CONSTRAINTS*

by

Jakša Cvitanić

Department of Statistics
Columbia University
New York, NY 10027
cj@stat.columbia.edu

Abstract

These are lecture notes on the techniques and results of the theory of optimal trading for a single agent under convex constraints on his portfolio process, in a continuous-time model. A similar methodology is applied to the case of policy dependent prices, different interest rates for borrowing and lending and transaction costs problems. We study the hedging problem and the portfolio optimization problem for the investor in this market. Mathematical tools involved are those of continuous-time martingales, convex duality, forward-backward SDE's and PDE's.

---

*Supported in part by the National Science Foundation under Grant NSF-DMS-95-03582, and C.I.M.E.

Key Words and Phrases: hedging, portfolio optimization, portfolio constraints, transaction costs, large investor forward-backward SDE's.

AMS-MOS (1991) Subject Classification: Primary 93E20, 90A09, 60H30. Secondary 60G44, 90A16.

CONTENTS

1. INTRODUCTION.

The main topic of these lecture notes is the problem of hedging (super-replication) and utility maximization for a single agent in a continuous-time financial market, under convex constraints on the proportions of wealth he/she invests in stocks. We present the model in Section 2; it is a standard, generalized Black-Scholes-Samuelson-Merton continuous-time diffusion model for several (d) risky assets (called "stocks") and one riskless "bank account". In Section 3 we introduce an agent who can trade in the assets, and describe his/her portfolio and wealth processes. We present the "equivalent martingale measure" approach to pricing financial contracts in the market in Section 4. The fair price of a contingent claim is obtained as its expectation under a change of probability measure that makes stocks prices martingales, after discounting by the interest rate of the bank account. In the special case of constant market parameters and the European call contract this leads to the famous Black-Scholes formula. Sections 5 and 6 generalize this approach to the case of a constrained market, in which the agent's hedging portfolio has to take values in a given closed convex set K. It is shown that the minimal hedging cost of a claim is now a supremum of Black-Scholes prices, taken over a family of auxiliary markets, parametrized by processes $\nu(\cdot)$, taking values in the domain of the support function of the set $-K$. These markets are chosen so that the wealth process becomes a super-martingale, under the appropriate change of measure. It is also shown that the supremum is attained if and only if the Black-Scholes (unconstrained) hedging portfolio happens to satisfy constraints. The latter result is used to prove, in the constant market parameters framework, that the minimal hedging cost under constraints can be calculated as the Black-Scholes price of an appropriately modified contingent claim, and that the corresponding hedging portfolio automatically satisfies the constraints. We end Section 6 by showing that there is no arbitrage in the constrained market if and only if a price of a claim is chosen in the interval determined by the minimal hedging price of the seller and the maximal hedging price of the buyer.

In Section 7 we show how the same methodology can be used to get analogous results in a market in which the drift of the wealth process is a concave function of the portfolio process. This can be regarded as a model in which the asset prices parameters depend on the trading strategy of the investor. More general model of this kind, in which both the drift and the diffusion terms of the prices depend on the portfolio and wealth of the investor is studied in Section 8, by different, Forward-Backward SDE's methods. An example is given in Section 9, providing a way of calculating the hedging price of options when the interest rate for borrowing is larger than the one for lending.

In Section 10 we introduce the concept of utility functions and prove existence of an optimal constrained portfolio strategy for maximizing expected utility from terminal wealth in Section 11. This is done indirectly, by first solving a dual problem, which is, loosely speaking, a problem to find an optimal change of probability measure associated to the constrained market. The optimal portfolio policy is the one that hedges the inverse of "marginal utility" (derivative of the utility function), evaluated at the Radon-Nikodym derivative corresponding to the optimal change of measure in the dual problem. Explicit solutions are provided for the case of logarithmic and power utilities. Next, in Section 12, we argue that it makes sense to price contingent claims in the constrained market by calculating the Black-Scholes price in the unconstrained auxiliary market that corresponds to the optimal dual change of measure. Although in general this price depends on the utility of the agent and his/her initial capital, in many cases it does not. In particular, if the contraints are given by a cone, and the market parameters are constant, the optimal dual process is independent of utility and initial capital.

In Sections 13-17 we study the hedging and utility maximization problem in the presence of proportional transaction costs. Similarly as in the case of constraints, we identify the family of (pairs of) changes of probability measure, under which the "wealth process" is a supermartingale, and the supremum over which gives the minimal hedging price of a claim in this market. In this case we do not know how to actually calculate this price, although it is known in some special cases. In particular, it is trivial for European call; namely, in order to hedge the call almost surely in the presence of positive transaction costs, one has to buy a whole share of the stock and hold it. Next, we consider the utility maximization problem in this setting, and its dual. We prove the existence in the primal problem, but, unfortunately, we do not know in general, whether an optimal solution exists for the dual problem. Under the assumption that the optimal dual solution does exist, the nature of the optimal terminal wealth in the primal problem is the same as in the case of constraints - it is equal to the inverse of the marginal utility evaluated at the optimal dual solution. This result is used to get sufficient conditions for the optimal policy to be the one of no trade at all - this is the case if the return rate of the stock is not very different from the interest rate of the bank account and the transaction costs are large relative to the time horizon.

In Section 18 we study the problem of maximizing the long-term growth rate of agent's wealth, under the constraint that the wealth never falls below a given fraction of its maximum up-to-date. A simple "trick" makes this problem equivalent to an unconstrained utility maximization problem, which is solved by the methods of previous sections.

We collect some basic results from stochastic calculus in Appendix.

. Finally, I would like to thank the PhD students at Columbia University who attended the course for which the first version of the lecture notes was prepared, and who gave a lot of useful remarks and suggestions: C. Hou, Y. Jin, Y. Lu, O. Mokliatchouk, H. Tang, X. Zhao, and, especially, Gennady Spivak. Thanks are also due to the finance group in CREST, Paris, in particular to Nizar Touzi and Huyen Pham. Moreover, big thanks are due to the participants and organizers of the CIME Summer School on Financial Mathematics, Bressanone 1996, for which the more expanded and polished version was prepared. In particular, without great enthusiasm and organizational skills of Professor Wolfgang Runggaldier, none of this would be possible.

2. THE MODEL

We consider a financial market $\mathcal{M}$ which consists of one *bank account* and several (d) *stocks*. The prices $P_0(t)$, $\{P_i(t)\}_{1 \le i \le d}$ of these financial instruments evolve according to the equations

$$dP_0(t) = P_0(t)r(t)dt , \quad P_0(0) = 1 \tag{2.1}$$

$$dP_i(t) = P_i(t)[b_i(t)dt + \sum_{j=1}^{d} \sigma_{ij}(t)dW^{(j)}(t)] , \quad P_i(0) = p_i \in (0, \infty) ; \quad i = 1, \dots, d .$$
$$\tag{2.2}$$

Here $W = (W^{(1)}, \dots, W^{(d)})^*$ is a standard Brownian motion in $\mathbb{R}^d$, defined on a complete probability space $(\Omega, \mathcal{F}, \mathbf{P})$, and we shall denote by $\{\mathcal{F}_t\}$ the **P**-augmentation of the filtration $\mathcal{F}_t^W = \sigma(W(s); 0 \le s \le t)$ generated by W. The *coefficients* (or parameters) of $\mathcal{M}$ - i.e., the processes $r(t)$ (scalar interest rate), $b(t) = (b_1(t), \dots, b_d(t))^*$ (vector of appreciation rates) and $\sigma(t) = \{\sigma_{ij}(t)\}_{1 \le i, j \le d}$ (volatility matrix) - are assumed to be progressively measurable with respect to $\{\mathcal{F}_t\}$ and *bounded* uniformly in $(t, \omega) \in [0, T] \times \Omega$. We shall also impose the following strong non-degeneracy condition on the matrix $a(t) := \sigma(t)\sigma^*(t)$:

$$\xi^* a(t)\xi \ge \varepsilon \|\xi\|^2, \quad \forall \, (t, \xi) \in [0, T] \times \mathbb{R}^d \tag{2.3}$$

almost surely, for a given real constant $\varepsilon > 0$.

We introduce also the "relative risk" process

$$\theta(t) := \sigma^{-1}(t)[b(t) - r(t)\mathbf{1}] , \tag{2.5}$$

where $\mathbf{1} = (1, \dots, 1)^*$. The exponential martingale

$$Z_0(t) := \exp[-\int_0^t \theta^*(s)dW(s) - \frac{1}{2}\int_0^t \|\theta(s)\|^2 ds] \tag{2.6}$$

and the discount process

$$\gamma_0(t) := \exp\{-\int_0^t r(s)ds\} \tag{2.7}$$

will be employed quite frequently.

2.1 Remark: It is a straightforward consequence of the strong non-degeneracy condition (2.3), that the matrices $\sigma(t)$, $\sigma^*(t)$ are invertible, and that the norms of $(\sigma(t))^{-1}, (\sigma^*(t))^{-1}$ are bounded above and below by δ and $1/\delta$, respectively, for some $\delta \in (1, \infty)$; see Karatzas & Shreve (1991), page 372. The boundedness of $b(\cdot), r(\cdot)$ and $(\sigma(\cdot))^{-1}$ implies that of $\theta(\cdot)$, and thus also the martingale property of the process $Z_0(\cdot)$ in (2.6).

3. PORTFOLIO AND WEALTH PROCESSES

Consider now an economic agent whose actions cannot affect market prices, and who can decide, at any time $t \in [0, T]$, what proportion $\pi_i(t)$ of his wealth $X(t)$ to invest in the ith stock $(1 \leq i \leq d)$. Of course these decisions can only be based on the current information $\mathcal{F}_t$, without anticipation of the future. With $\pi(t) = (\pi_1(t), \ldots, \pi_d(t))^*$ chosen, the amount $X(t)[1 - \sum_{i=1}^{d} \pi_i(t)]$ is invested in the bank. Thus, in accordance with the model set forth in (2.1), (2.2), the wealth process $X(t)$ satisfies the linear stochastic equation

$$
dX_t = \sum_{i=1}^{d} \pi_i(t) X_t \{b_i(t)dt + \sum_{j=1}^{d} \sigma_{ij}(t) dW^{(j)}(t)\} + \{1 - \sum_{i=1}^{d} \pi_i(t)\} X_t r_t dt
$$

$$
= r(t) X(t) dt + X(t) \pi^*(t) \sigma(t) dW_0(t) ; \quad X(0) = x > 0 ,
$$
(3.1)

where the real number $x > 0$ represents initial capital and

$$
W_0(t) := W(t) + \int_0^t \theta(s) ds , \quad 0 \leq t \leq T .
$$
(3.2)

We formalize the above discussion as follows.

3.1 Definition: (i) An $\mathbb{R}^d$ - valued, $\{\mathcal{F}_t\}$ - progressively measurable process $\pi = \{\pi(t), 0 \leq t \leq T\}$ with $\int_0^T X^2(t) \|\pi(t)\|^2 dt < \infty$, a.s., will be called a *portfolio process* (here, X is the corresponding wealth process defined in (ii) below).

(ii) Given the portfolio $\pi(\cdot)$ as above, the solution $X \equiv X^{x,\pi}$ of the equation (3.1) will be called the *wealth process* corresponding to the portfolio π and initial capital $x \in (0, \infty)$.

3.2 Definition: A portfolio process π is called *admissible* for the initial capital $x \in (0, \infty)$, if

$$
X^{x,\pi}(t) \geq 0, \quad \forall \, 0 \leq t \leq T
$$
(3.3)

holds almost surely. The set of admissible portfolios π will be denoted by $\mathcal{A}_0(x)$.
◊

In the notation of (2.6), (2.7), the equation (3.1) leads to

$$
M_0(t) := \gamma_0(t) X(t) = x + \int_0^t \gamma_0(s) X(s) \pi^*(s) \sigma(s) dW_0(s).
$$
(3.4)

In particular, the discounted wealth process $M_0(\cdot)$ of (3.4) is seen to be a continuous local martingale under the so-called "risk-neutral" probability measure (or "equivalent martingale measure")

$$
\mathbf{P}^0(A) := E[Z_0(T) 1_A], \quad A \in \mathcal{F}_T .
$$
(3.5)

If $\pi \in \mathcal{A}_0(x)$, the $\mathbf{P}^0$-local martingale $M_0(\cdot)$ of (3.4) is also nonnegative, thus a supermartingale. Consequently,

$$E^0[\gamma_0(T)X^{x,\pi,c}(T)] \leq x, \quad \forall \, \pi \in \mathcal{A}_0(x) . \qquad (3.6)$$

Here, E^0 denotes the expectation operator under the measure $\mathbf{P}^0$; under this measure, the process W_0 of (3.2) is standard Brownian motion by Girsanov theorem and the discounted stock prices $\gamma_0(\cdot)P_i(\cdot)$ are martingales, since

$$dP_i(t) = P_i(t)[r(t)dt + \sum_{j=1}^{d} \sigma_{ij}(t)dW_0^{(j)}(t)] , \;\; P_i(0) = p_i; \;\; i = 1,\ldots,d \qquad (3.7)$$

from (2.2) and (3.2).

3.3 Remark: For any given $\pi \in \mathcal{A}_0(x)$, let $X(\cdot) \equiv X^{x,\pi}(\cdot)$ and define the "bankruptcy time"

$$S := \inf\{t \in [0,T]; X(t) = 0\} \wedge T. \qquad (3.8)$$

Because the continuous process $M_0(\cdot)$ of (3.4) is a $\mathbf{P}^0$-supermartingale, the same is true of $\gamma_0(\cdot)X(\cdot)$, and thus (see Karatzas & Shreve (1991), Problem 1.3.29) for a.e. $\omega \in \{S < T\}$:

$$X(t,\omega) = 0, \quad \forall \, t \in [S(\omega), T]. \qquad (3.9)$$

In other words, if the wealth $X(\cdot)$ becomes equal to zero before the end T of the horizon, it stays there; further values of the portfolio $\pi(\cdot)$ become irrelevant. This remark seems to be unnecessary since the solution $X(\cdot)$ to the linear equation (3.1) is always positive. However, we shall, in fact, allow the possibility of bankruptcy; i.e., we shall allow continuous wealth processes modeled by (3.1) for $t < S$, where S is some stopping time, and $X(\cdot) \equiv 0$ for $S \leq t \leq T$.

3.4 Definition: We say that a portfolio strategy $\pi(\cdot)$ results in *arbitrage* if the initial investment $x = 0$, $X^{0,\pi}(T) \geq 0$, but $X^{0,\pi}(T) > 0$ with positive probability.

Notice that inequality (3.6) implies that an admissible portfolio $\pi \in \mathcal{A}_0(0)$ cannot result in arbitrage.

4. PRICING CONTINGENT CLAIMS IN A COMPLETE MARKET

Let us suppose now that an agent promises to pay a random amount $B(\omega) \geq 0$ at time $t = T$. *What is the value of this promise at time $t = 0$?* In other words, how much should the agent charge for selling a contractual obligation that entitles its holder to a payment of size $B(\omega)$ at $t = T$?

For instance, suppose that this obligation stipulates selling one share of the first stock at a contractually specified price q. If at time $t = T$ the price $P_1(T,\omega)$ of the stock is below q, the contract is worthless to its holder; if not, the holder can purchase the stock at the price q per share and then sell it at price $P_1(T,\omega)$, thus making a profit of $P_1(T,\omega) - q$. In other words, this contract entitles its

holder to a payment of $B(\omega) = (P_1(T, \omega) - q)^+$ at time $t = T$; it is called a
(European) *call option* with "exercise price" q and "maturity date" T.

To answer the question of the first paragraph, one argues as follows. Suppose
the agent sets aside an amount $x > 0$ at time $t = 0$; he/she invests in the market
$\mathcal{M}$ according to some portfolio $\pi(\cdot)$, but wants to be certain that at time $t = T$
he/she will be able to *cover his/her obligation*, i.e., that $X^{x,\pi}(T) \geq B$ will hold
almost surely. What is the smallest value of $x > 0$ for which such "hedging" is
possible? This smallest value will then be the "price" of the contract at time
$t = 0$.

4.1 Definition: A *Contingent Claim* is a nonnegative, $\mathcal{F}_T$-measurable random
variable B that satisfies

$$0 < E^0[\gamma_0(T)B] < \infty \ . \tag{4.1}$$

The *hedging price* of this contingent claim is defined by

$$u_0 := \inf\{x > 0; \ \exists \pi \in \mathcal{A}_0(x) \ \ s.t. \ \ X^{x,\pi}(T) \geq B \ \ a.s.\} \ . \tag{4.2}$$

The following classical result identifies u_0 as the expectation, under the risk-
neutral probability measure of (3.5), of the claim's discounted value (Harrison
& Kreps (1979), Harrison & Pliska (1981, 83)).

4.2 Proposition: *The infimum in (4.2) is attained, and we have*

$$u_0 = E^0[\gamma_0(T)B] \ . \tag{4.3}$$

*Furthermore, there exists a portfolio $\pi_0(\cdot)$ such that $X_0(\cdot) \equiv X^{u_0,\pi_0}(\cdot)$ is given
by*

$$X_0(t) = \frac{1}{\gamma_0(t)} E^0[\gamma_0(T)B|\mathcal{F}_t] \ , \ \ 0 \leq t \leq T \ . \tag{4.4}$$

Proof: Suppose $X^{x,\pi}(T) \geq B$ holds a.s. for some $x \in (0, \infty)$ and a suitable
$\pi \in \mathcal{A}_0(x)$. Then from (3.6) we have $x \geq z := E^0[\gamma_0(T)B]$ and thus $u_0 \geq z$.

On the other hand, from the martingale representation theorem, the process

$$X_0(t) := \frac{1}{\gamma_0(t)} E^0[\gamma_0(T)B|\mathcal{F}_t] \ , \ \ 0 \leq t \leq T$$

can be represented as

$$X_0(t) := \frac{1}{\gamma_0(t)} [z + \int_0^t \psi^*(s)dW_0(s)] \tag{4.5}$$

for a suitable $\{\mathcal{F}_t\}$-progressively measurable process $\psi(\cdot)$ with values in $\mathbb{R}^d$ and
$\int_0^T \|\psi(t)\|^2 dt < \infty$, a.s. Then $\pi_0(t) := \frac{1}{\gamma_0(t)X_0(t)}(\sigma^*(t))^{-1}\psi(t)$ is a well-defined,
portfolio process (recall Remarks 2.1 and 3.3), and a comparison of (4.5) with
(3.4) yields $X_0(\cdot) \equiv X^{z,\pi_0}(\cdot)$. Therefore, $z \geq u_0$.   ◇

Notice that

$$X_0(T) = X_0^{u_0, \pi_0}(T) = B \ , \quad a.s. \tag{4.6}$$

in Theorem 4.2; we express this by saying that the contingent claim is *attainable* (with initial capital u_0 and portfolio π_0).

4.3 Example. *Constant* $r(\cdot) \equiv r > 0, \sigma(\cdot) \equiv \sigma$ *nonsingular.* In this case, the solution $P(t) = (P_1(t), \ldots, P_d(t))^*$ is given by $P_i(t) = h_i(t - s, P(s), \sigma(W_0(t) - W_0(s)))$, $0 \le s \le t$ where $h : [0, \infty) \times \mathbb{R}_+^d \times \mathbb{R}^d \to \mathbb{R}_+^d$ is the function defined by

$$h_i(t, p, y; r) := p_i \exp[(r - \tfrac{1}{2} a_{ii})t + y_i] \ , \quad i = 1, \ldots, d. \tag{4.7}$$

Consider now a contingent claim of the type $B = \varphi(P(T))$, where $\varphi : \mathbb{R}_+^d \to [0, \infty)$ is a given continuous function, that satisfies polynomial growth conditions in both $\|p\|$ and $1/\|p\|$. Then the value process of this claim is given by

$$X_0(t) = e^{-r(T-t)} E^0[\varphi(P(T))|\mathcal{F}_t]$$

$$= e^{-r(T-t)} \int_{\mathbb{R}^d} \varphi(h(T - t, P(t), \sigma z)) \frac{1}{(2\pi(T-t))^{d/2}} \exp\{-\frac{\|z\|^2}{2(T-t)}\} dz$$

$$= U(T - t, P(t)), \tag{4.8}$$

where

$$U(t, p) := \left\{ \begin{array}{ll} e^{-rt} \int_{\mathbb{R}^d} \varphi(h(t, p, \sigma z; r)) \frac{e^{-\|z\|^2/2t}}{(2\pi t)^{d/2}} dz & ; \ t > 0, \ p \in \mathbb{R}_+^d \\ \varphi(p) & ; \ t = 0, \ p \in \mathbb{R}_+^d \end{array} \right\}. \tag{4.9}$$

In particular, the price u_0 of (4.3) is given, in terms of the function U of (4.9), by

$$u_0 = X_0(0) = U(T, P(0)) \ . \tag{4.10}$$

Moreover, function U is the unique solution to the Cauchy problem (by Feynman-Kac theorem)

$$\frac{1}{2} \sum_{n=1}^{d} \sum_{l=1}^{d} a_{nl} x_n x_l \frac{\partial^2 U}{\partial x_n \partial x_l} + \sum_{n=1}^{d} r(x_n \frac{\partial U}{\partial x_n} - U) = \frac{\partial U}{\partial t},$$

with the initial condition $U(0, x) = \varphi(x)$. Applying Ito's rule, we obtain

$$dU(T - t, P(t)) = rU(T - t, P(t)) + \sum_{n=1}^{d} \sum_{l=1}^{d} \sigma_{nl} P_n(t) \frac{\partial U}{\partial x_n}(T - t, P_n(t)) dW_0^{(l)}(t).$$

Comparing this with (3.1), we get that the hedging portfolio is given by

$$\pi_n(t) U(T - t, P(t)) = P_n(t) \frac{\partial U}{\partial x_n}(T - t, P(t)), \ n = 1, \ldots, d. \tag{4.11}$$

It should be noted that none of the above depends on vector $b(\cdot)$ of return rates.

A very explicit computation for the function U is possible for $d = 1$ in the case $\varphi(p) = (p - q)^+$ of a *European call option*: with $\sigma = \sigma_{11} > 0$, exercise price

$q > 0, \Phi(z) = \frac{1}{\sqrt{2\pi}} \int_{-\infty}^{z} e^{-u^2/2} du$ and $\nu_{\pm}(t, p) := \frac{1}{\sigma\sqrt{t}}\left[\log(\frac{p}{q}) + (r \pm \frac{\sigma^2}{2})t\right]$, we have the famous *Black & Scholes (1973) formula*

$$U(t, p) = \left\{ \begin{array}{ll} p\Phi(\nu_+(t, p)) - qe^{-rt}\Phi(\nu_-(t, p)) & ; \ t > 0, \ p \in (0, \infty) \\ (p - q)^+ & ; \ t = 0, \ p \in (0, \infty) \end{array} \right\}. \quad (4.12)$$

5. CONVEX SETS AND CONSTRAINED PORTFOLIOS

We shall fix throughout a nonempty, closed, convex set K in $\mathbb{R}^d$, and denote by

$$\delta(x) \equiv \delta(x|K) := \sup_{\pi \in K}(-\pi^* x) \ : \ \mathbb{R}^d \to \mathbb{R} \cup \{+\infty\} \quad (5.1)$$

the support function of the convex set $-K$. This is a closed, positively homogeneous, proper convex function on $\mathbb{R}^d$ (Rockafellar (1970), p.114). It is finite on its *effective domain*

$$\tilde{K} := \{x \in \mathbb{R}^d; \ \delta(x|K) < \infty\} = \{x \in \mathbb{R}^d; \ \exists \ \beta \in \mathbb{R} \ s.t. \ -\pi^* x \leq \beta, \ \forall \ \pi \in K\}, \quad (5.2)$$

which is a convex cone (called the "barrier cone" of $-K$). It will be assumed throughout this paper that the function

$$\delta(\cdot|K) \quad \text{is continuous on} \quad \tilde{K} \quad (5.3)$$

and that $0 \in K$, so that:

$$\delta(x|K) \geq 0, \quad \forall \ x \in \mathbb{R}^d. \quad (5.4)$$

5.1 Remark: Theorem 10.2, p.84 in Rockafellar (1970) guarantees that (5.3) is satisfied, in particular, if K is locally simplicial.

5.2 Examples: The role of the closed, convex set K that we just introduced, is to model reasonable constraints on portfolio choice. One may, for instance, consider the following examples, all of which satisfy conditions (5.3) and (5.4).

(i) *Unconstrained case:* $K = \mathbb{R}^d$. Then $\tilde{K} = \{0\}$, and $\delta \equiv 0$ on $\tilde{K}$.

(ii) *Prohibition of short-selling:* $K = [0, \infty)^d$. Then $\tilde{K} = K$, and $\delta \equiv 0$ on $\tilde{K}$.

(iii) *Incomplete Market:* $K = \{\pi \in \mathbb{R}^d; \pi_i = 0, \ \forall \ i = m + 1, \ldots, d\}$ for some fixed $m \in \{1, \ldots, d-1\}$. Then $\tilde{K} = \{x \in \mathbb{R}^d; \ x_i = 0, \ \forall \ i = 1, \ldots, m\}$ and $\delta \equiv 0$ on $\tilde{K}$.

(iv) *Incomplete Market with prohibition of short-selling:* $K = \{\pi \in \mathbb{R}^d; \ \pi_i \geq 0, \ \forall \ i = 1, \ldots, m \ and \ \pi_i = 0, \ \forall \ i = m+1, \ldots, d\}$ with m as in (iii). Then $\tilde{K} = \{x \in \mathbb{R}^d; \ x_i \geq 0, \ \forall \ i = 1, \ldots, m\}$ and $\delta \equiv 0$ on $\tilde{K}$.

(v) K *is a closed, convex cone in* $\mathcal{R}^d$. Then $\tilde{K} = \{x \in \mathbb{R}^d; \ \pi^* x \geq 0, \ \forall \ \pi \in K\}$ is the polar cone of $-K$, and $\delta \equiv 0$ on $\tilde{K}$. This case obviously generalizes (i) - (iv).

(vi) *Prohibition of borrowing:* $K = \{\pi \in \mathbb{R}^d; \sum_{i=1}^{d} \pi_i \leq 1\}$. Then $\tilde{K} = \{x \in \mathbb{R}^d; \ x_1 = \ldots = x_d \leq 0\}$, and $\delta(x) = -x_1$ on $\tilde{K}$.

(vii) *Rectangular constraints:* $K = \times_{i=1}^{d} I_i$, $I_i = [\alpha_i, \beta_i]$ for some fixed numbers $-\infty \leq \alpha_i \leq 0 \leq \beta_i \leq \infty$, with the understanding that the interval I_i is open to the right (left) if $b_i = \infty$ (respectively, if $\alpha_i = -\infty$). Then $\delta(x) = \sum_{i=1}^{d}(\beta_i x_i^- - \alpha_i x_i^+)$ and $\tilde{K} = \mathbb{R}^d$ if all the $\alpha_i's$, $\beta_i's$ are real. In general, $\tilde{K} = \{x \in \mathbb{R}^d; x_i \geq 0, \forall i \in S_+ \text{ and } x_j \leq 0, \forall j \in S_-\}$ where $S_+ := \{i = 1, \dots, d \mid \beta_i = \infty\}$, $S_- := \{i = 1, \dots, d \mid \alpha_i = -\infty\}$. ◇

From now on, we also allow our investor to spend some money for consumption. More precisely, we add the term $-dc(t)$ to the right-hand side of (3.1), where $c(\cdot)$ is a *cumulative consumption process*, a nondecreasing process, with $c(0) = 0$. The set of admissible policies $(\pi(\cdot), c(\cdot))$ is defined similarly as before, and still denoted by $\mathcal{A}_0(x)$. We consider only portfolios that take values in the given, convex, closed set $K \subset \mathbb{R}^d$, i.e., we replace the set of admissible policies $\mathcal{A}_0(x)$ with

$$\mathcal{A}'(x) := \{(\pi, c) \in \mathcal{A}_0(x); \pi(t, \omega) \in K \quad \text{for} \quad \ell \times \mathbf{P} - a.e. \ (t, \omega)\}. \quad (5.5)$$

Here, ℓ stands for Lebesgue measure on $[0, T]$. Consider the class $\mathcal{H}$ of $\tilde{K}$-valued, $\{\mathcal{F}_t\}$-progressively measurable processes $\nu = \{\nu(t), 0 \leq t \leq T\}$ which satisfy $E \int_0^T \|\nu(t)\|^2 dt + E \int_0^T \delta(\nu(t))dt < \infty$, and introduce for every $\nu \in \mathcal{H}$ the analogues

$$\theta_\nu(t) := \theta(t) + \sigma^{-1}(t)\nu(t), \quad (5.6)$$

$$\gamma_\nu(t) := \exp[-\int_0^t \{r(s) + \delta(\nu(s))\}ds], \quad (5.7)$$

$$Z_\nu(t) := \exp[-\int_0^t \theta_\nu^*(s)dW(s) - \frac{1}{2}\int_0^t \|\theta_\nu(s)\|^2 ds], \quad (5.8)$$

$$W_\nu(t) := W(t) + \int_0^t \theta_\nu(s)ds, \quad (5.9)$$

of the processes in (2.5)-(2.7), (3.2), as well as the measure

$$\mathbf{P}^\nu(A) := E[Z_\nu(T)1_A] = E^\nu[1_A], \quad A \in \mathcal{F}_T \quad (5.10)$$

by analogy with (3.5). Finally, denote by $\mathcal{D}$ the subset consisting of the processes $\nu \in \mathcal{H}$ which are bounded uniformly in (t, ω). Thus, for every $\nu \in \mathcal{D}$, the measure $\mathbf{P}^\nu$ of (5.10) is a probability measure and the process $W_\nu(\cdot)$ of (5.9) is a $\mathbf{P}^\nu$−Brownian motion.

In general, there are several interpretations for the processes $\nu \in \mathcal{D}$: they are stochastic "Lagrange multipliers" associated with the portfolio constraints; in economics jargon, they correspond to the shadow prices relevant to the incompleteness of the market introduced by constraints; they can also be considered as the dual processes appearing in the stochastic maximum principle corresponding to the stochastic control problems we shall be considering.

5.3 Definition: A contingent claim B will be called K-*hedgeable*, if it satisfies

$$V(0) := \sup_{\nu \in \mathcal{D}} E^{\nu}[\gamma_{\nu}(T)B] < \infty. \tag{5.11}$$

This definition will be justified in the next section; more precisely, it will be shown there that for any K-hedgeable contingent claim B, there exists a pair $(\pi, c) \in \mathcal{A}'(V(0))$ such that $X^{V(0),\pi,c}(T) = B$, and that $V(0)$ is the minimal initial wealth for which this can be achieved.

5.4 Remark: In the unconstrained case $K = \mathcal{R}^d$ we have $\tilde{K} = \{0\}$, and $V(0) = E^0[\gamma_0(T)B]$ is then the unconstrained hedging price for the contingent claim B, as in Proposition 4.2. The number $u_{\nu} := E^{\nu}[\gamma_{\nu}(T)B] = E[\gamma_{\nu}(T)Z_{\nu}(T)B]$ is the unconstrained hedging price for B in an *auxiliary market* $\mathcal{M}_{\nu}$; this market consists of a bank account with interest rate $r^{(\nu)}(t) := r(t) + \delta(\nu(t))$ and d stocks, with the same volatility matrix $\{\sigma_{ij}(t)\}_{1 \le i,j \le d}$ as before and appreciation rates $b_i^{(\nu)}(t) := b_i(t) + \nu_i(t) + \delta(\nu(t))$, $1 \le i \le d$, for any given $\nu \in \mathcal{D}$. We shall show that the price for hedging B with a constrained portfolio in the market $\mathcal{M}$, is given by the supremum of the unconstrained hedging prices $u_{\nu} = E^{\nu}[\gamma_{\nu}(T)B]$ in these auxiliary markets $\mathcal{M}_{\nu}$, $\nu \in \mathcal{D}$.

5.5 Remark: In terms of the $\mathbf{P}^{\nu}$-Brownian motion $W_{\nu}(\cdot)$ of (5.9), the stock price equations (2.2) can be re-written as

$$dP_i(t) = P_i(t) \left[(r(t) - \nu_i(t))dt + \sum_{j=1}^{d} \sigma_{ij}(t)dW_{\nu}^{(j)}(t) \right], \quad i = 1, \ldots, d \tag{5.12}$$

for any given $\nu \in \mathcal{D}$.

6. HEDGING WITH CONSTRAINED PORTFOLIOS

We introduce in this section the "hedging price" of a contingent claim B, with portfolios constrained to take values in the set K of section 5, and show that this price coincides with the number $V(0) = \sup_{\nu \in \mathcal{D}} E^{\nu}[\gamma_{\nu}(T)B]$ of (5.11).

6.1 Definition: The *hedging price with K-constrained portfolios* of a contingent claim B is defined by

$$h(0) := \left\{ \begin{array}{c} \inf\{x \in (0, \infty); \exists(\pi, c) \in \mathcal{A}'(x), \ s.t. \ X^{x,\pi,c}(T) \ge B \ a.s.\} \\ \infty \quad , \text{ if the above set is empty} \end{array} \right\}. \tag{6.1}$$

Let us denote by $\mathcal{S}$ the set of all $\{\mathcal{F}_t\}$-stopping times τ with values in $[0, T]$, and by $\mathcal{S}_{\rho,\sigma}$ the subset of $\mathcal{S}$ consisting of stopping times τ s.t. $\rho(\omega) \le \tau(\omega) \le \sigma(\omega)$, $\forall \omega \in \Omega$, for any two $\rho \in \mathcal{S}, \sigma \in \mathcal{S}$ such that $\rho \le \sigma$, a.s. For every $\tau \in \mathcal{S}$ consider also the $\mathcal{F}_{\tau}$-measurable random variable

$$V(\tau) := \operatorname*{ess\,sup}_{\nu \in \mathcal{D}} E^{\nu}[B\gamma_0(T) \exp\{-\int_{\tau}^{T} \delta(\nu(s))ds\}|\mathcal{F}_{\tau}]. \tag{6.2}$$

6.2 Proposition: *For any contingent claim that satisfies (5.11), the family (6.2) of random variables* $\{V(\tau)\}_{\tau \in S}$ *satisfies the equation of Dynamic Programming*

$$V(\tau) = ess \sup_{\nu \in \mathcal{D}_{\tau,\theta}} E^{\nu}[V(\theta)\exp\{-\int_{\tau}^{\theta}\delta(\nu(u))du\}|\mathcal{F}_{\tau}] ; \quad \forall\ \theta \in S_{\tau,T} , \quad (6.3)$$

where $\mathcal{D}_{\tau,\theta}$ *is the restriction of* $\mathcal{D}$ *to the stochastic interval* $[\![\tau, \theta]\!]$.

6.3 Proposition: *The process* $V = \{V(t), \mathcal{F}_t; 0 \le t \le T\}$ *of Proposition 6.2 can be considered in its RCLL modification and, for every* $\nu \in \mathcal{D}$,

$$\left\{ \begin{array}{c} Q_{\nu}(t) := V(t)e^{-\int_0^t \delta(\nu(u))du}, \mathcal{F}_t;\ 0 \le t \le T \\[2mm] is\ a\ \mathbf{P}^{\nu}\text{-}supermartingale\ with\ RCLL\ paths \end{array} \right\}. \quad (6.4)$$

Furthermore, V *is the smallest adapted, RCLL process that satisfies (6.4) as well as*

$$V(T) = B\gamma_0(T), \quad a.s. \quad (6.5)$$

Proof of Proposition 6.2: Let us start by observing that, for any $\theta \in \mathcal{S}$, the random variable

$$J_{\nu}(\theta) := E^{\nu}[V(T)e^{-\int_{\theta}^{T}\delta(\nu(s))ds}|\mathcal{F}_{\theta}]$$

$$= \frac{E[Z_{\nu}(\theta)Z_{\nu}(\theta,T)V(T)e^{-\int_{\theta}^{T}\delta(\nu(s))ds}|\mathcal{F}_{\theta}]}{E[Z_{\nu}(\theta)Z_{\nu}(\theta,T)|\mathcal{F}_{\theta}]}$$

$$= E[Z_{\nu}(\theta,T)V(T)e^{-\int_{\theta}^{T}\delta(\nu(s))ds}|\mathcal{F}_{\theta}]$$

depends only on the restriction of ν to $[\![\theta, T]\!]$ (we have used the notation $Z_{\nu}(\theta, T)$ $= \frac{Z_{\nu}(T)}{Z_{\nu}(\theta)}$). It is also easy to check that the family of random variables $\{J_{\nu}(\theta)\}_{\nu \in \mathcal{D}}$ is directed upwards; indeed, for any $\mu \in \mathcal{D}, \nu \in \mathcal{D}$ and with $A = \{(t,\omega); J_{\mu}(t,\omega) \ge J_{\nu}(t,\omega)\}$ the process $\lambda := \mu 1_A + \nu 1_{A^c}$ belongs to $\mathcal{D}$ and we have a.s. $J_{\lambda}(\theta)$ $= \min\{J_{\mu}(\theta), J_{\nu}(\theta)\}$; then from Neveu (1975), p.121, there exists a sequence $\{\nu_k\}_{k \in \mathbf{N}} \subseteq \mathcal{D}$ such that $\{J_{\nu_k}(\theta)\}_{k \in \mathbf{N}}$ is increasing and

(i) $$V(\theta) = \lim_{k \to \infty} \uparrow J_{\nu_k}(\theta), \quad a.s.$$

Returning to the proof itself, let us observe that

$$V(\tau) = ess \sup_{\nu \in \mathcal{D}_{\tau,T}} E^{\nu}[e^{-\int_{\tau}^{\theta}\delta(\nu(s))ds}E^{\nu}\{V(T)e^{-\int_{\theta}^{T}\delta(\nu(s))ds}|\mathcal{F}_{\theta}\}|\mathcal{F}_{\tau}]$$

$$\le ess \sup_{\nu \in \mathcal{D}_{\tau,T}} E^{\nu}[e^{-\int_{\tau}^{\theta}\delta(\nu(s))ds}V(\theta)|\mathcal{F}_{\tau}], \quad a.s.$$

To establish the opposite inequality, it certainly suffices to pick $\mu \in \mathcal{D}$ and show that

(ii) $$V(\tau) \ge E^{\mu}[V(\theta)e^{-\int_{\tau}^{\theta}\delta(\mu(s))ds}|\mathcal{F}_{\tau}]$$

holds almost surely.

Let us denote by $M_{\tau,\theta}$ the class of processes $\nu \in \mathcal{D}$ which agree with μ on $[\![\tau, \theta]\!]$. We have

$$V(\tau) \geq \underset{\nu \in M_{\tau,\theta}}{ess\ sup}\ E^\nu[e^{-\int_\tau^\theta \delta(\nu(s))ds - \int_\theta^T \delta(\nu(s))ds} V(T)|\mathcal{F}_\tau]$$

$$= \underset{\nu \in M_{\tau,\theta}}{ess\ sup}\ E^\nu[e^{-\int_\tau^\theta \delta(\nu(s))ds} E^\nu\{e^{-\int_\theta^T \delta(\nu(s))ds} V(T)|\mathcal{F}_\theta\}|\mathcal{F}_\tau].$$

Thus, for every $\nu \in M_{\tau,\theta}$, we have

$$V(\tau) \geq E^\nu[e^{-\int_\tau^\theta \delta(\nu(s))ds} J_\nu(\theta)|\mathcal{F}_\tau]$$

$$= \frac{E[Z_\nu(\tau)Z_\nu(\tau,\theta).E\{Z_\nu(\theta,T)|\mathcal{F}_\theta\}.e^{-\int_\tau^\theta \delta(\nu(s))ds} J_\nu(\theta)|\mathcal{F}_\tau]}{E[Z_\nu(\tau)Z_\nu(\tau,\theta).E\{Z_\nu(\theta,T)|\mathcal{F}_\theta\}|\mathcal{F}_\tau]}$$

$$= E[Z_\nu(\tau,\theta)e^{-\int_\tau^\theta \delta(\nu(s))ds} J_\nu(\theta)|\mathcal{F}_\tau]$$

$$= E[Z_\mu(\tau,\theta)e^{-\int_\tau^\theta \delta(\mu(s))ds} J_\nu(\theta)|\mathcal{F}_\tau]$$

$$= \ldots = E^\mu[e^{-\int_\tau^\theta \delta(\mu(s))ds} J_\nu(\theta)|\mathcal{F}_\tau].$$

Now clearly we may take $\{\nu_k\}_{k \in \mathbb{N}} \subseteq M_{\tau,\theta}$ in (i), as $J_\nu(\theta)$ depends only on the restriction of ν on $[\![\theta, T]\!]$; and from the above,

$$V(\tau) \geq \lim_{k \to \infty} \uparrow E^\mu[e^{-\int_\tau^\theta \delta(\mu(s))ds} J_{\nu_k}(\theta)|\mathcal{F}_\tau]$$

$$= E^\mu[e^{-\int_\tau^\theta \delta(\mu(s))ds} . \lim_{k \to \infty} \uparrow J_{\nu_k}(\theta)|\mathcal{F}_\tau]$$

$$= E^\mu[e^{-\int_\tau^\theta \delta(\mu(s))ds} V(\theta)|\mathcal{F}_\tau],\ a.s.$$

by Monotone Convergence. $\diamond$

It is an immediate consequence of this proposition that

(iii) $V(\tau)e^{-\int_0^\tau \delta(\nu(u))du} \geq E^\nu[V(\theta)e^{-\int_0^\theta \delta(\nu(u))du}|\mathcal{F}_\tau],\ a.s.$

holds for any given $\tau \in \mathcal{S}, \theta \in \mathcal{S}_{\tau,T}$ and $\nu \in \mathcal{D}$.

Proof of Proposition 6.3: Let us consider the positive, adapted process $\{V(t,\omega), \mathcal{F}_t;\ t \in [0,T] \cap \mathcal{Q}\}$ for $\omega \in \Omega$. From (iii), the process

$$\{V(t,\omega)e^{-\int_0^t \delta(\nu(s,\omega))ds},\ \mathcal{F}_t;\ t \in [0,T] \cap \mathcal{Q}\}\ \text{for}\ \omega \in \Omega$$

is a $\mathbf{P}^\nu$- supermartingale on $[0,T] \cap \mathcal{Q}$, where $\mathcal{Q}$ is the set of rational numbers, and thus has a.s. finite limits from the right and from the left (recall Proposition

1.3.14 in Karatzas & Shreve (1991), as well as the right-continuity of the filtration $\{\mathcal{F}_t\}$). Therefore,

$$V(t+,\omega) := \left\{ \begin{array}{cc} \lim_{\substack{s\downarrow t \\ s\in Q}} V(s,\omega) & ; \quad 0 \leq t < T \\ V(T,\omega) & ; \quad t = T \end{array} \right\}$$

$$V(t-,\omega) := \left\{ \begin{array}{cc} \lim_{\substack{s\uparrow t \\ s\in Q}} V(s,\omega) & ; \quad 0 < t \leq T \\ V(0) & ; \quad t = 0 \end{array} \right\}$$

are well-defined and finite for every $\omega \in \Omega^*$, $P(\Omega^*) = 1$, and the resulting processes are adapted. Furthermore (loc.cit.), $\{V(t+)e^{-\int_0^t \delta(\nu(s))ds}, \mathcal{F}_t; 0 \leq t \leq T\}$ is a RCLL, $\mathbf{P}^\nu$-supermartingale, for all $\nu \in \mathcal{D}$; in particular,

$$V(t+) \geq E^\nu[V(T)e^{-\int_t^T \delta(\nu(s))ds}|\mathcal{F}_t], \quad a.s.$$

holds for every $\nu \in \mathcal{D}$, whence $V(t+) \geq V(t)$ a.s. On the other hand, from Fatou's lemma we have for any $\nu \in \mathcal{D}$:

$$V(t+) = E^\nu[\lim_{n\to\infty} V(t+\frac{1}{n}) \, e^{-\int_t^{t+1/n} \delta(\nu(u))du}|\mathcal{F}_t]$$

$$\leq \lim_{n\to\infty} E^\nu[V(t+\frac{1}{n}) \, e^{-\int_t^{t+1/n} \delta(\nu(u))du}|\mathcal{F}_t] \leq V(t), \quad a.s.$$

and thus $\{V(t+), \mathcal{F}_t; 0 \leq t \leq T\}$, $\{V(t), \mathcal{J}_t; 0 \leq t \leq T\}$ are modifications of one another.

The remaining claims are immediate. ◇

6.4 Theorem: (Cvitanić & Karatzas (1993), El-Karoui & Quenez (1995)) *For an arbitrary contingent claim B, we have* $h(0) = V(0)$. *Furthermore, if* $V(0) < \infty$, *there exists a pair* $(\hat{\pi},\hat{c}) \in \mathcal{A}'(V(0))$ *such that* $X^{V(0),\hat{\pi},\hat{c}}(T) = B$, *a.s.*

Proof: We first want to show $h(0) \leq V(0)$. Clearly, we may assume $V(0) < \infty$. From (6.4), the martingale representation theorem and the Doob-Meyer decomposition, we have for every $\nu \in \mathcal{D}$:

$$Q_\nu(t) = V(0) + \int_0^t \psi_\nu^*(s)dW_\nu(s) - A_\nu(t), \quad 0 \leq t \leq T, \quad (6.6)$$

where $\psi_\nu(\cdot)$ is an $\mathbb{R}^d$-valued, $\{\mathcal{F}_t\}$-progressively measurable and a.s. square-integrable process and $A_\nu(\cdot)$ is adapted with increasing, RCLL paths and $A_\nu(0) = 0$, $EA_\nu(T) < \infty$ a.s. The idea then is to consider the positive, adapted, RCLL process

$$\hat{X}(t) := \frac{V(t)}{\gamma_0(t)} = \frac{Q_\nu(t)}{\gamma_\nu(t)}, \quad 0 \leq t \leq T \quad (\forall \nu \in \mathcal{D}) \quad (6.7)$$

with $\hat{X}(0) = V(0), \hat{X}(T) = B$ a.s., and to find a pair $(\hat{\pi},\hat{c}) \in \mathcal{A}'(V(0))$ such that $\hat{X}(\cdot) = X^{V(0),\hat{\pi},\hat{c}}(\cdot)$. This will prove that $h(0) \leq V(0)$.

In order to do this, let us observe that for any $\mu \in \mathcal{D}, \nu \in \mathcal{D}$ we have from (6.4)

$$Q_\mu(t) = Q_\nu(t) \exp\left[\int_0^t \{\delta(\nu(s)) - \delta(\mu(s))\}ds\right] ,$$

and from (6.6):

$$\begin{aligned}
dQ_\mu(t) &= \exp[\int_0^t \{\delta(\nu(s)) - \delta(\mu(s))\}ds] \cdot [Q_\nu(t)\{\delta(\nu(t)) - \delta(\mu(t))\}dt \\
&\quad + \psi_\nu^*(t)dW_\nu(t) - dA_\nu(t)] \\
&= \exp[\int_0^t \{\delta(\nu(s)) - \delta(\mu(s))\}ds] \cdot [\hat{X}(t)\gamma_\nu(t)\{\delta(\nu(t)) - \delta(\mu(t))\}dt \\
&\quad - dA_\nu(t) + \psi_\nu^*(t)\sigma^{-1}(t)(\nu(t) - \mu(t))dt + \psi_\nu^*(t)dW_\mu(t)] .
\end{aligned}$$

(6.8)

Comparing this decomposition with

$$dQ_\mu(t) = \psi_\mu^*(t)dW_\mu(t) - dA_\mu(t) , \qquad (6.6)'$$

we conclude that

$$\psi_\nu^*(t) \; e^{\int_0^t \delta(\nu(s))ds} = \psi_\mu^*(t) \; e^{\int_0^t \delta(\mu(s))ds}$$

and hence that this expression is independent of $\nu \in \mathcal{D}$:

$$\psi_\nu^*(t) \; e^{\int_0^t \delta(\nu(s))ds} = \hat{X}(t)\gamma_0(t)\hat{\pi}^*(t)\sigma(t); \quad \forall \, 0 \le t \le T, \nu \in \mathcal{D} , \qquad (6.9)$$

for some adapted, $\mathbb{R}^d$-valued, a.s. square integrable process $\hat{\pi}$ (we do not know yet that $\hat{\pi}$ takes values in K). If $X(t) = 0$, then $X(s) = 0$ for all $s \ge t$, and we can set, for example, $\pi(s) = 0$, $s \ge t$ (in fact, one can show that $\int_0^T 1_{\{\hat{X}(t)=0\}}\|\psi_\nu(t)\|^2 dt = 0$, a.s; see Karatzas & Kou (1994)).

Similarly, we conclude from (6.8),(6.9) and (6.6)':

$$e^{\int_0^t \delta(\nu(s))ds} dA_\nu(t) - \gamma_0(t)\hat{X}(t)[\delta(\nu(t)) + \hat{\pi}^*(t)\nu(t)]dt$$

$$= e^{\int_0^t \delta(\mu(s))ds} dA_\mu(t) - \gamma_0(t)\hat{X}(t)[\delta(\mu(t)) + \hat{\pi}^*(t)\mu(t)]dt$$

and hence this expression is also independent of $\nu \in \mathcal{D}$:

$$\hat{c}(t) := \int_0^t \gamma_\nu^{-1}(s)dA_\nu(s) - \int_0^t \hat{X}(s)[\delta(\nu(s)) + \nu^*(s)\hat{\pi}(s)]ds , \qquad (6.10)$$

for every $0 \le t \le T, \nu \in \mathcal{D}$. From (6.10) with $\nu \equiv 0$, we obtain $\hat{c}(t) = \int_0^t \gamma_0^{-1}(s)dA_0(s), 0 \le t \le T$ and hence

$$\left\{ \begin{array}{c} \hat{c}(\cdot) \text{ is an increasing, adapted, RCLL process} \\ \text{with } \hat{c}(0) = 0 \quad \text{and} \quad \hat{c}(T) < \infty, a.s. \end{array} \right\} . \qquad (6.11)$$

Next, we claim that

$$\delta(\nu) + \nu^* \hat{\pi}(t,\omega) \geq 0, \quad \ell \otimes \mathbf{P} - a.e. \tag{6.12}$$

holds for every $\nu \in \tilde{K}$. Then Theorem 13.1 of Rockafellar (1970) (together with continuity of $\delta(\cdot)$ and closedness of K) leads to the fact that

$$\hat{\pi}(t,\omega) \in K \quad \text{holds} \quad \ell \otimes \mathbf{P} - a.e. \text{ on } [0,T] \times \Omega \tag{6.12}'$$

In order to verify (6.12), notice that from (6.10) we obtain

$$\int_0^t \gamma_\nu^{-1}(s) A_\nu(s) ds = \hat{c}(t) + \int_0^t \hat{X}(s)\{\delta(\nu_s) + \nu_s^* \hat{\pi}_s\} ds; \; 0 \leq t \leq T, \; \nu \in \mathcal{D}.$$

Fix $\nu \in \tilde{K}$ and define the set $F_\nu := \{(t,\omega) \in [0,T] \times \Omega; \; \delta(\nu)) + \nu^* \hat{\pi}(t,\omega) < 0\}$. Let $\mu(t) := [\nu 1_{F_\nu^c} + n\nu 1_{F_\nu}]$, $n \in \mathbf{N}$; then $\mu \in \mathcal{D}$, and assuming that (6.12) does not hold, we get for n large enough

$$E[\int_0^T \gamma_\mu^{-1}(s) A_\mu(s) ds] = E\left[\hat{c}(T) + \int_0^T \hat{X}(t) 1_{F_\nu^c} \{\delta(\nu) + \nu^* \hat{\pi}(t)\} dt\right]$$

$$+ nE\left[\int_0^T \hat{X}(t) 1_{F_\nu} \{\delta(\nu) + \nu^* \hat{\pi}(t)\} dt\right] < 0 \; ,$$

a contradiction.

Now we can put together (6.6)-(6.10) to deduce

$$d(\gamma_\nu(t)\hat{X}(t)) = dQ_\nu(t) = \psi_\nu^*(t) dW_\nu(t) - dA_\nu(t)$$
$$= \gamma_\nu(t)[-d\hat{c}(t) - \hat{X}(t)\{\delta(\nu(t)) + \nu^*(t)\hat{\pi}(t)\} dt \tag{6.13}$$
$$+ \hat{X}(t)\hat{\pi}^*(t)\sigma(t) dW_\nu(t)] \; ,$$

for any given $\nu \in \mathcal{D}$. As a consequence, the process

$$\hat{M}_\nu(t) := \gamma_\nu(t)\hat{X}(t) + \int_0^t \gamma_\nu(s) d\hat{c}(s) + \int_0^t \gamma_\nu(s)\hat{X}(s)[\delta(\nu(s)) + \nu^*(s)\hat{\pi}(s)] ds$$

$$= V(0) + \int_0^t \gamma_\nu(s)\hat{X}(s)\hat{\pi}^*(s)\sigma(s) dW_\nu(s) \; , \quad 0 \leq t \leq T$$

$$\tag{3.4}'$$

is a nonnegative, $\mathbf{P}^\nu$-local martingale.

In particular, for $\nu \equiv 0$, (6.13) gives:

$$d(\gamma_0(t)\hat{X}(t)) = -\gamma_0(t) d\hat{c}(t) + \gamma_0(t)\hat{X}(t)\hat{\pi}^*(t)\sigma(t) dW_0(t),$$
$$\hat{X}(0) = V(0) \; , \quad \hat{X}(T) = B \; , \tag{6.13}'$$

which is equation (3.1) (plus the consumption term) for the process $\hat{X}(\cdot)$ of (6.7). This shows $\hat{X}(\cdot) \equiv X^{V(0),\hat{\pi},\hat{c}}(\cdot)$, and hence $h(0) \leq V(0) < \infty$.

To complete the proof, it thus suffices to show $h(0) \geq V(0)$. Clearly, we may assume $h(0) < \infty$, and then there exists a number $x \in (0,\infty)$ such that

$X^{x,\pi,c}(T) \geq B$, a.s., for some $(\pi, c) \in \mathcal{A}'(x)$. Then the analogue of (6.13) holds, and it follows from the supermartingale property that

$$x \geq E^\nu[\gamma_\nu(T)X^{x,\pi,c}(T) + \int_0^T \gamma_\nu(t)dc(t)$$

$$+ \int_0^T \gamma_\nu(t)X^{x,\pi,c}(t)\{\delta(\nu(t)) + \nu^*(t)\pi(t)\}dt] \tag{6.14}$$

$$\geq E^\nu[B\gamma_\nu(T)] ,$$

$\forall \nu \in \mathcal{D}$. Therefore, $x \geq V(0)$, and thus $h(0) \geq V(0)$.   ⋄

6.5 Definition: We say that a K-hedgeable contingent claim B is $K-attainable$ if there exists a portfolio process π with values in K such that $(\pi, 0) \in \mathcal{A}'(V(0))$ and $X^{V(0),\pi,0}(T) = B$, a.s.

6.6 Theorem: *For a given K-hedgeable contingent claim B, and any given $\lambda \in \mathcal{D}$, the conditions*

$$\{Q_\lambda(t) = V(t)e^{-\int_0^t \delta(\lambda(u))du}, \mathcal{F}_t; \ 0 \leq t \leq T\} \quad \text{is a } \mathbf{P}^\lambda\text{-martingale} \tag{6.15}$$

$$\lambda \text{ achieves the supremum in } V(0) = \sup_{\nu\in\mathcal{D}} E^\nu[B\gamma_\nu(T)] \tag{6.16}$$

$$\left\{ \begin{array}{l} B \text{ is } K\text{-attainable (by a portfolio } \pi\text{), and the} \\ \text{corresponding } \gamma_\lambda(\cdot)X^{V(0),\pi,0}(\cdot) \text{ is a } \mathbf{P}^\lambda\text{-martingale} \end{array} \right\} \tag{6.17}$$

are equivalent, and imply

$$\hat{c}(t,\omega) = 0, \ \delta(\lambda(t,\omega)) + \lambda^*(t,\omega)\hat{\pi}(t,\omega) = 0; \ \ell \otimes P - a.e. \tag{6.18}$$

for the pair $(\hat{\pi}, \hat{c}) \in \mathcal{A}'(V(0))$ of Theorem 6.4.

Proof: The $\mathbf{P}^\lambda$-supermartingale $Q_\lambda(\cdot)$ is a $\mathbf{P}^\lambda$-martingale, if and only if $Q_\lambda(0) = E^\lambda Q_\lambda(T) \Leftrightarrow V(0) = E^\lambda[B\gamma_\lambda(T)] \Leftrightarrow$ (6.16).

On the other hand, (6.15) implies $A_\lambda(\cdot) \equiv 0$, and so from (6.10): $\hat{c}(t) = -\int_0^t \hat{X}(s)[\delta(\lambda(s)) + \lambda^*(s)\hat{\pi}(s)]ds$. Now (6.18) follows from the increase of $\hat{c}(\cdot)$ and the nonnegativity of $\delta(\lambda) + \lambda^*\hat{\pi}$, since $\hat{\pi}$ takes values in K.

¿From (6.16) (and its consequences (6.15), (6.18)), the process $\hat{X}(\cdot)$ of (6.7) and (6.13) coincides with $X^{V(0),\hat{\pi},0}(\cdot)$, and we have: $\hat{X}(T) = B$ almost surely, $\gamma_\lambda(\cdot)\hat{X}(\cdot)$ is a $\mathbf{P}^\lambda$-martingale; thus (6.17) is satisfied with $\pi \equiv \hat{\pi}$. On the other hand, suppose that (6.17) holds; then $V(0) = E^\lambda[B\gamma_\lambda(T)]$, so (6.16) holds.

6.7 Theorem: *Let B be a K-hedgeable contingent claim. Suppose that, for any $\nu \in \mathcal{D}$ with $\delta(\nu) + \nu^*\hat{\pi} \equiv 0$,*

$$Q_\nu(\cdot) \text{ in (6.4) is of class } DL[0,T], \text{ under } \mathbf{P}^\nu. \tag{6.19}$$

Then, for any given $\lambda \in \mathcal{D}$, *the conditions (6.15), (6.16), (6.18) are equivalent, and imply*

$$\left\{ \begin{array}{l} B \text{ is } K\text{-attainable (by a portfolio } \pi), \text{ and the} \\ \text{corresponding } \gamma_0(\cdot) X^{V(0),\pi,0}(\cdot) \text{ is a } \mathbf{P}^0\text{-martingale} \end{array} \right\}. \qquad (6.17)^o$$

Proof: We have already shown the implications (6.15) $\Leftrightarrow$ (6.16) $\Rightarrow$ (6.18). To prove that these three conditions are actually *equivalent* under (6.19), suppose that (6.18) holds; then from (6.10): $A_\lambda(\cdot) \equiv 0$, whence the $\mathbf{P}^\lambda$-local martingale $Q_\lambda(\cdot)$ is actually a $\mathbf{P}^\lambda$-martingale (from (6.6) and the assumption (6.19)); thus (6.15) is satisfied.

Clearly then, if (6.15), (6.16), (6.18) are satisfied for some $\lambda \in \mathcal{D}$, they are satisfied for $\lambda \equiv 0$ as well; and from Theorem 6.6, we know then that $(6.17)^o$ (i.e., (6.17) with $\lambda \equiv 0$) holds.

6.8 Remark: (i) Roughly speaking, Theorems 6.6, 6.7 say that *the supremum in (6.16) is attained if and only if it is attained by $\lambda \equiv 0$, if and only if the Black-Scholes (unconstrained) portfolio happens to satisfy constraints.*

(ii) It can be shown that the conditions $V(0) < \infty$ and (6.19) are satisfied (the latter, in fact, for every $\nu \in \mathcal{D}$) in the case of the simple European call option $B = (P_1(T) - q)^+$, provided

the function $\quad x \mapsto \delta(x) + x_1 \quad$ is bounded from below on $\tilde{K}$. $\qquad$ (6.20)

The same is true for any contingent claim B that satisfies $B \le \alpha P_1(T)$ a.s., for some $\alpha \in (0, \infty)$.

6.9 Remark: Note that the condition (6.20) is indeed satisfied, if the convex set

K *contains both the origin and the point* $(1, 0, \ldots, 0)$ $\qquad$ (6.20)'

(and thus also the line-segment adjoining these points); for then $x_1 + \delta(x) \ge x_1 + \sup_{0 \le \alpha \le 1}(-\alpha x_1) = x_1^+ \ge 0$, $\forall x \in \tilde{K}$. This is the case in the Examples 5.2 (i)-(iv), (vi), and (vii) with $1 \le \beta_1 \le \infty$.

6.10 Remark: *If the condition (6.20) is not satisfied, we have* $V(0) = \infty$ *for the European call option* $B = (P_1(T) - q)^+$ *with* $\delta(\cdot) \ge 0$, $r(\cdot) \ge 0$. In other words, such constraints make impossible the hedging of this contingent claim , starting with a finite initial capital.

6.11 Remark: A slight modification of the proof of Theorem 6.4 shows that

$$V(0) = \sup_{\nu \in \mathcal{D}} E^\nu[B\gamma_\nu(T)] = \sup_{\nu \in \mathcal{H}} E[B\gamma_\nu(T)Z_\nu(T)] \qquad (6.23)$$

holds for an arbitrary contingent claim B. The straightforward details are left to the diligence of the reader.

We would like now to get a method for calculating the price $h(0)$. In order to do that, we assume constant market coefficients r, b, σ and consider only the claims of the form $B = b(P(T))$, for a given function b. Similarly as in the no-constraints case, the minimal hedging process will be given as $X(t) = U(t, P(t))$, for some function $U(t, p)$, depending on the constraints. Introduce also, for a given process $\nu(\cdot)$ in $\mathbb{R}^d$, the auxiliary, shadow economy vector of stock prices $P^\nu(\cdot)$ (in analogy with (5.12)) by

$$dP_i^\nu(t) = P_i^\nu(t) \left[(r - \nu_i(t))dt + \sum_{j=1}^d \sigma_{ij} dW^{(j)}(t) \right] \qquad (6.24)$$

Then, by standard results from stochastic control theory, we can restate Theorem 6.4 as follows:

6.12 Theorem: *With the above notation and assumptions, we have*

$$U(t, p) = \sup_{\nu \in \mathcal{D}} E\left[b(P^\nu(T)) e^{-\int_t^T (r + \delta(\nu(s)))ds} \,\Big|\, P^\nu(t) = p \right]. \qquad (6.25)$$

We will show that this complex looking stochastic control problem has a simple solution. First, we modify the value of the claim by considering the following function:

$$\hat{b}(p) = \sup_{\nu \in \tilde{K}} b(pe^{-\nu}) e^{-\delta(\nu)}. \qquad (6.26)$$

Here, $pe^{-\nu} = (p_1 e^{-\nu_1}, \dots, p_d e^{-\nu_d})^*$, and we use the same notation for the componentwise product of two vectors throughout. We use below the term Feynman-Kac assumptions, with the understanding that those are the assumptions under which relevant expected values satisfy corresponding PDE's. A set of such assumptions is given in Duffie (1992). Here is the result which gives a way of calculating $U(t, p)$:

6.13 Theorem: *(Broadie, Cvitanić & Soner (1996)) The minimal K-hedging price function $U(t, p)$ of the claim $b(P(T))$ is the Black-Scholes cost function for replicating $\hat{b}(P(T))$. In particular, under Feynman-Kac assumptions, it is the solution to the PDE*

$$V_t + \frac{1}{2} \sum_{i=1}^d \sum_{j=1}^d a_{ij} p_i p_j V_{p_i p_j} + r \left(\sum_{i=1}^d p_i V_{p_i} - V \right) = 0, \qquad (6.27)$$

with the terminal condition

$$V(T, p) = \hat{b}(p), \quad p \in \mathbb{R}_+^d. \qquad (6.28)$$

Moreover, the corresponding hedging strategy π satisfies the constraints. Under Feynman-Kac assumptions, it is given by

$$\pi_i(t) = P_i(t) V_{p_i}(t, P(t)) / V(t, P(t)), \quad i = 1, \dots, d \qquad (6.29)$$

and $\pi(\cdot)$ takes values in K.

Proof: (a) We first show that portfolio π satisfies the constraints. Let $\nu \in \mathcal{D}$ and observe that, from the properties of the support function and the cone property of $\tilde{K}$,

$$(i) \quad \hat{b} = \hat{b}$$

$$(ii) \quad \int_t^T \delta(\nu_s)ds \geq \delta(\int_t^T \nu_s ds),$$

$$(iii) \quad \int_t^T \nu_s ds \text{ is an element of } \tilde{K},$$

where $\int_t^T \nu(s)ds := (\int_t^T \nu_1(s)ds, ..., \int_t^T \nu_d(s)ds)^*$. Moreover, we have

$$(iv) \quad P_i^\nu(t) = P_i^0(t)e^{-\int_0^t \nu_i(s)ds},$$

because the processes on the left-hand side and the right-hand side satisfy the same linear SDE. Then, for every $\nu \in \mathcal{D}$ we have

$$E[\hat{b}(P^\nu(T))e^{-\int_0^T (r+\delta(\nu(s)))ds}] \leq E[\hat{b}(P^0(T)e^{-\int_0^T \nu(s)ds})e^{-\delta(\int_0^T \nu(s)ds)}e^{-rT}]$$

$$\leq E[\sup_{\nu \in \tilde{K}} \hat{b}(P^0(T)e^{-\nu})e^{-\delta(\nu)}e^{-rT}]$$

$$= E[\hat{\hat{b}}(P^0(T))e^{-rT}] = E[\hat{b}(P^0(T))e^{-rT}].$$

$$(6.30)$$

Therefore the supremum (over $\mathcal{D}$) of the initial expression is obtained for $\nu = 0$. Similarly for conditional expectations of (6.25). It follows then from Theorems 6.6, 6.7 that $\hat{b}(P(T))$ can be attained by a portfolio which satisfies the constraints. Moreover, under Feynman-Kac assumptions, its value function is the solution to (6.27)-(6.28), and the portfolio is given by (6.29).

(b) To conclude we have to show that to hedge $b(P(T))$ we have to hedge at least $\hat{b}(P(T))$. It is sufficient to prove that the left limit of $U(t,p)$ at $t = T$ is larger than $\hat{b}(p)$. For this, let $\{\nu^k\}$ be the maximizing sequence in the cone $\tilde{K}$ attaining $\hat{b}(p)$, i.e., such that $b(pe^{-\nu^k})e^{-\delta(\nu^k)}$ converges to $\hat{b}(p)$ as k goes to infinity. Then, using (for fixed $t < T$) constant deterministic controls $\nu^k/(T-t)$ in (6.25), we get

$$U(t,p) \geq E[b(P^0(T)e^{-\nu^k})e^{-\delta(\nu^k)}e^{-r(T-t)} \mid P^0(t) = p],$$

hence

$$\lim_{t \to T} U(t,p) \geq b(pe^{-\nu^k})e^{-\delta(\nu^k)}$$

and letting k to infinity, we finish the proof.

Here is a sketch of a PDE proof for part (a) in the proof above: Let V be the solution to (6.27)-(6.28). For a given $\nu \in \tilde{K}$, consider the function $W_\nu = (pV_p)^*\nu + \delta(\nu)V$, where V_p is the vector of partial derivatives of V with respect to p_i, $i = 1, ..., d$. By Theorem 13.1 in Rockafellar (1970), to prove that portfolio π of (6.29) takes values in K, it is sufficient to prove that W_ν is non-negative, for all $\nu \in \tilde{K}$. It is not difficult to see (assuming enough smoothness) that W_ν solves PDE (6.27), too. Moreover, it is also straightforward to check that $W_\nu(p, T) \geq 0$. So, by the maximum principle, $W_\nu \geq 0$ everywhere.

6.14 Examples: We restrict ourselves to the case of only one stock, $d = 1$, and the constraints of the type

$$K = [-l, u], \tag{6.31}$$

with $0 \leq l, u \leq +\infty$, with the understanding that the interval K is open to the right (left) if $u = +\infty$ (respectively, if $l = +\infty$). It is straightforward to see that

$$\delta(\nu) = l\nu^+ + u\nu^-,$$

and $\tilde{K} = \mathbb{R}$ if both l and u are finite. In general,

$$\tilde{K} = \{x \in \mathbb{R} \ : \ x \geq 0 \text{ if } u = +\infty, \ x \leq 0 \text{ if } l = +\infty\}. \tag{6.32}$$

For the European call $b(p) = (p - k)^+$, one easily gets that $\hat{b}(p) \equiv \infty$, if $u < 1$, $\hat{b}(p) = p$ if $u = 1$ (no-borrowing) and $\hat{b}(p) = b(p)$ if $u = \infty$ (short-selling constraints don't matter). For $1 < u < \infty$ we have (by ordinary calculus)

$$\hat{b}(p) = \begin{cases} p - k & ; \ p \geq \frac{ku}{u-1} \\ \frac{k}{u-1}\left(\frac{(u-1)p}{ku}\right)^u & ; \ p < \frac{ku}{u-1} \end{cases}. \tag{6.33}$$

For the European put $b(p) = (k - p)^+$, one gets $\hat{b} = b$ if $l = \infty$ (borrowing constraints don't matter), $\hat{b} \equiv k$ if $l = 0$ (no short-selling), and otherwise

$$\hat{b}(p) = \begin{cases} k - p & ; \ p \leq \frac{kl}{l+1} \\ \frac{k}{l+1}\left(\frac{ku}{(l+1)p}\right)^l & ; \ p > \frac{kl}{l+1} \end{cases}. \tag{6.34}$$

We finish this section by a theorem on connections with arbitrage. First, let us remark that one can, analogously to $h(0)$, define the hedging price $\tilde{h}(0)$ for the buyer of the claim B, as the maximal amount of money the buyer can borrow at $t = 0$ and have more than $-B$ at time $t = T$. We have

Theorem 6.15: (Karatzas & Kou (1994)) *There is no arbitrage with constrained portfolios if the price $B(0)$ of B satisfies*

$$\tilde{h}(0) < B(0) < h(0). \tag{6.35}$$

Conversely, if $B(0)$ is strictly larger than $h(0)$ or strictly smaller then $\tilde{h}(0)$, then there is arbitrage.

Proof: First the converse: suppose, for example, that $B(0) > h(0)$. Then one can sell the claim for $B(0)$, put $B(0) - h(0) > 0$ in the bank, and replicate B with $h(0)$ - arbitrage!

Suppose now that (6.35) is satisfied, and suppose, for example, that arbitrage can be obtained by selling the claim for $B(0)$ and investing $x < B(0)$ in the market, using the policy (π, c) such that $X^{x,\pi,c}(T) \geq B$, a.s.. But this is in contradiction with $x < B(0) < h(0)$.

7. NONLINEAR PORTFOLIO DYNAMICS - LARGE INVESTOR.

We consider an investor whose investment policy influences the behavior of the prices $P_0, \{P_i\}_{1 \leq i \leq d}$ of the financial instruments. More precisely, these prices evolve according to the equations

$$dP_0(t) = P_0(t)[r(t) + f_0(\pi_t)]dt , \qquad P_0(0) = 1 \qquad (7.1)$$

$$dP_i(t) = P_i(t)[(b_i(t)+f_i(\pi_t))dt+\sum_{j=1}^{d} \sigma_{ij}(t)dW^{(j)}(t)] , \quad P_i(0) = p_i \in (0,\infty) \ (7.2)$$

for $i = 1, \ldots, d$. Here $f_i : \mathbb{R}^d \mapsto \mathbb{R}$, $i = 0, \ldots, d$ are some given functions that describe the effect of the investor's strategy on the prices.

Let us remark that the interpretation of policy-dependent prices is not the only possible one; one could simply start with the economy in which the wealth process of the investor is the one whose dynamics are given in (7.4) below, and forget about the prices.

For a given $\{\mathcal{F}_t\}$−progressively measurable process $\nu(\cdot)$ with values in $\mathbb{R}^d$ and $\mu(\cdot)$ with values in $\mathbb{R}$, we introduce the discount process

$$\beta_\mu(t) := \beta_\mu(0,t), \quad \beta_\mu(u,t) := \exp\{-\int_u^t \mu(s)ds\}. \qquad (7.3)$$

The wealth process $X(t)$ satisfies the stochastic differential equation

$$dX(t) = X(t)g(t, \pi_t)dt + X(t)\pi^*(t)\sigma(t)dW(t) - dc(t) , \quad X(0) = x > 0, \quad (7.4)$$

where

$$g(t, \pi) := r(t) + f_0(\pi) + \sum_{i=1}^{d} \pi_i[b_i(t) + f_i(\pi) - r(t) - f_0(\pi)] . \qquad (7.5)$$

We will restrict ourselves by imposing the following assumption:

7.1 Assumption: The function $g(t, \cdot)$ is concave for all $t \in [0, T]$, and uniformly (with respect to t) Lipschitz:

$$|g(t, x) - g(t, y)| \leq k\|x - y\|, \quad \forall \, t \in [0, T]; \ x, y \in \mathbb{R}^d,$$

for some $0 < k < \infty$.

In analogy with the case of constraints we define the convex conjugate function $\tilde{g}$ of g by

$$\tilde{g}(t, \nu) := \sup_{\pi \in \mathbb{R}^d} \{g(t, \pi) + \pi^*\nu\}, \qquad (7.5)'$$

on its *effective domain* $\mathcal{D}_t := \{\nu : \tilde{g}(\nu, t) < \infty\}$. Introduce also the class $\mathcal{D}$ of processes $\nu(t)$ taking values in $\mathcal{D}_t$, for all t. We shall also assume that

$$\mathcal{D} \text{ is not empty;} \tag{7.6}$$

The function $\tilde{g}(t, \cdot)$ is bounded on its effective domain, uniformly in t. (7.7)

Denote

$$\gamma_\nu(t, u) := \exp\{-\int_t^u \tilde{g}(s, \nu_s)ds\}, \quad \gamma_\nu(t) := \gamma_\nu(0, t),$$

$$dZ_\nu(t) := -\sigma^{-1}(t)\nu(t)Z_\nu(t)dW(t), \quad Z_\nu(0) = 1, \quad H_\nu(t) := Z_\nu(t)\gamma_\nu(t) . \tag{7.8}$$

For every $\nu \in \mathcal{D}$ we have (by Ito's rule)

$$H_\nu(t)X(t) + \int_0^t H_\nu(s)\left[X(s)(\tilde{g}(s, \nu_s) - g(s, \pi_s) - \pi^*(s)\nu(s))ds + dc(s)\right]$$

$$\tag{7.9}$$

$$= x + \int_0^t H_\nu(s)X(s)\left[\pi^*(s)\sigma(s) + \sigma^{-1}(s)\nu(s)\right]dW(s).$$

In particular, the process on the right-hand side is a nonnegative local martingale, hence a supermartingale. Therefore we get the following *necessary condition for π to be admissible*:

$$\sup_{\nu \in \mathcal{D}} E\left[H_\nu(T)X(T) + \int_0^T H_\nu(s)X(s)\{\tilde{g}(s, \nu_s) - g(s, \pi_s) - \pi^*(s)\nu(s)\}ds\right] \le x .$$

$$\tag{7.10}$$

7.2 Remark: The supermartingale property excludes arbitrage opportunities from this market: if $x = 0$, then necessarily $X(t) = 0$, $\forall\ 0 \le t \le T$, a.s.. If $f_i \equiv 0$, $i = 0, \ldots, d$, then $\mathcal{D}$ consists of only one process $\hat{\nu}(\cdot)$, given by $\hat{\nu}_i(t) = r(t) - b_i(t)$, $i = 1, \ldots, d$, i.e., we are in the standard Black-Scholes-Merton complete market model with the unique "equivalent martingale risk neutral measure" $\mathbf{P}^{\hat{\nu}}$, defined below.

Next, for a given $\nu \in \mathcal{D}$, introduce the process

$$W_\nu(t) := W(t) - \int_0^t \sigma^{-1}(s)\nu(s)ds , \tag{7.11}$$

as well as the measure

$$\mathbf{P}^\nu(A) := E[Z_\nu(T)1_A] = E^\nu[1_A], \quad A \in \mathcal{F}_T . \tag{7.12}$$

Notice that Assumption 7.1 implies that the sets $\mathcal{D}_t$ are uniformly bounded. Therefore, if $\nu \in \mathcal{D}$, then Z_ν *is a martingale*. Thus, for every $\nu \in \mathcal{D}$, the measure $\mathbf{P}^\nu$ is a probability measure and the process $W_\nu(\cdot)$ is a $\mathbf{P}^\nu$−Brownian motion, by Girsanov theorem.

Given a contingent claim B, consider, for every stopping time τ, the $\mathcal{F}_\tau$-measurable random variable

$$V(\tau) := ess\sup_{\nu \in \mathcal{D}} E^\nu[B\gamma_\nu(\tau, T)|\mathcal{F}_\tau]. \tag{7.13}$$

The proof of the following theorem is similar to the corresponding theorem in the case of constraints.

7.3 Theorem: (El-Karoui, Peng & Quenez (1994)) *For an arbitrary contingent claim B, we have $h(0) = V(0)$. Furthermore, there exists a pair $(\hat{\pi}, \hat{c}) \in \mathcal{A}_0(V(0))$ such that $X^{V(0),\hat{\pi},\hat{c}}(\cdot) = V(\cdot)$.*

The theorem gives the minimal hedging price for a claim B; in fact, it is easy to see (using the same supermartingale argument) that the process $V(\cdot)$ is the minimal wealth process that hedges B. There remains the question of whether consumption is necessary. We show that, in fact, $\hat{c}(\cdot) \equiv 0$.

7.4 Theorem: *Every contingent claim B is attainable, namely the process $\hat{c}(\cdot)$ from Theorem 7.3 is a zero-process.*
Proof : Let $\{\nu_n; n \in \mathcal{N}\}$ be a maximizing sequence for achieving $V(0)$, i.e., $\lim_{n\to\infty} E^{\nu_n} B\gamma_{\nu_n}(T) = V(0)$. The necessary condition (7.11) (with $X(\cdot) = V(\cdot)$) implies $\lim_{n\to\infty} E^{\nu_n} \int_0^T \gamma_{\nu_n}(t)d\hat{c}(t) = 0$ and, since the processes $\gamma_{\nu_n}(\cdot)$ are bounded away from zero (uniformly in n), $\lim_{n\to\infty} E[Z_{\nu_n}(T)\hat{c}(T)] = 0$. Using weak compactness arguments as in Cvitanić & Karatzas (1993, Theorem 9.1) we can show that there exists $\nu \in \mathcal{D}$ such that $\lim_{n\to\infty} E[Z_{\nu_n}\hat{c}(T)] = E[Z_\nu(T)\hat{c}(T)] = 0$ (along a subsequence). It follows that $\hat{c}(\cdot) \equiv 0$. ◇

8. GENERAL NONLINEARITIES AND FORWARD-BACKWARD SDE's.

We denote by π the vector of amounts of money invested in stocks and change the model (2.1), (2.2) for the asset prices to

$$dP_0(t) = P_0(t)r(t, X_t, \pi_t)dt, \qquad P_0(0) = 1 \tag{8.1}$$

$$dP_i(t) = b_i(t, P_t, X_t, \pi_t)dt + \sum_{j=1}^d \sigma_{ij}(t, P_t, X_t, \pi_t)dW^{(j)}(t), \qquad P_i(0) = p_i \in (0, \infty) \tag{8.2}$$

for $i = 1, \ldots, d$. We require that the wealth replicates, at time T, the contingent claim with value $l(P(T))$, for a given function l, the assumptions on which are specified below. The wealth equation becomes

$$dX(t) = \hat{b}(t, P_t, X_t, \pi_t)dt + \hat{\sigma}(t, P_t, X_t, \pi_t)dW(t); \quad X(T) = l(P(T)), \tag{8.3}$$

where

$$\hat{b}(t, p, x, \pi) = \left(x - \sum_{i=1}^{d} \pi_i\right) r(t, x, \pi) + \sum_{i=1}^{d} \frac{\pi_i}{p_i} b_i(t, p, x, \pi);$$

$$\hat{\sigma}_j(t, p, x, \pi) = \sum_{i=1}^{d} \frac{\pi_i}{p_i} \sigma_{ij}(t, p, x, \pi), \quad j = 1, \cdots, d,$$

(8.4)

for $(t, p, x, \pi) \in [0, T] \times \mathbb{R}^d \times \mathbb{R} \times \mathbb{R}^d$. The system of SDE's (8.2) and (8.3) is called a *Forward-Backward* SDE; the forward component is the price process, having been assigned an initial value, whereas the backward component is the wealth process, having been assigned the terminal value $X(T) = l(P(T))$. An existence theory for such equations has been developed by Ma, Protter and Yong (1994).

The main differences compared to the model of Sections 2 and 7 are: (i) more general, nonconvex nonlinearities, including the volatility term; (ii) the contingent claim value $l(P(T))$ is not given in advance, but depends on the portfolio strategy and wealth of the investor through P; (iii) Markovian structure of the model.

In this section we shall use the following notations throughout: we denote $\mathbb{R}^d_+ = \{(x_1, \cdots, x_d) \in \mathbb{R}^d | x_i > 0, i = 1, \cdots, d\}$; the inner product in $\mathbb{R}^d$ by $\langle \cdot, \cdot \rangle$; the norm in $\mathbb{R}^d$ by $| \cdot |$ and that of $\mathbb{R}^{d \times d}$, the space of all $d \times d$ matrices, by $\| \cdot \|$ and the transpose of a matrix $A \in \mathbb{R}^{d \times d}$ (resp. a vector $x \in \mathbb{R}^d$) by A^T (resp. x^T). We also denote $\underline{1}$ to be the vector $\underline{1} := (1, \cdots, 1) \in \mathbb{R}^d$, and define a (diagonal) matrix-valued function $\Lambda : \mathbb{R}^d \mapsto \mathbb{R}^{d \times d}$ by

$$\Lambda(x) := \begin{bmatrix} x_1 & 0 & \cdots & 0 \\ 0 & x_2 & \cdots & 0 \\ \vdots & \vdots & \ddots & \vdots \\ 0 & 0 & \cdots & x_d \end{bmatrix}, \quad x = (x_1, \cdots, x_d) \in \mathbb{R}^d. \quad (8.5)$$

It is obvious that $\|\Lambda(x)\| = |x|$ for any $x \in \mathbb{R}^d$, and whenever $x \notin \partial \mathbb{R}^d_+$, $\Lambda(x)$ is invertible and $[\Lambda(x)]^{-1}$ is of the same form as $\Lambda(x)$ with $x_1, \cdots, x_d$ being replaced by $x_1^{-1}, \cdots, x_d^{-1}$. We can then rewrite functions $\hat{b}$ and $\hat{\sigma}$ in (8.4) as

$$\widehat{b}(t, p, x, \pi) = xr(t, x, \pi) + \langle \pi, b^1(t, p, x, \pi) - r(t, x, \pi)\underline{1} \rangle;$$

$$\widehat{\sigma}(t, p, x, \pi) = \langle \pi, \sigma^1(t, p, x, \pi) \rangle,$$

(8.6)

where

$$b^1(t, p, x, \pi) := [\Lambda(p)]^{-1} b(t, p, x, \pi) = \left(\frac{b_1}{p_1}, \cdots, \frac{b_d}{p_d}\right)(t, p, x, \pi);$$

$$\sigma^1(t, p, x, \pi) := [\Lambda(p)]^{-1} \sigma(t, p, x, \pi) = \left\{\frac{\sigma_{ij}}{p_i}\right\}_{i,j=1}^{d} (t, p, x, \pi).$$

(8.7)

To be consistent with the standard model, we henceforth call b^1 the *appreciation rate* and σ^1 the *volatility matrix* of the stock market. We restrict ourselves to the portfolios for which $E \int_0^T |\pi(t)|^2 dt < \infty$ and $X(t) \geq 0$, $\forall t \in [0, T]$, a.s.. Let us impose the following *Standing Assumptions*:

(A1) The functions b, $\sigma : [0, T] \times \mathbb{R}^d \times \mathbb{R} \times \mathbb{R}^d \mapsto \mathbb{R}$ and $l : \mathbb{R}^d \mapsto \mathbb{R}$ are twice continuously differentiable, such that $b(t, 0, x, \pi) = \sigma(t, 0, x, \pi) = 0$, for all $(t, x, \pi) \in [0, T] \times \mathbb{R} \times \mathbb{R}^d$. The functions b^1 and σ^1, together with their first order partial derivatives in p, x and π are bounded, uniformly in (t, p, x, π). Further, we assume that partial derivatives of b^1 and σ^1 in p satisfy

$$\sup_{(t,p,x,\pi)} \left\{ \left| p_k \frac{\partial b^1}{\partial p_k} \right|, \left| p_k \frac{\partial \sigma^1_{ij}}{\partial p_k} \right| \right\} < \infty, \qquad i, j, k = 1, \cdots, d. \qquad (8.8)$$

(A2) The function σ satisfies $\sigma \sigma^T(t, p, x, \pi) > 0$, for all (t, p, x, π) with $p \notin \partial \mathbb{R}^d_+$; and there exists a positive constant $\mu > 0$, such that

$$a^1(t, p, x, \pi) \geq \mu I, \qquad \text{for all} \quad (p, t, x, \pi), \qquad (8.9)$$

where $a^1 = \sigma^1 (\sigma^1)^T$.

(A3) The function r is twice continuously differentiable and such that the following conditions are satisfied:

(a) For $(t, x, \pi) \in [0, T] \times \mathbb{R} \times \mathbb{R}^d$, $0 < r(t, x, \pi) \leq K$, for some constant $K > 0$.

(b) The partial derivatives of r in x and π, denoted by a generic function ψ, satisfy

$$\varlimsup_{|x|,|\pi| \to \infty,} (|x| + |\pi|)^2 |\psi(t, x, \pi)| < \infty. \qquad (8.10)$$

Either

(A4.a) The function l is bounded, C^2 and nonnegative; Its partial derivatives up to second order are all bounded;

or

(A4.b) The function l is nonnegative and $\lim_{|p| \to \infty} l(p) = \infty$; moreover, l has bounded, continuous partial derivatives up to third order, and there exist constants $K, M > 0$ such that

$$\begin{cases} |\Lambda(p) l_p(p)| \leq K(1 + l(p)); \\ \sup_{p \in \mathbb{R}^d_+} \|\Lambda^2(p) l_{pp}\| = M < \infty. \end{cases}$$

(A5) The partial derivatives of σ^1 in x and π satisfy

$$\sup_{(t,p,x,\pi)} \left\{ \left| x \frac{\partial \sigma^1_{ij}}{\partial x} \right| + \left| x \frac{\partial \sigma^1_{ij}}{\partial \pi_k} \right| \right\} < \infty, \qquad i, j, k = 1, \cdots, d. \qquad (8.11)$$

Remark 8.1. The conditions here are quite restrictive, which is largely due to the generality of the setting of this section. The PDE method described below often works even if the assumptions are far from being satisfied. In particular, it works in the case of the model used in the previous section, if we restrict ourselves to the Markovian setting and to standard European options. We note that the assumptions (A1) and (A2) obviously contain those cases in which $b(t, p, x, \pi) = \Lambda(p)b_1(t, x, \pi)$ and $\sigma(t, p, x, \pi) = \Lambda(p)\sigma_1(t, x, \pi)$ where b_1 and σ_1 are bounded, continuously differentiable functions with bounded first order partial derivatives; and $\sigma_1 \sigma_1^T$ is positive definite and bounded away from zero, as is the case in the standard model. The second conditions on l restricts it to have at most quadratic growth. An example of a function σ satisfying (A1), (A2) and (A4.b) could be $\sigma(t, p, x, \pi) = p(\sigma(t) + \arctan(x^2 + |\pi|^2))$ with $\sigma(\cdot)$ satisfying (A2).

All the results below are proved in Cvitanić & Ma (1996):

Lemma 8.2. *Suppose that (A1), (A2) hold. Then for any portfolio π and initial wealth x, the price process P satisfies: $P_i(t) > 0$, $i = 1, \cdots, d$ for all $t \in [0, T]$, almost surely, provided the initial prices $p_1, \cdots, p_d$ are positive.*

The Four Step Scheme of Ma, Protter & Yong (1994), in our setting, consists of the following (and consists of three steps only):

Step 1: Solve the Black-Scholes type (but nonlinear) PDE

$$\begin{cases} 0 = \theta_t + \frac{1}{2}\text{tr}\{\sigma\sigma^T(t, p, \theta, \Lambda(p)\theta_p)\theta_{pp}\} + (\langle p, \theta_p \rangle - \theta)r(t, \theta, \Lambda(p)\theta_p), \\ \theta(T, p) = l(p), \qquad p \in \mathbb{R}_+^d. \end{cases} \tag{8.12}$$

Step 2: Setting

$$\begin{cases} \tilde{b}(t, p) = b(t, p, \theta(t, p), \Lambda(p)\theta_p(t, p)) \\ \tilde{\sigma}(t, p) = \sigma(t, p, \theta(t, p), \Lambda(p)\theta_p(t, p)), \end{cases} \tag{8.13}$$

solve the forward SDE

$$P(t) = p + \int_0^t \tilde{b}(s, P(s))ds + \int_0^t \tilde{\sigma}(s, P(s))dW(s). \tag{8.14}$$

Step 3: Set

$$\begin{cases} X(t) = \theta(t, P(t)) \\ \pi(t) = \Lambda(P(t))\theta_p(t, P(t)), \end{cases} \tag{8.15}$$

Theorem 8.3. *Suppose that the standing assumptions (A1)—(A3), (A4.b) and (A5) hold. Then the PDE (8.12) and the SDE (8.14) admit unique solutions. Moreover, for any given $p \in \mathbb{R}_+^d$, the FBSDE (8.2), (8.3) admits a unique adapted solution (P, X, π), given by (8.15) with θ being the solution of (8.12).*

The theorem implies that the initial value $X(0)$ of the backward process provides an upper bound for the minimal hedging price $h(0)$ of the contingent claim $B = l(P(T))$, since the claim can indeed be hedged starting with $X(0)$. The next result shows that $X(0) = h(0)$, and that π given by (8.15) is the least expensive hedging portfolio.

Theorem 8.4. *(Comparison Theorem): Suppose that (A1) - (A5) hold. Let initial prices $p \in \mathbb{R}_+^d$ be given, and let π be any admissible portfolio such that the corresponding price/wealth process (P, Y) satisfies $Y(T) \geq l(P(T))$. Then $Y(\cdot) \geq \theta(\cdot, P(\cdot))$, where θ is the solution to (8.12). In particular, $Y(0) \geq$*

$\theta(0,p) = X(0)$, where X is the solution to the FBSDE (8.2), (8.3), starting from $p \in \mathbb{R}^d_+$, constructed by the Four-Step Scheme.

We conclude by examples in which a model like the one of this section would be appropriate:

(i) *Large investor.* If the investor keeps too much capital in the bank, the government (or the market) decides to decrease the bank interest rate. For example, we can assume that $r(t, x, \pi)$ is a decreasing function of $x - \pi$, for $x - \pi$ large.

(ii) Borrowing rate could be decreasing in wealth.

(iii) *Several agents - equilibrium model.* In Platen & Schweizer (1994), an SDE for the stock price is obtained from equilibrium considerations; both its drift and volatility coefficients depend on the hedging strategy of the agents in the market in a rather complex fashion. As the authors mention, " it is not clear at all how one should compute option prices in an economy where agents' strategies affect the underlying stock price process". Our results provide the price that would enable the seller to hedge against all the risk, i.e., the upper bound for the price.

9. EXAMPLE: HEDGING CLAIMS WITH HIGHER INTEREST RATE FOR BORROWING

We have studied so far a model in which one is allowed to borrow money, at an interest rate $R(\cdot)$ equal to the bank rate $r(\cdot)$. In this section we consider the more general case of a financial market $\mathcal{M}^*$ in which $R(\cdot) \geq r(\cdot)$, without constraints on portfolio choice. We assume that the progressively measurable process $R(\cdot)$ is also bounded.

In this market $\mathcal{M}^*$ it is not reasonable to borrow money and to invest money in the bank at the same time. Therefore, we restrict ourselves to policies for which the relative amount borrowed at time t is equal to $\left(1 - \sum_{i=1}^{d} \pi_i(t)\right)^-$. Then, the wealth process $X = X^{x,\pi,c}$ corresponding to initial capital $x > 0$ and portfolio/cumulative consumption pair (π, c), satisfies

$$
\begin{aligned}
dX(t) =& r(t)X(t)dt - dc(t) \\
&+ X(t)\left[\pi^*(t)\sigma(t)dW_0(t) - (R(t) - r(t))\left(1 - \sum_{i=1}^{d} \pi_i(t)\right)^- dt\right].
\end{aligned} \tag{9.1}
$$

In the notation of Section 7, we get $\tilde{g}(\nu(t)) = r(t) - \nu_1(t)$ for $\nu \in \mathcal{D}$, where

$$
\begin{aligned}
\mathcal{D} := \{&\nu; \ \nu \text{ progressively measurable, } \mathcal{R}^d - \text{valued process with} \\
&r - R \leq \nu_1 = \ldots = \nu_d \leq 0, \ \ell \otimes \mathbf{P} - a.e.\}.
\end{aligned} \tag{9.2}
$$

We also have

$$\tilde{g}(\nu(t)) - g(t,\pi(t)) - \pi^*(t)\nu(t) = [R(t) - r(t) + \nu_1(t)]\Big(1 - \sum_{i=1}^{d} \pi_i(t)\Big)^{-}$$

$$- \nu_1(t)\Big(1 - \sum_{i=1}^{d} \pi_i(t)\Big)^{+}, \tag{9.3}$$

for $0 \le t \le T$. It can be shown, in analogy to the case of constraints, that the optimal dual process $\hat{\lambda}(\cdot) \in \mathcal{D}$ can be taken as the one that attains zero in (9.3), namely as

$$\hat{\lambda}(t) = \hat{\lambda}_1(t)\mathbf{1}, \quad \hat{\lambda}_1(t) := [r(t) - R(t)]\, 1_{\{\sum_{i=1}^{d} \hat{\pi}_i(t) > 1\}}. \qquad\qquad \diamond \tag{9.4}$$

Consider the case $d = 1$, $B = \varphi(P_1(T))$ with $\varphi : \mathbb{R}_+ \to [0,\infty)$, and with constant $R > r$, If $p\varphi'(p) \ge \varphi(p)$ holds everywhere on $\mathbb{R}_+$ and strictly on a set of positive measure, then we may take $\hat{\lambda} \equiv r - R$, and the Black-Scholes formulae, remain valid if we replace in them r by R. This follows as in the following example.

9.1 Example: Let us consider the case of constant coefficients $r, R, \{\sigma_{ij}\} = \sigma$. Then the vector $P(t) = (P_1(t), \ldots, P_d(t))$ of stock price processes satisfies the equations

$$dP_i(t) = P_i(t)[b_i(t)dt + \sum_{i=1}^{d} \sigma_{ij} dW^{(j)}(t)]$$

$$= P_i(t)[(r - \nu_1(t))dt + \sum_{i=1}^{d} \sigma_{ij} dW_\nu^{(j)}(t)], \quad 1 \le i \le d, \tag{9.5}$$

for every $\nu \in \mathcal{D}$. Consider now a contingent claim of the form $B = \varphi(P(T))$, for a given continuous function $\varphi : \mathbb{R}_+^d \to [0,\infty)$ that satisfies a polynomial growth condition, as well as the *value function*

$$Q(t,p) := \sup_{\nu \in \mathcal{D}} E^\nu[\varphi(P(T))e^{-\int_t^T (r - \nu_1(s))ds}|P(t) = p] \tag{9.6}$$

on $[0,T] \times \mathbb{R}_+^d$. Clearly, the processes $\hat{X}, V$ are given as

$$\hat{X}(t) = Q(t, P(t)), \quad V(t) = e^{-rt}\hat{X}(t); \quad 0 \le t \le T,$$

where Q solves the semilinear parabolic partial differential equation of Hamilton-Jacobi-Bellman (HJB) type

$$\frac{\partial Q}{\partial t} + \frac{1}{2}\sum_i \sum_j a_{ij} p_i p_j \frac{\partial^2 Q}{\partial p_i \partial p_j} + \max_{r-R \le \nu_1 \le 0}\left[(r - \nu_1)\{\sum_i p_i \frac{\partial Q}{\partial p_i} - Q\}\right] = 0, \tag{9.7}$$

for $0 \le t < T$, $p \in \mathbb{R}^d_+$,

$$Q(T,p) = \varphi(p); \quad p \in \mathbb{R}^d_|$$

associated with the control problem of (9.6) and the dynamics (9.5) (cf. Ladyženskaja, Solonnikov & Ural'tseva (1968) for the basic theory of such equations, and Fleming & Rishel (1975) for the connections with stochastic control). Clearly, the maximization in (9.7) is achieved by $\nu^*_1 = -(R - r).1_{\{\sum_i p_i \frac{\partial Q}{\partial p_i} \ge Q\}}$; the portfolio $\hat{\pi}(\cdot)$ and the process $\hat{\lambda}_1(\cdot)$ are then given, respectively, by

$$\hat{\pi}_i(t) = \frac{P_i(t) \cdot \frac{\partial}{\partial p_i} Q(t, P(t))}{Q(t, P(t))}, \qquad i = 1, \dots, d \qquad (9.8)$$

and

$$\hat{\lambda}_1(t) = (r - R)1_{\{\sum_i \hat{\pi}_i(t) \ge 1\}}. \qquad (9.9)$$

Suppose now that the function φ satisfies $\sum_i p_i \frac{\partial \varphi(p)}{\partial p_i} \ge \varphi(p)$, $\forall p \in \mathbb{R}^d_+$. Then the solution Q also satisfies this inequality:

$$\sum_i p_i \frac{\partial Q(t,p)}{\partial p_i} \ge Q(t,p), \qquad 0 \le t \le T \qquad (9.10)$$

for all $p \in \mathbb{R}^d_+$, and is actually given explicitly as

$$Q(t,p) = E^{(r-R)1}[e^{-R(T-t)}\varphi(P(T))|P(t) = p]$$
$$= \begin{cases} e^{-R(T-t)} \int_{R^d} \varphi(h(T-t,p,\sigma z; R))(2\pi t)^{-d/2} e^{-\frac{\|z\|^2}{2t}} dz & ; \quad t < T, p > 0 \\ \varphi(p) & ; \quad t = T, p > 0 \end{cases}$$
$$(9.11)$$

in the notation of previous sections. This is because, in this case, the PDE (9.7) becomes the Black-Scholes PDE

$$\frac{\partial Q}{\partial t} + \frac{1}{2}\sum_i \sum_j p_i p_j a_{ij} \frac{\partial^2 Q}{\partial p_i \partial p_j} + R\left(\sum_i p_i \frac{\partial Q}{\partial p_i} - Q\right) = 0; t < T, p > 0$$

$$Q(T,p) = \varphi(p); p > 0$$

In this case the portfolio $\hat{\pi}(\cdot)$ always borrows: $\sum^d_{i=1} \hat{\pi}_i(t) \ge 1$, $0 \le t \le T$ (a.s.), and thus $\hat{\lambda}_1(t) = r - R$, $0 \le t \le T$.

10. UTILITY FUNCTIONS.

A function $U : (0, \infty) \to \mathbb{R}$ will be called a *utility function* if it is strictly increasing, strictly concave, of class C^1, and satisfies

$$U'(0+) := \lim_{x \downarrow 0} U'(x) = \infty, \quad U'(\infty) := \lim_{x \to \infty} U'(x) = 0. \qquad (10.1)$$

We shall denote by I the (continuous, strictly decreasing) inverse of the function U'; this function maps $(0, \infty)$ onto itself, and satisfies $I(0+) = \infty, I(\infty) = 0$. We also introduce the Legendre-Fenchel transform

$$\tilde{U}(y) := \max_{x>0}[U(x) - xy] = U(I(y)) - yI(y), \quad 0 < y < \infty \qquad (10.2)$$

of $-U(-x)$; this function $\tilde{U}$ is strictly decreasing and strictly convex, and satisfies

$$\tilde{U}'(y) = -I(y), \quad 0 < y < \infty, \qquad (10.3)$$

$$U(x) = \min_{y>0}[\tilde{U}(y) + xy] = \tilde{U}(U'(x)) + xU'(x) , \quad 0 < x < \infty . \qquad (10.4)$$

The useful inequalities

$$U(I(y)) \geq U(x) + y[I(y) - x] \qquad (10.5)$$

$$\tilde{U}(U'(x)) + x[U'(x) - y] \leq \tilde{U}(y) , \qquad (10.6)$$

valid for all $x > 0, y > 0$, are direct consequences of (10.2), (10.4). It is also easy to check that

$$\tilde{U}(\infty) = U(0+), \quad \tilde{U}(0+) = U(\infty) \qquad (10.7)$$

hold; cf. KLSX (1991), Lemma 4.2.

10.1 Remark: We shall have occasion, in the sequel, to impose the following conditions on our utility functions:

$$c \mapsto cU'(c) \quad \text{is nondecreasing on } (0, \infty) , \qquad (10.8)$$

for some $\alpha \epsilon (0, 1), \gamma \epsilon (1, \infty)$ we have: $\quad \alpha U'(x) \geq U'(\gamma x), \quad \forall \, x \epsilon (0, \infty) .$
$$\qquad (10.9)$$

10.2 Remark: Condition (10.8) is equivalent to

$$y \mapsto yI(y) \quad \text{is nonincreasing on } (0, \infty) , \qquad (10.8)'$$

and implies that

$$x \mapsto \tilde{U}(e^x) \quad \text{is convex on } \mathcal{R} . \qquad (10.8)''$$

(If U is of class C^2, then condition (10.8) amounts to the statement that $\frac{-cU''(c)}{U'(c)}$, the so-called "Arrow-Pratt measure of relative risk - aversion", does not exceed 1.)

Similarly, condition (10.9) is equivalent to having

$$I(\alpha y) \leq \gamma I(y), \quad \forall \, y \epsilon (0, \infty) \quad \text{for some } \alpha \epsilon (0, 1), \quad \gamma > 1 . \qquad (10.9)'$$

Iterating (10.9)', we obtain the apparently stronger statement

$$\forall \, \alpha \epsilon (0, 1), \, \exists \, \gamma \epsilon (1, \infty) \text{ such that } \quad I(\alpha y) \leq \gamma I(y), \quad \forall \, y \epsilon (0, \infty) . \quad (10.9)''$$

11. PORTFOLIO OPTIMIZATION PROBLEM.

In this section we consider the optimization problem of maximizing utility from terminal wealth for our investor, i.e., we want to maximize

$$J(x; \pi) := EU(X^{x,\pi}(T)) , \tag{11.1}$$

over the constrained portfolios $\pi \in \mathcal{A}'(x)$, provided that the expectation is well-defined. More precisely, we have

11.1 Definition: The *utility maximization problem* is to maximize the expression of (11.1) over the class $\mathcal{A}'(x)$ of processes $\pi \in \mathcal{A}'(x)$ that satisfy

$$EU^-(X^{x,\pi}(T)) < \infty. \tag{11.2}$$

(x^- denotes the negative part of x : $x^- = max\{-x, 0\}$.) The value function of this problem will be denoted by

$$V(x) := \sup_{\pi \in \mathcal{A}'_0(x)} J(x; \pi) , \quad x \in (0, \infty) . \tag{11.3}$$

11.2 Assumption: $V(x) < \infty, \ \forall \, x \in (0, \infty) .$

It is fairly straightforward that the function $V(\cdot)$ is increasing and concave on $(0, \infty)$.

11.3 Remark: It can be checked that the Assumption 11.2 is satisfied if the function U is nonnegative and satisfies the growth condition

$$0 \le U(x) \le \kappa(1 + x^\alpha) ; \quad \forall \, (x) \in (0, \infty) \tag{11.7}$$

for some constants $\kappa \in (0, \infty)$ and $\alpha \in (0, 1)$ - cf. KLSX (1991) for details.

Denote

$$H_\nu(t) = \gamma_\nu(t) Z_\nu(t),$$

in the notation of (5.7) and (5.8).

11.4 Definition: We introduce the function

$$\mathcal{X}_\nu(y) := E\Big[H_\nu(T)I(yH_\nu(T))\Big] , \quad 0 < y < \infty \tag{11.9}$$

and consider the subclass D' of $\mathcal{H}$ (in the notation of Section 5) given by

$$D' := \{\nu \in \mathcal{H}; \ \mathcal{X}_\nu(y) < \infty, \ \forall \, y \in (0, \infty)\} . \tag{11.10}$$

For every $\nu \in D'$, the function $\mathcal{X}_\nu(\cdot)$ of (11.9) is continuous and strictly decreasing, with $\mathcal{X}_\nu(0+) = \infty$ and $\mathcal{X}_\nu(\infty) = 0$; we denote its inverse by $\mathcal{Y}_\nu(\cdot)$.

11.5 Remark: Suppose that $U(\cdot)$ satisfies condition (11.9). It is then easy to see, using (11.9)$''$, that $\mathcal{X}_\nu(y) < \infty$ for some $y \in (0, \infty)$ implies: $\nu \in D'$.

Next, we prove a crucial lemma, which provides sufficient conditions for optimality in the problem of (11.1). The duality approach of the lemma and subsequent analysis was implicitly used in Pliska (1986), Karatzas, Lehoczky & Shreve (1987), Cox & Huang (1989) in the case of no constraints, and explicitly in He & Pearson (1991), Karatzas et al. (1991), Xu & Shreve (1992) for special types of constraints.

11.6 Lemma: For any given $x > 0$, $y > 0$ and $\pi \in \mathcal{A}'(x)$, we have

$$EU(X^{x,\pi}(T)) \leq E\tilde{U}(yH_\nu(T)) + yx, \quad \forall \, \nu \in \mathcal{H}. \tag{11.11}$$

In particular, if $\hat{\pi} \in \mathcal{A}'(x)$ is such that *equality* holds in (11.11), for some $\lambda \in \mathcal{H}$ and $\hat{y} > 0$, then $\hat{\pi}$ is optimal for our (primal) optimization problem, while λ is optimal for the *dual problem*

$$\tilde{V}(\hat{y}) = \inf_{\nu \in \mathcal{H}} E\tilde{U}(\hat{y}H_\nu(T)). \tag{11.12}$$

Furthermore, equality holds in (11.11) if and only if

$$X^{x,\pi}(T) = I(yH_\nu(T)) \ a.s., \tag{11.13}$$

$$\delta(\nu_t) = -\nu^*(t)\pi(t) \ a.e., \tag{11.14}$$

$$E[H_\nu(T)X^{x,\pi}(T)] = x \tag{11.15}$$

(the latter being equivalent to $\nu \in \mathcal{D}'$ and $y = \mathcal{Y}_\nu(x)$, if (11.13) holds).

Proof: By definitions of $\tilde{U}, \delta$ we get

$$U(X(T)) \leq \tilde{U}(yH_\nu(T)) + yH_\nu(T)X(T) + \int_0^T H_\nu(t)X(t)[\delta(\nu_t) + \nu^*(t)\pi(t)]dt. \tag{11.16}$$

The upper bound of (11.11) follows from the supermartingale property (6.14); condition (11.13) follows from (10.2), condition (11.14) is obvious, and condition (11.15) corresponds to equality holding in (6.14).   ◇

Remark 11.7: Lemma 11.6 suggests the following strategy for solving the optimization problem:

(i) show that the dual problem (11.12) has an optimal solution $\lambda_y \in \mathcal{D}'$ for all $y > 0$;

(ii) using Theorem 6.4, find the minimal hedging price $h_y(0)$ and a corresponding portfolio $\hat{\pi}_y$ for hedging $B := I(yH_{\lambda_y}(T))$;

(iii) prove (11.14) for the pair $(\hat{\pi}_y, \lambda_y)$;

(iv) show that, for every $x > 0$, you can find $y = y_x > 0$ such that $x = h_y(0) = E[H_{\lambda_y}(T)I(yH_{\lambda_y}(T))]$.

Then (i)-(iv) would imply that $\hat{\pi}_{y_x}$ is the optimal portfolio process for the utility maximization problem of an investor starting with initial capital equal to x.

To verify that step (i) can be accomplished, we impose the following condition:

$$\forall \, y \, \in \, (0,\infty), \quad \exists \, \nu \in \mathcal{H} \text{ such that } \tilde{J}(y;\nu) := E\tilde{U}(yH_\nu(T)) < \infty \qquad (11.17)$$

We shall also impose the assumption

$$U(0+) > -\infty \, , \qquad U(\infty) = \infty \, . \qquad (11.18)$$

11.8 Remark: Under the conditions of Remark 11.3, the requirement (11.17) is satisfied. Indeed, the condition (11.7) leads to

$$0 \le \tilde{U}(y) \le \tilde{\kappa}(1 + y^{-\rho}) \, ; \quad \forall \, y \, \in \, (0,\infty)$$

for some $\tilde{\kappa} \, \in \, (0,\infty)$ and $\rho = \frac{\alpha}{1-\alpha}$.

Even though the log function does not satisfy (11.18), we solve that case directly in examples below.

11.9 Theorem: (Cvitanić & Karatzas (1992)) *Assume that (10.8), (10.9), (11.17) and (11.18) are satisfied. Then condition (i) of Remark 11.7 is true, i.e. the dual problem admits a solution in the set $\mathcal{D}'$, for every $y > 0$.*

The fact that the dual problem admits a solution under the conditions of Theorem 11.9 follows almost immediately (by standard weak compactness arguments) from Proposition 11.10 below. The details, as well as a relatively straightforward proof of Proposition 11.10, can be found in CK(1992). Denote by $\mathcal{H}'$ the Hilbert space of progressively measurable processes ν with norm $[\![\nu]\!] = E \int_0^T \nu^2(s)ds < \infty$.

11.10 Proposition: *Under the assumptions of Theorem 11.9, the functional $\tilde{J}(y;\cdot) : \mathcal{H}' \to \mathbb{R}\cup\{+\infty\}$ of (11.12) is (i) convex, (ii) coercive:* $\lim_{[\![\nu]\!]\to\infty} \tilde{J}(y;\nu) = \infty$, *and (iii) lower-semicontinuous: for every $\nu\epsilon\mathcal{H}'$ and $\{\nu_n\}_{n\epsilon\mathbb{N}} \subseteq \mathcal{H}'$ with $[\![\nu_n - \nu]\!] \to 0$ as $n \to \infty$, we have*

$$\tilde{J}(y;\nu) \le \lim_{n\to\infty} \tilde{J}(y;\nu_n) \, . \qquad (11.19)$$

11.11 Remark: optimal λ_y is an element of $\mathcal{D}$. It can be shown that the optimal dual process λ_y satisfies $\lambda_y \in \mathcal{D}'$; see, for example, Karatzas, Lehoczky, Shreve & Xu (1991), proof of Theorem 12.3.

We move now to step (ii) of Remark 11.7. We have the following useful fact:

11.12 Lemma: For every $\nu \in \mathcal{H}$, $0 < y < \infty$, we have

$$E[H_\nu(T)B_{\lambda_y}] \le E[H_{\lambda_y}(T)B_{\lambda_y}] \, . \qquad (11.20)$$

11.13 Remark: In fact, (11.20) is equivalent to λ_y being optimal for the dual problem, but we shall not need that result; its proof is quite lengthy and technical (see CK (1992, Theorem 10.1). We are going to provide a simpler proof for Lemma 11.12, but under the additional assumption that

$$E[H_{\lambda_y}(T)I(yH_\nu(T))] < \infty, \ \forall \nu \in \mathcal{H}, y > 0. \tag{11.21}$$

Proof of Lemma 11.12: Fix $\varepsilon \in (0,1), \nu \in \mathcal{H}$ and define (supressing dependence on t)

$$
\begin{aligned}
G_\varepsilon &:= (1-\varepsilon)H_{\lambda_y} + \varepsilon H_\nu, \quad \mu_\varepsilon := G_\varepsilon^{-1}((1-\varepsilon)H_{\lambda_y}\lambda_y + \varepsilon H_\nu \nu), \\
\tilde{\mu}_\varepsilon &:= G_\varepsilon^{-1}((1-\varepsilon)H_{\lambda_y}\delta(\lambda_y) + \varepsilon H_\nu \delta(\nu)).
\end{aligned}
\tag{11.22}
$$

Then $\mu_\varepsilon \in \mathcal{H}$, because of the convexity of $\tilde{K}$. Moreover, we have

$$dG_\varepsilon = (\theta + \sigma^{-1}\mu_\varepsilon)G_\varepsilon dW - \tilde{\mu}_\varepsilon G_\varepsilon dt,$$

and convexity of δ implies $\delta(\mu_\varepsilon) \le \tilde{\mu}_\varepsilon$, and therefore, comparing the solutions to the respective (linear) SDE's, we get

$$G_\varepsilon(\cdot) \le H_{\mu_\varepsilon}(\cdot), \ a.s.. \tag{11.23}$$

Since λ_y is optimal and $\tilde{U}$ is decreasing, (11.23) implies

$$\varepsilon^{-1}\left(E[\tilde{U}(yH_{\lambda_y}(T)) - \tilde{U}(yG_\varepsilon(T))]\right) \le 0. \tag{11.24}$$

Next, recall that $I = -\tilde{U}'$ and denote by V_ε the random variable inside the expectation operator in (11.24). Fix $\omega \in \Omega$, and assume, supressing the dependence on ω and T, that $H_\nu \ge H_{\lambda_y}$. Then $\varepsilon^{-1}V_\varepsilon = I(F)y(H_\nu - H_{\lambda_y})$, where $yH_{\lambda_y} \le F \le yH_{\lambda_y} + \varepsilon y(H_\nu - H_{\lambda_y})$. Since I is decreasing we get $\varepsilon^{-1}V_\varepsilon \ge yI(yH_\nu)(H_\nu - H_{\lambda_y})$. We get the same result when assuming $H_\nu \le H_{\lambda_y}$. This and assumption (11.21) imply that we can use Fatou's lemma when taking the limit as $\varepsilon \downarrow 0$ in (11.24), which gives us (11.20).   ⋄

Now, given $y > 0$ and the optimal λ_y for the dual problem, let π_y be the portfolio of Theorem 6.4 for hedging the claim $B_{\lambda_y} = I(yH_{\lambda_y}(T))$. Lemma 11.12 implies that, in the notation of Section 6,

$$h_y(0) = V_y(0) = E[H_{\lambda_y}(T)I(yH_{\lambda_y}(T))] = \ \text{initial capital for portfolio } \pi_y,$$

so (11.15) is satisfied for $x = h_y(0)$. It also implies, by (6.18), that (11.14) holds for the pair (π_y, λ_y). Therefore we have completed both steps (ii) and (iii). Step (iv) is a corollary of the following result.

11.13 Proposition: *Under the assumptions of Theorem 11.9, for any given* $x > 0$, *there exists* $y_x > 0$ *that achieves* $\inf_{y>0}[\tilde{V}(y) + xy]$ *and satisfies*

$$x = \mathcal{X}_{\lambda_{y_x}}(y_x). \tag{11.25}$$

For the (straightforward) proof see CK (1992, Proposition 12.2). We now put together the results of this section:

11.14 Theorem: *Under the assumptions of Theorem 11.9, for any given* $x > 0$ *there exists an optimal portfolio process* $\hat{\pi}$ *for the utility maximization problem of Definition 11.1.* $\hat{\pi}$ *is equal to the portfolio of Theorem 6.4 for minimaly hedging* the claim $I(y_x H_{\lambda_{y_x}}(T))$, where y_x is given by Proposition 11.13 and λ_{y_x} is the optimal process for the dual problem (11.12).

12. EXAMPLES

12.1 Example: *Logarithmic utility.* If $U(x) = \log x$, for $x \epsilon\ (0, \infty)$, we have $I(y) = \frac{1}{y}$, $\tilde{U}(y) = -(1 + \log y)$ and

$$\mathcal{X}_\nu(y) = \frac{1}{y}, \quad \mathcal{Y}_\nu(x) = \frac{1}{x} \tag{12.1}$$

and therefore, the optimal terminal wealth is

$$X_\lambda(T) = x\frac{1}{H_\lambda(T)} \tag{12.2}$$

for $\lambda \epsilon \mathcal{H}$ optimal. In particular $\mathcal{D}' = \mathcal{H}$ in this case. Therefore,

$$E\left[\tilde{U}(\mathcal{Y}_\lambda(x)H_\nu(T))\right] = = -1 - \log\frac{1}{x} + E\left(\log\frac{1}{H_\nu(T)}\right) . \tag{12.3}$$

But

$$E\left(\log\frac{1}{H_\nu(T)}\right) = E\int_0^T \left[r(s) + \delta(\nu(s)) + \frac{1}{2}\|\theta(s) + \sigma^{-1}(s)\nu(s)\|^2\right]ds ,$$

and thus the dual problem amounts to a point-wise minimization of the convex function
$\delta(x) + \frac{1}{2}\|\theta(t) + \sigma^{-1}(t)x\|^2$ over $x \epsilon \tilde{K}$, for every $t \epsilon\ [0, T]$:

$$\lambda(t) = \arg\min_{x\epsilon\tilde{K}}\ [\ 2\delta(x) + \|\theta(t) + \sigma^{-1}(t)x\|^2\ ] . \tag{12.4}$$

Furthermore, (12.2) gives

$$H_\lambda(t)X_\lambda(t) = x; \quad 0 \leq t \leq T ,$$

and using Ito's rule to get the SDE for $H_\lambda X_\lambda$ we get, by equating the integrand in the stochastic integral term to zero, $\sigma^*(t)\hat{\pi}(t) = \theta_\lambda(t)$, $\ell \otimes P$ - a.e.

We conclude that the optimal portfolio is given by

$$\hat{\pi}(t) = (\sigma(t)\sigma^*(t))^{-1}[\lambda(t) + b(t) - r(t)\underline{1}] \tag{12.5}$$

in terms of the market coëfficients and the process λ of (12.4).

12.2 Example: *(Constraints on borrowing)* From the point of view of applications, an interesting example is the one in which the total proportion $\sum_{i=1}^d \pi_i(t)$ of wealth invested in stocks is bounded from above by some real constant $a > 0$. For example, if we take $a = 1$, we exclude borrowing; with $a \in (1, 2)$, we allow borrowing up to a fraction $1-a$ of wealth. If we take $a = 1/2$, we have to invest at least half of the wealth in the bank.

To illustrate what happens in this situation, let again $U(x) = \log x$,, and, for the sake of simplicity, $d = 2$, $\sigma =$ unit matrix, and the constraints on the portfolio be given by

$$K = \{x\epsilon\mathbb{R}^2; \ x_1 \geq 0, x_2 \geq 0, \ x_1 + x_2 \leq a\}$$

for some $a\epsilon(0, 1]$ (obviously, we also exclude short-selling with this K). We have here $\delta(x) \equiv a\max\{x_1^-, x_2^-\}$, and thus $\tilde{K} = \mathbb{R}^2$. By some elementary calculus and/or by inspection, and omitting the dependence on t, we can see that the optimal dual process λ that minimizes $\frac{1}{2}\|\theta_t + \nu_t\|^2 + \delta(\nu_t)$, and the optimal portfolio $\pi_t = \theta_t + \lambda_t$, are given respectively by

$$\lambda = -\theta \ ; \ \pi = (0,0)^* \ \text{if} \ \theta_1, \theta_2 \leq 0$$

(do not invest in stocks if the interest rate is larger than the stocks appreciation rates),

$$\lambda = (0, -\theta_2)^*; \quad \pi = (\theta_1, 0)^* \quad \text{if} \ \theta_1 \geq 0, \theta_2 \leq 0, \ a \geq \theta_1 \ ,$$
$$\lambda = (a - \theta_1, -\theta_2)^*; \quad \pi = (a, 0)^* \quad \text{if} \ \theta_1 \geq 0, \theta_2 \leq 0, \ a < \theta_1 \ ,$$
$$\lambda = (-\theta_1, 0)^*; \quad \pi = (0, \theta_2)^* \quad \text{if} \ \theta_1 \leq 0, \theta_2 \geq 0, \ a \geq \theta_2 \ ,$$
$$\lambda = (-\theta_1, a - \theta_2)^*; \quad \pi = (0, a)^* \quad \text{if} \ \theta_1 \leq 0, \theta_2 \geq 0, \ a < \theta_2 \ ,$$

(do not invest in the stock whose rate is less than the interest rate, invest $X \min\{a, \theta_i\}$ in the i-th stock whose rate is larger than the interest rate),

$$\lambda = (0,0)^*; \quad \pi = \theta \ \text{if} \ \theta_1, \theta_2 \geq 0, \ \theta_1 + \theta_2 \leq a$$

(invest $\theta_i X$ in the respective stocks-as in the no constraints case-whenever the optimal portfolio of the no constraints case happens to take values in K),

$$\lambda = (a - \theta_1, -\theta_2)^*; \quad \pi = (a, 0)^* \ \text{if} \ \theta_1, \theta_2 \geq 0, \ a \leq \theta_1 - \theta_2 \ ,$$
$$\lambda = (-\theta_1, a - \theta_2)^*; \quad \pi = (0, a)^* \ \text{if} \ \theta_1, \theta_2 \geq 0, \ a \leq \theta_2 - \theta_1$$

(with both $\theta_1, \theta_2 \geq 0$ and $\theta_1 + \theta_2 > a$ do not invest in the stock whose rate is smaller, invest aX in the other one if the absolute value of the difference of the stocks rates is larger than a),

$$\lambda_1 = \lambda_2 = \frac{a - \theta_1 - \theta_2}{2} \ ; \quad \pi_1 = \frac{a + \theta_1 - \theta_2}{2} \ , \ \pi_2 = \frac{a + \theta_2 - \theta_1}{2}$$

if $\theta_1, \theta_2 \geq 0, \theta_1 + \theta_2 > a > |\theta_1 - \theta_2|$ (if none of the previous conditions is satisfied, invest the amount $\frac{a}{2}X$ in the stocks, corrected by the difference of their rates).

Let us consider now the case, where the coëfficients $r(\cdot), b(\cdot), \sigma(\cdot)$ of the market model are deterministic functions on $[0, T]$, which we shall take for simplicity to be bounded and continuous. Then there is a formal HJB (Hamilton-Jacobi-Bellman) equation associated with the dual optimization problem, namely,

$$Q_t + \inf_{x \in \tilde{K}} [\frac{1}{2} y^2 Q_{yy} ||\theta(t) + \sigma^{-1}(t)x||^2 - yQ_y\delta(x)] - yQ_y r(t) = 0 , \qquad (12.6)$$

in $[0,T) \times (0,\infty)$;

$$Q(T,y) = \tilde{U}(y) ; \quad y \in (0,\infty) . \qquad (12.7)$$

If there exists a classical solution $Q \in C^{1,2}([0,T) \times (0,\infty))$ of this equation, that satisfies appropriate growth conditions, then standard verification theorems in stochastic control (e.g. Fleming & Soner (1993)) lead to the representation

$$\tilde{V}(y) = Q(0,y), \quad 0 < y < \infty \qquad (12.8)$$

for the dual value function.

12.3 Example: *(Cone constraints)* Suppose that $\delta \equiv 0$ on $\tilde{K}$. Then

$$\lambda(t) = arg \min_{x \in \tilde{K}} ||\theta(t) + \sigma^{-1}(t)x||^2 \qquad (12.9)$$

is *deterministic*, the same for all $y \in (0,\infty)$, and the equation (12.6) becomes

$$Q_t + \frac{1}{2}||\theta_\lambda(t)||^2 y^2 Q_{yy} - r(t)yQ_y + \tilde{U}_1(t,y) = 0 ; \quad \text{in } [0,T) \times (0,\infty) . \qquad (12.10)$$

Standard theory guarantees then the existence and uniqueness of a classical solution for this equation.

In the case of constant coëfficients, this solution can even be computed explicitly; see Cvitanić & Karatzas (1992).

12.4 Problem: *(Power utility)* Consider the case $U(x) = \frac{x^\alpha}{\alpha}$, $x \in (0,\infty)$ for some $\alpha \in (0,1)$. Then $\tilde{U}(y) = \frac{1}{\rho}y^{-\rho}$, $0 < y < \infty$ with $\rho := \frac{\alpha}{1-\alpha}$. Again, the process $\lambda(\cdot)$ is *deterministic*, namely

$$\lambda(t) = arg \min_{x \in \tilde{K}} [||\theta(t) + \sigma^{-1}(t)x||^2 + 2(1-\alpha)\delta(x)] ,$$

and is the same for all $y \in (0,\infty)$. ◇

Show that, in this case,

$$\pi_\lambda(t) = \frac{1}{1-\alpha}(\sigma(t)\sigma^*(t))^{-1}[b(t) - r(t)1 + \lambda(t)] .$$

12.5 Example: *(Different interest rates for borrowing and lending)* We consider the example of Section 9, with different interest rates for borrowing R, and lending r, $R(\cdot) \geq r(\cdot)$. The methodology of the previous section can still be used in the context of the model in Section 7, of which the different interest rates case is just one example. See Cvitanić & Karatzas (1992) for details. We are looking for an optimal process $\lambda_y \in \mathcal{H}$ for the corresponding dual problem, and, for any given $x \in (0, \infty)$, for an optimal portfolio $\hat{\pi}$ for the original primal control problem. In the case of logaritmic utility $U(x) = \log x$, we see as in (12.4) that $\lambda(t) = \lambda_1(t)\mathbf{1}$, where

$$\lambda_1(t) = arg \min_{r(t) - R(t) \le x \le 0} (-2x + \|\theta(t) + \sigma^{-1}(t)\mathbf{1}x\|^2) .$$

With $A(t) := tr[(\sigma^{-1}(t))^*(\sigma^{-1}(t))]$, $B(t) := \theta^*(t)\sigma^{-1}(t)\mathbf{1}$, this minimization is achieved as follows:

$$\lambda_1(t) = \left\{ \begin{array}{lll} \frac{1 - B(t)}{A(t)} & ; \quad if & 0 < B(t) - 1 < A(t)(R(t) - r(t)) \\ 0 & ; \quad if & B(t) \le 1 \\ r(t) - R(t) & ; \quad if & B(t) - 1 \ge A(t)(R(t) - r(t)) \end{array} \right\} .$$

¿From (12.5), the optimal portfolio is then computed as

$$\hat{\pi}_t = \left\{ \begin{array}{lll} (\sigma_t \sigma_t^*)^{-1}[b_t - (r_t + \frac{B_t - 1}{A_t})\mathbf{1}] & ; & 0 < B_t - 1 \le A_t(R_t - r_t) \\ (\sigma_t \sigma_t^*)^{-1}[b_t - r_t\mathbf{1}] & ; & B_t \le 1 \\ (\sigma_t \sigma_t^*)^{-1}[b_t - R_t\mathbf{1}] & ; & B_t - 1 \ge A_t(R_t - r_t) \end{array} \right.$$

In the case $U(x) = \frac{x^\alpha}{\alpha}$, for some $\alpha \in (0, 1)$, we get $\lambda(t) = \lambda_1(t)\mathbf{1}$ with

$$\lambda_1(t) = arg \min_{r(t) - R(t) \le x \le 0} \left[-2(1 - \alpha)x + \|\theta(t) + \sigma^{-1}(t)\mathbf{1}x\|^2 \right]$$

$$= \left\{ \begin{array}{lll} \frac{1 - \alpha - B(t)}{A(t)} & ; \quad if & 0 < B(t) - 1 + \alpha < A(t)(R(t) - r(t)) \\ 0 & ; \quad if & B(t) \le 1 - \alpha \\ r(t) - R(t) & ; \quad if & B(t) - 1 + \alpha \ge A(t)(R(t) - r(t)). \end{array} \right\}$$

The optimal portfolio is given as

$$\hat{\pi}_t = \left\{ \begin{array}{lll} \frac{(\sigma_t \sigma_t^*)^{-1}}{A_t}[b_t - (r_t + \frac{B_t - 1 + \alpha}{A_t})\mathbf{1}] & ; & 0 < B_t - 1 + \alpha < A_t(R_t - r_t) \\ \frac{(\sigma_t \sigma_t^*)^{-1}}{1 - \alpha}[b_t - r_t\mathbf{1}] & ; & B_t \le 1 - \alpha \\ \frac{(\sigma_t \sigma_t^*)^{-1}}{1 - \alpha}[b_t - R_t\mathbf{1}] & ; & B_t - 1 + \alpha \ge A_t(R_t - r_t) \end{array} \right.$$

13. UTILITY BASED PRICING

How to choose a price of a contingent claim B in the no-arbitrage pricing interval $[\tilde{h}(0), h(0)]$ of Theorem 6.15, in the case of incomplete markets, i.e., when the interval is non-degenerate (consists of more than just the Black-Scholes price)? There have been many attempts to provide a satisfactory answer to this question. We describe one suggested by Davis (1994), as presented in Karatzas & Kou (1996). It is based on the following "zero marginal rate of substitution" principle: Given the agent's utility function U and initial wealth x, the price $\hat{p}$ is the one that makes the agent neutral with respect to diversion of a small amount of funds into the contingent claim at time zero, while maximizing the utility from total wealth at the exercise time T. We will show that

$$\hat{p} = E[H_{\lambda_x}(T)B], \tag{13.1}$$

where λ_x is the associated optimal dual process. In particular, this price can be calculated in the context of examples of Section 12, and *does not depend on U and x*, in the case of cone constraints ($\delta \equiv 0$) and constant coefficients (Example 12.3). It can also be shown that, in this case, it gives the probability measure $\mathbf{P}^{\lambda_x}$ which minimizes the relative entropy with respect to the original measure $\mathbf{P}$.

In order to show (13.1), for a given $-x < \delta < x$ and price p of the claim, we introduce the value function

$$Q(\delta, p, x) := \sup_{\pi \in \mathcal{A}'(x - \delta)} EU(X^{x-\delta}(T) + \frac{\delta}{p}B). \tag{13.2}$$

In other words, the agent acquires δ/p units of the claim B at price p at time zero, and maximizes his/her terminal wealth at time T. Davis (1994) suggests to use price $\hat{p}$ for which

$$\frac{\partial Q}{\partial \delta}(\delta, \hat{p}, x)\bigg|_{\delta=0} = 0, \tag{13.3}$$

so that this diversion of funds has a neutral effect on the expected utility. Since the derivative in (13.3) need not exist, we have the following

13.1 Definition: For a given $x > 0$, we call $\hat{p}$ a *weak solution* of (13.3) if, for every function $\varphi : (-x, x) \mapsto \mathbb{R}$ of class C^1 which satisfies

$$\varphi(\delta) \geq Q(\delta, \hat{p}, x), \forall \delta \in (-x, x), \quad \varphi(0) = Q(0, p, x) = V(x), \tag{13.4}$$

we have $\varphi'(0) = 0$. If it is unique, then we call it the *utility based price* of B.

We have

13.2 Theorem: *Under the conditions of Theorem 11.14, the utility based price of B is given as in (13.1).*

Proof: Denote by $\hat{X}^x(T)$ the optimal terminal wealth for the utility maximization problem of the agent, starting with $\hat{X}(0) = x$. It is not difficult to see that $\hat{X}^x(T)$ is a.s. increasing in x. Also, for the concave function U, one has

$$U(z) + (y - z)U'(y) \leq U(y) \leq U(z) + (y - z)U'(z), \quad 0 < z < y. \tag{13.5}$$

We get, for $0 < \delta < x$, and a given function φ as in the theorem,

$$\varphi(\delta) \geq Q(\delta, p, x) \geq EU(\hat{X}^{x-\delta}(T) + \frac{\delta}{p}B)$$

$$\geq EU(\hat{X}^{x-\delta}(T)) + \frac{\delta}{p}E[U'(\hat{X}^{x-\delta}(T) + \frac{\delta}{p}B)B].$$

Since $\varphi(0) = EU(\hat{X}^0(T))$, we obtain, by monotone convergence,

$$\varphi'(0) \geq \frac{1}{p}E[U'(\hat{X}^x(T))B] - V'(x). \tag{13.6}$$

Similarly, with $-x < \delta < 0$, setting formally $U(z) = U'(z) = \infty$ for $z < 0$, we get

$$\varphi'(0) \le \frac{1}{p} E[U'(\hat{X}^x(T))B] - V'(x). \tag{13.7}$$

Now, (13.6), (13.7) imply that $\varphi'(0) = 0$ iff

$$p = \hat{p}(x) := \frac{E[U'(\hat{X}^x(T))B]}{V'(x)}.$$

But from previous sections we know that $U'(\hat{X}^x(T)) = y_x H_{\lambda_x}(T)$, for a suitable $y_x > 0$. One can also easily show (see Cvitanić & Karatzas (1992)) that $V'(x) = y_x$. This completes the proof.

14. HEDGING AND PORTFOLIO OPTIMIZATION IN THE PRESENCE OF TRANSACTION COSTS: THE MODEL

We consider a financial market consisting of one riskless asset, *bank account* with price $B(\cdot)$ given by

$$dB(t) = B(t)r(t)dt, \quad B(0) = 1; \tag{14.1}$$

and of one risky asset, *stock*, with price-per-share $S(\cdot)$ governed by the stochastic equation

$$dS(t) = S(t)[b(t)dt + \sigma(t)dW(t)], \quad S(0) = p \in (0, \infty), \tag{14.2}$$

for $t \in [0, T]$. Here $W = \{W(t), 0 \le t \le T\}$ a standard, one-dimensional Brownian motion. The coefficients of the model $r(\cdot)$, $b(\cdot)$ and $\sigma(\cdot) > 0$ are assumed to be bounded and $\mathbf{F}$−progressively measurable processes; furthermore, $\sigma(\cdot)$ is also assumed to be bounded away from zero (uniformly in (t, ω)).

Now, a *trading strategy* is a pair (L, M) of $\mathbf{F}$−adapted processes on $[0, T]$, with left-continuous, nondecreasing paths and $L(0) = M(0) = 0$; $L(t)$ (respectively, $M(t)$) represents the total amount of funds transferred from bank-account to stock (respectively, from stock to bank-account) by time t. Given *proportional transaction costs* $0 < \lambda, \mu < 1$ for such transfers, and initial holdings x, y in bank and stock, respectively, the portfolio holdings $X(\cdot) = X^{x,L,M}(\cdot), Y(\cdot) = Y^{y,L,M}(\cdot)$ corresponding to a given trading strategy (L, M), evolve according to the equations:

$$X(t) = x - (1 + \lambda)L(t) + (1 - \mu)M(t) + \int_0^t X(u)r(u)du, \quad 0 \le t \le T \tag{14.3}$$

$$Y(t) = y + L(t) - M(t) + \int_0^t Y(u)[b(u)du + \sigma(u)dW(u)], \quad 0 \le t \le T. \tag{14.4}$$

14.1 Definition: A *contingent claim* is a pair (C_0, C_1) of $\mathcal{F}(T)$−measurable random variables. We say that a trading strategy (L, M) *hedges the claim* (C_0, C_1)

starting with (x, y) as initial holdings, if $X(\cdot), Y(\cdot)$ of (14.3), (14.4) satisfy

$$X(T) + (1 - \mu)Y(T) \geq C_0 + (1 - \mu)C_1 \qquad (14.5)$$

$$X(T) + (1 + \lambda)Y(T) \geq C_0 + (1 + \lambda)C_1. \qquad (14.6)$$

Interpretation: Here C_0 (respectively, C_1) is understood as a target-position in the bank-account (resp., the stock) at the terminal time $t = T$: for example

$$C_0 = -q\mathbf{1}_{\{S(T)>q\}}, \ C_1 = S(T)\mathbf{1}_{\{S(T)>q\}} \qquad (14.7)$$

in the case of a European call-option; and

$$C_0 = q\mathbf{1}_{\{S(T)<q\}}, \ C_1 = -S(T)\mathbf{1}_{\{S(T)<q\}} \qquad (14.8)$$

for a European put-option (both with exercise price $q \geq 0$).

"Hedging", in the sense of (14.5) and (14.6), simply means that "one is able to cover these positions at $t = T$". Indeed, assume that we have both $Y(T) \geq C_1$ and (14.5), in the form

$$X(T) + (1 - \mu)[Y(T) - C_1] \geq C_0 \ ; \qquad (14.5)'$$

then (14.6) holds too, and (14.5)' shows that we can cover the position in the bank-account as well, by transferring the amount $Y(T) - C_1 \geq 0$ to it. Similarly for the case $Y(T) < C_1$.

14.3 Remark: The equations (14.3), (14.4) can be written in the equivalent form

$$d\left(\frac{X(t)}{B(t)}\right) = \left(\frac{1}{B(t)}\right)[(1 - \mu)dM(t) - (1 + \lambda)dL(t)], \quad X(0) = x \qquad (14.9)$$

$$d\left(\frac{Y(t)}{S(t)}\right) = \left(\frac{1}{S(t)}\right)[dL(t) - dM(t)], \quad Y(0) = y \qquad (14.10)$$

in terms of "number-of-shares" (rather than amounts) held.

15. AUXILIARY MARTINGALES.

Consider the class $\mathcal{D}$ of pairs of strictly positive **F**-martingales $(Z_0(\cdot), Z_1(\cdot))$ with

$$Z_0(0) = 1 \ , \quad z := Z_1(0) \in [p(1 - \mu), p(1 + \lambda)] \qquad (15.1)$$

and

$$1 - \mu \leq R(t) := \frac{Z_1(t)}{Z_0(t)P(t)} \leq 1 + \lambda, \ \forall \ 0 \leq t \leq T, \qquad (15.2)$$

where

$$P(t) := \frac{S(t)}{B(t)} = p + \int_0^t P(u)[(b(u) - r(u))du + \sigma(u)dW(u)] \ , \ \ 0 \leq t \leq T \quad (15.3)$$

is the *discounted stock price*.

The martingales $Z_0(\cdot), Z_1(\cdot)$ are the feasible *state-price densities* for holdings in bank and stock, respectively, in this market with transaction costs; as such, they reflect the "constraints" or "frictions" inherent in this market, in the form of condition (15.2). ¿From the martingale representation theorem there exist **F**−progressively measurable processes $\theta_0(\cdot), \theta_1(\cdot)$ with $\int_0^T (\theta_0^2(t) + \theta_1^2(t))dt < \infty$ a.s. and

$$Z_i(t) = Z_i(0) \exp \left\{ \int_0^t \theta_i(s)dW(s) - \frac{1}{2} \int_0^t \theta_i^2(s)ds \right\} , \quad i = 0,1; \qquad (15.4)$$

thus, the process $R(\cdot)$ of (15.2) has the dynamics

$$\begin{aligned} dR(t) =& R(t)[\sigma^2(t) + r(t) - b(t) - (\theta_1(t) - \theta_0(t))(\sigma(t) + \theta_0(t))]dt \\ &+ R(t)(\theta_1(t) - \sigma(t) - \theta_0(t))dW(t), \quad R(0) = z/p. \end{aligned} \qquad (15.5)$$

15.1 Remark: A rather "special" pair $(Z_0^*(\cdot), Z_1^*(\cdot)) \in \mathcal{D}$ is obtained, if we take in (15.4) the processes $(\theta_0(\cdot), \theta_1(\cdot))$ to be given as

$$\theta_0^*(t) := \frac{r(t) - b(t)}{\sigma(t)} , \quad \theta_1^*(t) := \sigma(t) + \theta_0^*(t) , \quad 0 \le t \le T, \qquad (15.6)$$

and let $Z_0^*(0) = 1$, $p(1 - \mu) \le Z_1^*(0) = z \le p(1 + \lambda)$. Because then, from (15.5), $R^*(\cdot) := \frac{Z_1^*(\cdot)}{Z_0^*(\cdot)P(\cdot)} \equiv \frac{z}{p}$; in fact, the pair of (15.6) and $z = p$ provide the *only* member $(Z_0^*(\cdot), Z_1^*(\cdot))$ of $\mathcal{D}$, if $\lambda = \mu = 0$. Notice that the processes $\theta_0^*(\cdot)$, $\theta_1^*(\cdot)$ of (15.6) are *bounded*.

15.2 Remark: Let us observe also that the martingales $Z_0(\cdot), Z_1(\cdot)$ play the role of *adjoint processes* to the "number-of-share holdings" processes $X(\cdot)/B(\cdot)$, $Y(\cdot)/S(\cdot)$, respectively, in the sense that

$$\begin{aligned} Z_0(t)\frac{X(t)}{B(t)} &+ Z_1(t)\frac{Y(t)}{S(t)} + \int_0^t \frac{Z_0(s)}{B(s)}[(1 + \lambda) - R(s)]dL(s) \\ &+ \int_0^t \frac{Z_0(s)}{B(s)}[R(s) - (1 - \mu)]dM(s) \\ &= x + \frac{yz}{p} + \int_0^t \frac{Z_0(s)}{B(s)}[X(s)\theta_0(s) + R(s)Y(s)\theta_1(s)]dW(s), \quad t \in [0,T] \end{aligned} \qquad (15.7)$$

is a **P**−local martingale, for any $(Z_0(\cdot), Z_1(\cdot)) \in \mathcal{D}$ and any trading strategy (L, M); this follows directly from (14.9), (14.10), (15.4) and the product rule. Equivalently, (15.7) can be re-written as

$$\begin{aligned} \frac{X(t) + R(t)Y(t)}{B(t)} &+ \int_0^t \frac{(1 + \lambda) - R(s)}{B(s)}dL(s) + \int_0^t \frac{R(s) - (1 - \mu)}{B(s)}dM(s) \\ &= x + \frac{yz}{p} + \int_0^t \frac{R(s)Y(s)}{B(s)}(\theta_1(s) - \theta_0(s))dW_0(s) = \mathbf{P_0} - \text{local martingale}, \end{aligned} \qquad (15.8)$$

where

$$W_0(t) := W(t) - \int_0^t \theta_0(s)ds, \quad 0 \le t \le T \tag{15.9}$$

is (by Girsanov's theorem), a Brownian motion under the equivalent probability measure

$$\mathbf{P}_0(A) := E[Z_0(T)\mathbf{1}_A], \quad A \in \mathcal{F}(T). \tag{15.10}$$

15.3 Remark: We shall denote by $Z_0^*(\cdot), W_0^*(\cdot)$ and $\mathbf{P}_0^*$ the processes and probability measure, respectively, corresponding to the process $\theta_0^*(\cdot)$ of (15.6), via the equations (15.4) (with $Z_0^*(0) = 1$), (15.9) and (15.10). With this notation, (15.3) becomes $dP(t) = P(t)\sigma(t)dW_0^*(t)$, $P(0) = p$.

15.4 Definition: Let $\mathcal{D}_\infty$ be the class of positive martingales $(Z_0(\cdot), Z_1(\cdot)) \in \mathcal{D}$, for which the random variable

$$\frac{Z_0(T)}{Z_0^*(T)}, \quad \text{and thus also} \quad \frac{Z_1(T)}{Z_0^*(T)P(T)}, \tag{15.11}$$

is essentially bounded.

15.5 Definition: We shall say that a given trading strategy (L, M) is *admissible* for (x, y), and write $(L, M) \in \mathcal{A}(x, y)$, if

$$\frac{X(\cdot) + R(\cdot)Y(\cdot)}{B(\cdot)} \text{ is a } \mathbf{P}_0 - \text{supermartingale}, \quad \forall\ (Z_0(\cdot), Z_1(\cdot)) \in \mathcal{D}_\infty. \tag{15.12}$$

Consider, for example, a trading strategy (L, M) that satisfies the no-bankruptcy conditions

$$X(t) + (1 + \lambda)Y(t) \ge 0 \text{ and } X(t) + (1 - \mu)Y(t) \ge 0, \quad \forall\ 0 \le t \le T.$$

Then $X(\cdot) + R(\cdot)Y(\cdot) \ge 0$ for every $(Z_0(\cdot), Z_1(\cdot)) \in \mathcal{D}$ (recall (15.2), and note Remark 15.6 below); this means that the $\mathbf{P}_0$−local martingale of (15.8) is nonnegative, hence a $\mathbf{P}_0$−supermartingale. But the second and the third terms

$$\int_0^\cdot \frac{1 + \lambda - R(s)}{B(s)}dL(s), \quad \int_0^\cdot \frac{R(s) - (1 - \mu)}{B(s)}dM(s)$$

in (15.8) are increasing processes, thus the first term $\frac{X(\cdot)+R(\cdot)Y(\cdot)}{B(\cdot)}$ is also a $\mathbf{P}_0$−supermartingale, for every pair $(Z_0(\cdot), Z_1(\cdot))$ in $\mathcal{D}$. The condition (15.12) is actually weaker, in that it requires this property only for pairs in $\mathcal{D}_\infty$. This provides a motivation for Definition 15.4, namely, *to allow for as wide a class of trading strategies as possible, and still exclude arbitrage·opportunities*. This is usually done by imposing a lower bound on the wealth process; however, that excludes simple strategies of the form "trade only once, by buying a fixed number of shares of the stock at a specified time t", which may require (unbounded) borrowing. We shall have occasion, to use such strategies in the sequel; see, for example, (16.20).

15.6 Remark: Here is a trivial (but useful) observation: if $x + (1 - \mu)y \ge a + (1 - \mu)b$ and $x + (1 + \lambda)y \ge a + (1 + \lambda)b$, then $x + ry \ge a + rb, \forall\ 1 - \mu \le r \le 1 + \lambda$.

16. HEDGING PRICE.

Suppose that we are given an initial holding $y \in \mathbb{R}$ in the stock, and want to hedge a given contingent claim (C_0, C_1) with strategies which are admissible (in the sense of Definitions 14.1, 15.4). What is the smallest amount of holdings in the bank

$$h(C_0, C_1; y) := \inf\{x \in \mathbb{R} / \exists (L, M) \in \mathcal{A}(x, y) \text{ and } (L, M) \text{ hedges } (C_0, C_1)\} \tag{16.1}$$

that allows to do this? We call $h(C_0, C_1; y)$ the *hedging price* of the contingent claim (C_0, C_1) for initial holding y in the stock, and with the convention that $h(C_0, C_1; y) = \infty$ if the set in (16.1) is empty.

Suppose this is not the case, and let $x \in \mathbb{R}$ belong to the set of (16.1); then for any $(Z_0(\cdot), Z_1(\cdot)) \in \mathcal{D}_\infty$ we have from (15.12), the Definition 14.1 of hedging, and Remark 15.6:

$$x + \frac{y}{p} E Z_1(T) = x + \frac{y}{p} z \geq E_0 \left[\frac{X(T) + R(T)Y(T)}{B(T)} \right]$$

$$\geq E_0 \left[\frac{C_0 + R(T)C_1}{B(T)} \right] = E \left[\frac{Z_0(T)}{B(T)} (C_0 + R(T)C_1) \right],$$

so that $x \geq E\left[\frac{Z_0(T)}{B(T)} (C_0 + R(T)C_1) - \frac{y}{p} Z_1(T) \right]$. Therefore

$$h(C_0, C_1; y) \geq \sup_{\mathcal{D}_\infty} E \left[\frac{Z_0(T)}{B(T)} (C_0 + R(T)C_1) - \frac{y}{p} Z_1(T) \right], \tag{16.2}$$

and this inequality is clearly also valid if $h(C_0, C_1; y) = \infty$.

16.1 Lemma: *If the contingent claim (C_0, C_1) is bounded from below, in the sense*

$$C_0 + (1 + \lambda)C_1 \geq -K \text{ and } C_0 + (1 - \mu)C_1 \geq -K, \text{ for some } 0 \leq K < \infty \tag{16.3}$$

then

$$\sup_{\mathcal{D}_\infty} E\left[\frac{Z_0(T)}{B(T)} (C_0 + R(T)C_1) - \frac{y}{p} Z_1(T) \right]$$

$$= \sup_{\mathcal{D}} E\left[\frac{Z_0(T)}{B(T)} (C_0 + R(T)C_1) - \frac{y}{p} Z_1(T) \right]. \tag{16.4}$$

Proof: Start with arbitrary $(Z_0(\cdot), Z_1(\cdot)) \in \mathcal{D}$ and define the sequence of stopping times $\{\tau_n\} \uparrow T$ by

$$\tau_n := \inf\{t \in [0, T] / \frac{Z_0(t)}{Z_0^*(t)} \geq n\} \wedge T, \ n \in \mathbb{N}.$$

Consider also, for $i = 0, 1$ and in the notation of (15.6):

$$\theta_i^{(n)}(t) := \left\{ \begin{array}{ll} \theta_i(t), & 0 \leq t < \tau_n \\ \theta_i^*(t), & \tau_n \leq t \leq T \end{array} \right\}$$

and

$$Z_i^{(n)}(t) = z_i \exp\{ \int_0^t \theta_i^{(n)}(s)dW(s) - \frac{1}{2} \int_0^t (\theta_i^{(n)}(s))^2 ds\}$$

with $z_0 = 1$, $z_1 = Z_1(0) = EZ_1(T)$. Then, for every $n \in \mathbb{N}$, both $Z_0^{(n)}(\cdot)$ and $Z_1^{(n)}(\cdot)$ are positive martingales, $R^{(n)}(\cdot) = \frac{Z_1^{(n)}(\cdot)}{Z_0^{(n)}(\cdot)P(\cdot)} = R(\cdot \wedge \tau_n)$ takes values in $[1 - \mu, 1 + \lambda]$ (by (15.2) and Remark 15.1), and $Z_0^{(n)}(\cdot)/Z_0^*(\cdot)$ is bounded by n (in fact, constant on $[\tau_n, T]$). Therefore, $(Z_0^{(n)}(\cdot), Z_1^{(n)}(\cdot)) \in \mathcal{D}_\infty$. Now let κ denote an upper bound on $K/B(T)$, and observe, from Remark 15.6, (16.3) and Fatou's lemma:

$$E\left[\frac{Z_0(T)}{B(T)}(C_0 + R(T)C_1) - \frac{y}{p}Z_1(T) \right] + \frac{y}{p}Z_1(0) + \kappa$$

$$= E\left[Z_0(T)\left\{ \frac{C_0 + R(T)C_1}{B(T)} + \kappa \right\} \right]$$

$$= E\left[\lim_n Z_0^{(n)}(T)\left\{ \frac{C_0 + R^{(n)}(T)C_1}{B(T)} + \kappa \right\} \right] \qquad (16.5)$$

$$\leq \lim_n E\left[Z_0^{(n)}(T)\left\{ \frac{C_0 + R^{(n)}(T)C_1}{B(T)} + \kappa \right\} \right]$$

$$= \lim_n E\left[\frac{Z_0^{(n)}(T)}{B(T)}(C_0 + R^{(n)}(T)C_1) - \frac{y}{p}Z_1^{(n)}(T) \right] + \frac{y}{p}Z_1(0) + \kappa.$$

This shows that the left-hand-side dominates the right-hand-side in (16.4); the reverse inequality is obvious. ⋄

Remark: Formally taking $y = 0$ in (16.5), we deduce

$$E_0\left(\frac{C_0 + R(T)C_1}{B(T)} \right) \leq \lim_{n \to \infty} E_0^{(n)}\left(\frac{C_0 + R^{(n)}(T)C_1}{B(T)} \right), \qquad (16.6)$$

where $E_0, E_0^{(n)}$ denote expectations with respect to the probability measures $\mathbf{P_0}$ of (15.10) and $\mathbf{P}_0^{(n)}(\cdot) = E[Z_0^{(n)}(T)\mathbf{1}.]$, respectively.

Here is the main result of this section.

16.2 Theorem: (Cvitanić & Karatzas (1996)) *Under the conditions (16.3) and*

$$E_0^*(C_0^2 + C_1^2) < \infty , \qquad (16.7)$$

we have

$$h(C_0, C_1; y) = \sup_{\mathcal{D}} E\left[\frac{Z_0(T)}{B(T)}(C_0 + R(T)C_1) - \frac{y}{p}Z_1(T) \right]. \qquad (16.8)$$

In (16.7), E_0^* denotes expectation with respect to the probability measure $\mathbf{P}_0^*$. The conditions (16.3), (16.7) are both easily verified for a European call or put. In fact, one can show that if a pair of admissible terminal holdings $(X(T), Y(T))$ hedges a pair $(\tilde{C}_0, \tilde{C}_1)$ satisfying (16.7) (for example, $(\tilde{C}_0, \tilde{C}_1) \equiv$

$(0,0))$, then necessarily the pair $(X(T), Y(T))$ also satisfies (16.7) – and so does any other pair of random variables (C_0, C_1) which are bounded from below and are hedged by $(X(T), Y(T))$. In other words, any strategy which satisfies the "no-bankruptcy" condition of hedging $(0,0)$, necessarily results in a square-integrable final wealth. In this sense, the condition (16.7) is consistent with the standard "no-bankruptcy" condition, hence not very restrictive.

It would be of significant interest to be able to describe the least expensive hedging strategy associated with a general hedgeable contingent claim; Our *functional-analytic proof*, which takes up the remainder of this section and was inspired by similar arguments in Kusuoka (1995), does not provide the construction of such a strategy.

Proof: In view of Lemma 16.1 and the inequality (16.2), it suffices to show

$$h(C_0, C_1; y) \leq \sup_{\mathcal{D}} E\left[Z_0(T)\frac{C_0}{B(T)} + Z_1(T)\left(\frac{C_1}{S(T)} - \frac{y}{p}\right)\right] =: R. \qquad (16.9)$$

And in order to alleviate somewhat the (already rather heavy) notation, we shall take $p = 1$, $r(\cdot) \equiv 0$, thus $B(\cdot) \equiv 1$, for the remainder of the section; the reader will verify easily that this entails no loss of generality.

We start by taking an arbitrary $b < h(C_0, C_1; y)$ and considering the sets

$$A_0 := \{(U, V) \in (\mathbf{L}_2^*)^2 : \exists (L, M) \in \mathcal{A}(0,0) \text{ that hedges } (U, V) \text{ starting with}$$
$$x = 0, y = 0\}$$
$$\qquad (16.10)$$
$$A_1 := \{(C_0 - b, C_1 - yS(T))\}, \qquad (16.11)$$

where $\mathbf{L}_2^* = \mathbf{L}_2(\Omega, \mathcal{F}(T), \mathbf{P}_0^*)$. It is not hard to prove (see below) that

$$A_0 \text{ is a convex cone, and contains the origin } (0,0), \text{ in } (\mathbf{L}_2^*)^2, \qquad (16.12)$$
$$A_0 \cap A_1 = \emptyset. \qquad (16.13)$$

It is, however, considerably harder to establish that

$$A_0 \text{ is closed in } (\mathbf{L}_2^*)^2. \qquad (16.14)$$

The proof can be found in the appendix of Cvitanić & Karatzas(1996). From (16.12)-(16.14) and the Hahn-Banach theorem there exists a pair of random variables $(\rho_0^*, \rho_1^*) \in (\mathbf{L}_2^*)^2$, not equal to $(0,0)$, such that

$$E_0^*[\rho_0^* V_0 + \rho_1^* V_1] = E[\rho_0 V_0 + \rho_1 V_1] \leq 0, \quad \forall (V_0, V_1) \in A_0 \qquad (16.15)$$
$$E_0^*[\rho_0^*(C_0 - b) + \rho_1^*(C_1 - yS(T))] = E[\rho_0(C_0 - b) + \rho_1(C_1 - yS(T))] \geq 0, \qquad (16.16)$$

where $\rho_i := \rho_i^* Z_0^*(T)$, $i = 0, 1$. It is also not hard to check (see below) that

$$(1 - \mu)E[\rho_0|\mathcal{F}(t)] \leq \frac{E[\rho_1 S(T)|\mathcal{F}(t)]}{S(t)} \leq (1 + \lambda)E[\rho_0|\mathcal{F}(t)], \forall\, 0 \leq t \leq T \qquad (16.17)$$

$$\rho_1 \geq 0, \ \rho_0 \geq 0 \text{ and } E\rho_0 > 0, \ E(\rho_1 S(T)) > 0. \qquad (16.18)$$

In view of (16.18), we may take $E\rho_0 = 1$, and then (16.16) gives

$$b \leq E[\rho_0 C_0 + \rho_1(C_1 - yS(T))]. \qquad (16.19)$$

Consider now *arbitrary* $0 < \varepsilon < 1$, $(Z_0(\cdot), Z_1(\cdot)) \in \mathcal{D}$, and define

$$\tilde{Z}_0(t) := \varepsilon Z_0(t) + (1 - \varepsilon)E[\rho_0|\mathcal{F}(t)], \quad \tilde{Z}_1(t) := \varepsilon Z_1(t) + (1 - \varepsilon)E[\rho_1 S(T)|\mathcal{F}(t)],$$

for $0 \leq t \leq T$. Clearly these are positive martingales, and $\tilde{Z}_0(0) = 1$; on the other hand, multiplying in (16.17) by $1 - \varepsilon$, and in $(1 - \mu)Z_0(t) \leq Z_1(t)/S(t) \leq (1+\lambda)Z_0(t)$, $0 \leq t \leq T$ by ε, and adding up, we obtain $(\tilde{Z}_0(\cdot), \tilde{Z}_1(\cdot)) \in \mathcal{D}$. Thus, in the notation of (16.9),

$$R \geq E\left[\tilde{Z}_0(T)C_0 + \tilde{Z}_1(T)\left(\frac{C_1}{S(T)} - y\right)\right]$$

$$= (1 - \varepsilon)E[\rho_0 C_0 + \rho_1(C_1 - yS(T))] + \varepsilon E\left[Z_0(T)C_0 + Z_1(T)\left(\frac{C_1}{S(T)} - y\right)\right]$$

$$\geq b(1 - \varepsilon) + \varepsilon E\left[Z_0(T)C_0 + Z_1(T)\left(\frac{C_1}{S(T)} - y\right)\right]$$

from (16.19); letting $\varepsilon \downarrow 0$ and then $b \uparrow h(C_0, C_1; y)$, we obtain (16.9), as required to complete the proof of Theorem 16.2.

Proof of (16.13): Suppose that $A_0 \cap A_1$ is not empty, i.e., that there exists $(L, M) \in \mathcal{A}(0, 0)$ such that, with $X(\cdot) = X^{0,L,M}(\cdot)$ and $Y(\cdot) = Y^{0,L,M}(\cdot)$, the process $X(\cdot) + R(\cdot)Y(\cdot)$ is a $\mathbf{P}_0$−supermartingale for every $(Z_0(\cdot), Z_1(\cdot)) \in \mathcal{D}_\infty$, and we have:

$$X(T) + (1 - \mu)Y(T) \geq (C_0 - b) + (1 - \mu)(C_1 - yS(T)),$$

$$X(T) + (1 + \lambda)Y(T) \geq (C_0 - b) + (1 + \lambda)(C_1 - yS(T)).$$

But then, with

$$\tilde{X}(\cdot) := X^{b,L,M}(\cdot) = b + X(\cdot), \quad \tilde{Y}(\cdot) := Y^{y,L,M}(\cdot) = Y(\cdot) + yS(\cdot)$$

we have, from above, that $\tilde{X}(\cdot) + R(\cdot)\tilde{Y}(\cdot) = X(\cdot) + R(\cdot)Y(\cdot) + b + yZ_1(\cdot)/Z_0(\cdot)$ is a $\mathbf{P}_0$−supermartingale for every $(Z_0(\cdot), Z_1(\cdot)) \in \mathcal{D}_\infty$, and that

$$\tilde{X}(T) + (1 - \mu)\tilde{Y}(T) \geq C_0 + (1 - \mu)C_1,$$

$$\tilde{X}(T) + (1 + \lambda)\tilde{Y}(T) \geq C_0 + (1 + \lambda)C_1.$$

In other words, (L, M) belongs to $\mathcal{A}(b, y)$ and hedges (C_0, C_1) starting with (b, y) – a contradiction to the definition (16.1), and to the fact $h(C_0, C_1; y) > b$.

Proof of (16.17), (16.18): Fix $t \in [0, T]$ and let ξ be an arbitrary bounded, nonnegative, $\mathcal{F}(t)$−measurable random variable. Consider the strategy of starting with $(x, y) = (0, 0)$ and buying ξ shares of stock at time $s = t$, otherwise doing nothing ("buy-and-hold strategy"); more explicitly, $M^\xi(\cdot) \equiv 0$,

$L^\xi(s) = \xi S(t)\mathbf{1}_{(t,T]}(s)$ and thus

$$
\begin{aligned}
X^\xi(s) &:= X^{0,L^\xi,M^\xi}(\cdot) = -\xi(1+\lambda)S(t)\mathbf{1}_{(t,T]}(s), \\
Y^\xi(s) &:= Y^{0,L^\xi,M^\xi}(s) = \xi S(s)\mathbf{1}_{(t,T]}(s),
\end{aligned}
\tag{16.20}
$$

for $0 \le s \le T$. Consequently, $Z_0(s)[X^\xi(s) + R(s)Y^\xi(s)] = \xi[Z_1(s) - (1 + \lambda)S(t)Z_0(s)]\mathbf{1}_{(t,T]}(s)$ is a $\mathbf{P}$–supermartingale for every $(Z_0(\cdot), Z_1(\cdot)) \in \mathcal{D}$, since, for instance with $t < s \le T$:

$$
\begin{aligned}
E[Z_0(s)(X_s^\xi + R_s Y_s^\xi)|\mathcal{F}_t] &= \xi\,(E[Z_1(s)|\mathcal{F}_t] - (1+\lambda)S_t E[Z_0(s)|\mathcal{F}_t]) \\
&= \xi[Z_1(t) - (1+\lambda)S(t)Z_0(t)] = \xi S(t)Z_0(t)[R(t) - (1+\lambda)] \\
&\le 0 = Z_0(t)[X^\xi(t) + R(t)Y^\xi(t)].
\end{aligned}
$$

Therefore, $(L^\xi, M^\xi) \in \mathcal{A}(0,0)$, thus $(X^\xi(T), Y^\xi(T))$ belongs to the set A_0 of (16.10), and, from (16.15):

$$
\begin{aligned}
0 &\ge E[\rho_0 X^\xi(T) + \rho_1 Y^\xi(T)] = E[\xi(\rho_1 S(T) - (1 + \lambda)\rho_0 S(t))] \\
&= E\big[\xi\,(E[\rho_1 S(T)|\mathcal{F}(t)] - (1 + \lambda)S(t)E[\rho_0|\mathcal{F}(t)])\big].
\end{aligned}
$$

¿From the arbitrariness of $\xi \ge 0$, we deduce the inequality of the right-hand side in (16.17), and a dual argument gives the inequality of the left-hand side, for given $t \in [0,T)$. Now all three processes in (16.17) have continuous paths; consequently, (16.17) is valid for *all* $t \in [0,T]$.

Next, we notice that (16.17) with $t = T$ implies $(1 - \mu)\rho_0 \le \rho_1 \le (1 + \lambda)\rho_0$, so that ρ_0, hence also ρ_1, is nonnegative. Similarly, (16.17) with $t = 0$ implies $(1 - \mu)E\rho_0 \le E[\rho_1 S(T)] \le (1 + \lambda)E\rho_0$, and therefore, since (ρ_0, ρ_1) is not equal to $(0,0)$, $E\rho_0 > 0$, hence also $E[\rho_1 S(T)] > 0$. This proves (16.18).   ◇

16.3 Example: Consider the *European call option* of (14.7). ¿From (16.8) with $y = 0$, we have

$$
h(C_0, C_1) \equiv h(C_0, C_1; 0) = \sup_{\mathcal{D}} E\left[Z_1(T)\mathbf{1}_{\{S(T)>q\}} - q\frac{Z_0(T)}{B(T)}\mathbf{1}_{\{S(T)>q\}} \right],
\tag{16.21}
$$

and therefore, $h(C_0, C_1) \le \sup_{\mathcal{D}} EZ_1(T) = \sup_{\mathcal{D}} Z_1(0) \le (1 + \lambda)p$. The number $p(1+\lambda)$ corresponds to the cost of the "buy-and-hold strategy", of acquiring one share of the stock at $t = 0$ (at a price $p(1 + \lambda)$, due to the transaction cost), and holding on to it until $t = T$. Davis & Clark (1993) conjectured that this hedging strategy is actually the cheapest:

$$
h(C_0, C_1) = (1 + \lambda)p.
\tag{16.22}
$$

The conjecture (16.22) was proved by Soner, Shreve & Cvitanić (1995), as well as by Leventhal & Skorohod (1995). It is an open question *to derive (16.22)* directly from the representation (16.21); in other words, to find a sequence $\{(Z_0^{(n)}(\cdot), Z_1^{(n)}(\cdot))\}_{n\in\mathbb{N}}$ with

$$\mathbf{P}_0^{(n)}[S(T) > q] \to 0, \quad E[Z_1^{(n)}(T)\mathbf{1}_{\{S(T)>q\}}] \to 1, \quad Z_1^{(n)}(0) \to 1 + \lambda,$$

as $n \uparrow \infty$.

17. MAXIMIZING EXPECTED UTILITY FROM TERMINAL WEALTH.

Consider now a small investor, who can make decisions in the context of the market model of (13.1), (13.2) as described in section 13, and who derives utility $U(X(T+))$ from his *terminal wealth*

$$X(T+) := X(T) + f(Y(T)), \quad \text{where} \quad f(u) := \left\{ \begin{array}{l} (1 + \lambda)u \; ; \; u \le 0 \\ (1 - \mu)u \; ; \; u > 0 \end{array} \right\}. \quad (17.1)$$

In other words, this agent liquidates at the end of the day his position in the stock, incurs the appropriate transaction cost, and collects all the money in the bank-account. For a given initial holding $y \ge 0$ in the stock, his optimization problem is to find an admissible pair $(\hat{L}, \hat{M}) \in \mathcal{A}^+(x, y)$ that *maximizes expected utility from terminal wealth*, i.e., attains the supremum

$$V(x;y) := \sup_{(L,M)\in\mathcal{A}^+(x,y)} EU\left(X^{x,L,M}(T) + f(Y^{y,L,M}(T))\right), \quad 0 < x < \infty, \quad (17.2)$$

where $\mathcal{A}^+(x, y)$ is the class of processes $(L, M) \in \mathcal{A}(x, y)$ for which $X^{x,L,M}(T) + f(Y^{y,L,M}(T)) \ge 0$. It can be shown, using standard convex/functional analysis arguments, that the supremum of (17.2) is attained, i.e., that there exists an optimal pair $(\hat{L}, \hat{M})$ for this problem, and that $V(x, y) < \infty$. Our purpose in this section is to describe the nature of this optimal pair, by using results of section 16 in the context of the *dual problem*

$$\tilde{V}(\zeta;y) := \inf_{(Z_0,Z_1)\in\mathcal{D}} E\left[\tilde{U}\left(\zeta\frac{Z_0(T)}{B(T)}\right) + \frac{y}{p}\zeta Z_1(T)\right], \quad 0 < \zeta < \infty, \quad (17.3)$$

under the following assumption.

17.1 Assumption: There exists a pair $(\hat{Z}_0(\cdot), \hat{Z}_1(\cdot)) \in \mathcal{D}$, that attains the infimum in (17.3), and does so for all $0 < \zeta < \infty$. Moreover, for all $0 < \zeta < \infty$, we have

$$\tilde{V}(\zeta;y) < \infty \quad \text{and} \quad E\left[\frac{\hat{Z}_0(T)}{B(T)}I\left(\frac{\zeta Z_0^*(T)}{B(T)}\right)\right] < \infty.$$

17.2 Remark: The assumption that the infimum of (17.3) is attained is a big one; we have not yet been able to obtain a general existence result to this effect, only very simple examples that can be solved explicitly. The assumption that the minimization in (17.3) can be carried out for all $0 < \zeta < \infty$ simultaneously, is made only for simplicity; it can be dispensed with using methods analogous to those in previous sections. Note, however, that this latter assumption is satisfied if $y = 0$ and either $U(x) = \log x$ or $U(x) = \frac{1}{\delta}x^\delta$ for $0 < \delta < 1$. It should also be mentioned that the optimal pair $(\hat{Z}_0(\cdot), \hat{Z}_1(\cdot))$ of Assumption 17.1 need not be unique; thus, in the remainder of this section, $(\hat{Z}_0(\cdot), \hat{Z}_1(\cdot))$ will denote *any* pair that attains the infimum in (17.3), as in Assumption 17.1.

For any such pair, we have then the following property, proved similarly as the corresponding result in the utility maximization under constraints.

17.3 Lemma: *Under the Assumption 17.1 and the condition*

$$xU'(x) \leq a + (1-b)U(x), \quad 0 < x < \infty,$$

for some $a \geq 0$, $0 < b \leq 1$, we have

$$E\left[\frac{Z_0(T)}{B(T)}I\left(\zeta\frac{\hat{Z}_0(T)}{B(T)}\right) - \frac{y}{p}Z_1(T)\right] \leq E\left[\frac{\hat{Z}_0(T)}{B(T)}I\left(\zeta\frac{\hat{Z}_0(T)}{B(T)}\right) - \frac{y}{p}\hat{Z}_1(T)\right] < \infty,$$

(17.4)

for all $0 < \zeta < \infty$, for every $(Z_0(\cdot), Z_1(\cdot))$ in $\mathcal{D}$.

Now, because the function $\zeta \mapsto E\left[\frac{\hat{Z}_0(T)}{B(T)}I(\zeta\frac{\hat{Z}_0(T)}{B(T)})\right] : (0,\infty) \to (0,\infty)$ is continuous and strictly decreasing, there exists a unique $\hat{\zeta} = \hat{\zeta}(x; y, U) \in (0,\infty)$ that satisfies

$$E\left[\frac{\hat{Z}_0(T)}{B(T)}I\left(\hat{\zeta}\frac{\hat{Z}_0(T)}{B(T)}\right)\right] = x + \frac{y}{p}E\hat{Z}_1(T).$$

(17.5)

And with

$$\hat{C}_0 := I\left(\hat{\zeta}\frac{\hat{Z}_0(T)}{B(T)}\right), \quad \hat{C}_1 := 0,$$

(17.6)

it follows from (17.4) that

$$\sup_{(Z_0, Z_1) \in \mathcal{D}} E\left[Z_0(T)\frac{\hat{C}_0}{B(T)} + Z_1(T)\left(\frac{\hat{C}_1}{S(T)} - \frac{y}{p}\right)\right]$$

$$= E\left[\hat{Z}_0(T)\frac{\hat{C}_0}{B(T)} + \hat{Z}_1(T)\left(\frac{\hat{C}_1}{S(T)} - \frac{y}{p}\right)\right] \qquad (17.7)$$

$$= x.$$

Consequently, if in addition we have $\hat{C}_0 \in \mathbf{L}_2^*$, then Theorem 16.2 gives $h(\hat{C}_0, \hat{C}_1; y) = x$. Now it can be shown (see Cvitanić & Karatzas (1996)) that *the infimum in (16.1) is actually attained*; in other words, there exists a pair $(\hat{L}, \hat{M}) \in \mathcal{A}(x, y)$ such that, with $\hat{X}(\cdot) \equiv X^{x, \hat{L}, \hat{M}}(\cdot)$, $\hat{Y}(\cdot) \equiv Y^{y, \hat{L}, \hat{M}}(\cdot)$, we have

$$\hat{X}(T) + (1-\mu)\hat{Y}(T) \geq \hat{C}_0, \quad \hat{X}(T) + (1+\lambda)\hat{Y}(T) \geq \hat{C}_0. \qquad (17.8)$$

17.4 Theorem: (Cvitanić & Karatzas (1996)) *Under assumptions of Lemma 17.3 and the condition*

$$E_0^*[\hat{C}_0^2] = E_0^*\left[I^2(\hat{\zeta}\hat{Z}_0(T)/B(T))\right] < \infty, \qquad (17.9)$$

the above pair $(\hat{L}, \hat{M}) \in \mathcal{A}(x, y)$ is optimal for the problem of (17.2), and satisfies

$$\hat{X}(T+) := \hat{X}(T) + f(\hat{Y}(T)) = I(\hat{\zeta}\hat{Z}_0(T)/B(T)) = \hat{C}_0 \qquad (17.10)$$

$$\hat{L}(\cdot) \text{ is flat off the set } \{0 \leq t \leq T / \hat{R}(t) = 1 + \lambda\} \qquad (17.11)$$

$$\hat{M}(\cdot) \text{ is flat off the set } \{0 \leq t \leq T / \hat{R}(t) = 1 - \mu\} \qquad (17.12)$$

$$\frac{\hat{X}(t) + \hat{R}(t)\hat{Y}(t)}{B(t)} = \hat{E}_0\left[\frac{I(\hat{\zeta}\hat{Z}_0(T)/B(T))}{B(T)}\bigg|\mathcal{F}(t)\right], \ 0 \leq t \leq T, \qquad (17.13)$$

where $\hat{R}(\cdot) := \frac{\hat{Z}_1(\cdot)}{\hat{Z}_0(\cdot)P(\cdot)}$. Furthermore, we have $\tilde{V}(\hat{\zeta}; y) = V(x; y) - x\hat{\zeta} < \infty$.

Proof: As we just argued, (17.9) and Theorem 16.2 imply the existence of a pair $(\hat{L}, \hat{M}) \in \mathcal{A}(x, y)$, so that (17.8) is satisfied; and from (17.8), we know that both

$$\hat{X}(T) + \hat{R}(T)\hat{Y}(T) \geq \hat{C}_0, \quad \hat{X}(T) + f(\hat{Y}(T)) \geq \hat{C}_0 \qquad (17.14)$$

hold. On the other hand, (15.12) implies that the process

$$\frac{\hat{X}(\cdot) + \hat{R}(\cdot)\hat{Y}(\cdot)}{B(\cdot)} \text{ is a } \hat{P}_0 - \text{supermartingale.} \qquad (17.15)$$

Therefore, from (17.5), (17.14) and (17.15) we have

$$x + \frac{y}{p}E\hat{Z}_1(T) = E\left[\frac{\hat{Z}_0(T)}{B(T)}I\left(\hat{\zeta}\frac{\hat{Z}_0(T)}{B(T)}\right)\right] = \hat{E}_0\left(\frac{\hat{C}_0}{B(T)}\right)$$

$$\leq \hat{E}_0\left(\frac{\hat{X}(T) + \hat{R}(T)\hat{Y}(T)}{B(T)}\right) \leq x + \frac{y}{p}E\hat{Z}_1(T), \qquad (17.16)$$

whence

$$\hat{X}(T) + \hat{R}(T)\hat{Y}(T) = \hat{C}_0. \qquad (17.17)$$

But now from (17.8), (17.14) we deduce $\hat{R}(T) = 1 - \mu$ on $\{\hat{Y}(T) > 0\}$, and $\hat{R}(T) = 1 + \lambda$ on $\{\hat{Y}(T) < 0\}$; thus

$$\hat{C}_0 = \hat{X}(T) + \hat{R}(T)\hat{Y}(T)$$
$$= \hat{X}(T) + \hat{Y}(T)[(1 + \lambda)1_{\{\hat{Y}(T) \leq 0\}} + (1 - \mu)1_{\{\hat{Y}(T) > 0\}}] = \hat{X}(T) + f(\hat{Y}(T)),$$

and (17.10) follows.

It develops from (17.15), (17.16) that the process $\frac{\hat{X}(\cdot) + \hat{R}(\cdot)\hat{Y}(\cdot)}{B(\cdot)}$ is a $\hat{P}_0$−super-martingale with constant expectation, thus a $\hat{P}_0$−martingale; from this and (17.17), we obtain (17.13), as well as the fact that this process is nonnegative, hence that the $\hat{P}_0$−local martingale

$$\frac{\hat{X}(t) + \hat{R}(t)\hat{Y}(t)}{B(t)} + \int_0^t \frac{1 + \lambda - \hat{R}(s)}{B(s)}d\hat{L}(s) + \int_0^t \frac{\hat{R}(s) - (1 - \mu)}{B(s)}d\hat{M}(s), \ 0 \leq t \leq T \qquad (17.18)$$

is also nonnegative. Hence, the process of (17.18) is a $\hat{\mathbf{P}}_0$−supermartingale with $\hat{\mathbf{P}}_0$−expectation at most $x + \frac{y}{p}E\hat{Z}_1(T)$ at $t = T$; but this is equal to the $\hat{\mathbf{P}}_0$−expectation of $\frac{\hat{X}(T)+\hat{R}(T)\hat{Y}(T)}{B(T)}$ by (17.16), whence the nonnegative terms

$$\int_0^T \frac{1+\lambda-\hat{R}(s)}{B(s)}d\hat{L}(s) \ , \quad \int_0^T \frac{\hat{R}(s)-(1-\mu)}{B(s)}d\hat{M}(s)$$

must have $\hat{\mathbf{P}}_0$−expectation equal to zero. The claims (17.11), (17.12) follow.

Now for the optimality of the pair $(\hat{L}, \hat{M})$: we have from (17.10) and (17.5)

$$EU(\hat{X}(T) + f(\hat{Y}(T)))) = EU(\hat{C}_0) = EU\left(I\left(\hat{\zeta}\frac{\hat{Z}_0(T)}{B(T)}\right)\right)$$

$$= E\tilde{U}\left(\hat{\zeta}\frac{\hat{Z}_0(T)}{B(T)}\right) + \hat{\zeta}E\left[\frac{\hat{Z}_0(T)}{B(T)}I\left(\hat{\zeta}\frac{\hat{Z}_0(T)}{B(T)}\right)\right] \qquad (17.19)$$

$$= E\tilde{U}\left(\hat{\zeta}\frac{\hat{Z}_0(T)}{B(T)}\right) + \hat{\zeta}x + \hat{\zeta}\frac{y}{p}E\hat{Z}_1(T) = \tilde{V}(\hat{\zeta};y) + x\hat{\zeta}.$$

Consider also the holdings processes $X(\cdot) \equiv X^{x,L,M}(\cdot)$, $Y(\cdot) \equiv Y^{y,L,M}(\cdot)$ corresponding to an *arbitrary* strategy $(L, M) \in \mathcal{A}(x, y)$. We have

$$U(X(T) + (1 - \mu)Y(T)) \le \tilde{U}\left(\hat{\zeta}\frac{\hat{Z}_0(T)}{B(T)}\right) + \hat{\zeta}\frac{\hat{Z}_0(T)}{B(T)}[X(T) + (1 - \mu)Y(T)]$$

$$U(X(T) + (1 + \lambda)Y(T)) \le \tilde{U}\left(\hat{\zeta}\frac{\hat{Z}_0(T)}{B(T)}\right) + \hat{\zeta}\frac{\hat{Z}_0(T)}{B(T)}[X(T) + (1 + \lambda)Y(T)]$$

and thus, in conjunction with Remark 15.6, (16.6) and (15.12),

$$EU(X(T)+f(Y(T))) \le E\tilde{U}\left(\hat{\zeta}\frac{\hat{Z}_0(T)}{B(T)}\right) + \hat{\zeta}\hat{E}_0\left(\frac{X(T) + \hat{R}(T)Y(T)}{B(T)}\right)$$

$$\le E\tilde{U}\left(\hat{\zeta}\frac{\hat{Z}_0(T)}{B(T)}\right) + \hat{\zeta}\lim_{n\to\infty}\hat{E}_0^{(n)}\left(\frac{X(T) + \hat{R}^{(n)}(T)Y(T)}{B(T)}\right)$$

$$\le E\tilde{U}\left(\hat{\zeta}\frac{\hat{Z}_0(T)}{B(T)}\right) + \hat{\zeta}(x + \frac{y}{p}E\hat{Z}_1(T))$$

$$= \tilde{V}(\hat{\zeta};y) + x\hat{\zeta}.$$

$$(17.20)$$

The optimality of $(\hat{L}, \hat{M}) \in \mathcal{A}(x, y)$ for the problem of (17.2), as well as the equality $V(x; y) = \tilde{V}(\hat{\zeta}; y) + x\hat{\zeta}$, follow now directly from (17.19) and (17.20).

◇

Notice that, if $r(\cdot)$ is deterministic, then Jensen's inequality gives

$$E\left[\tilde{U}\left(\varsigma\frac{Z_0(T)}{B(T)}\right) + \frac{y}{p}\varsigma Z_1(T)\right] \geq \tilde{U}\left(\frac{\varsigma}{B(T)}EZ_0(T)\right) + \frac{y}{p}\varsigma Z_1(0)$$

$$\geq \tilde{U}\left(\frac{\varsigma}{B(T)}\right) + y\varsigma(1-\mu),$$

(17.21)

for all $(Z_0(\cdot), Z_1(\cdot)) \in \mathcal{D}$. We shall use this observation to find examples, in which the optimal strategy $(\hat{L}, \hat{M})$ of Theorem 17.4 never trades.

17.5 Example: $r(\cdot)$ *deterministic*, $y = 0$. In this case we see from (17.21) that

$$\tilde{V}(\varsigma; 0) = \inf_{(Z_0, Z_1)\in\mathcal{D}} E\tilde{U}\left(\varsigma\frac{Z_0(T)}{B(T)}\right) \geq \tilde{U}(\varsigma/B(T)),$$

and the infimum is achieved by taking $\hat{Z}_0(\cdot) \equiv 1$, i.e., by any pair $(1, \hat{Z}_1(\cdot)) \in \mathcal{D}$ that satisfies $1 - \mu \leq \hat{R}(\cdot) = \hat{Z}_1(\cdot)/P(\cdot) \leq 1 + \lambda$, if such exists. In particular, one can take $\hat{Z}_1(0) = (1 + \lambda)p$ and $\hat{\theta}_1(\cdot) \equiv \sigma(\cdot)$, in which case $(1, \hat{Z}_1(\cdot)) \in \mathcal{D}$ if and only if

$$0 \leq \int_0^t (b(s) - r(s))ds \leq \log\frac{1+\lambda}{1-\mu}, \forall\, 0 \leq t \leq T.$$

(17.22)

Furthermore, from (17.10) and (17.5),(17.6) we have

$$\hat{X}(T) + f(\hat{Y}(T)) = I(\hat{\varsigma}/B(T)) = \hat{C}_0 = xB(T).$$

All the conditions (17.4), (17.9) and the Assumption 17.1 are satisfied rather trivially; and the *no-trading-strategy* $\hat{L} \equiv 0$, $\hat{M} \equiv 0$ is optimal, from Theorem 17.4 (and gives $\hat{X}(T) = xB(T)$, $\hat{Y}(T) = 0$). The condition (17.22) is satisfied, for instance, if

$$r(\cdot) \leq b(\cdot) \leq r(\cdot) + \rho, \text{ for some } 0 \leq \rho \leq \frac{1}{T}\log\frac{1+\lambda}{1-\mu}.$$

(17.23)

If $b(\cdot) = r(\cdot)$ the result is not surprising – even without transaction costs, it is then optimal not to trade. However, for $b(\cdot) > r(\cdot)$ the optimal portfolio always invests a positive amount in the stock, if there are no transaction costs; the same is true even in the presence of transaction costs, if one is maximizing expected discounted utility from consumption over an infinite time-horizon, and if the market coefficients are constant – see Shreve & Soner (1994), Theorem 11.6.

The situation here, on the finite time-horizon $[0, T]$, is quite different: if the excess rate of return $b(\cdot) - r(\cdot)$ is positive but small relative to the transaction costs, and/or if the time-horizon is small, in the sense of (17.23), then it is optimal not to trade.

Remark: In the infinite time-horizon case with constant market coefficients, as in Davis & Norman (1990), Shreve & Soner (1994), the ratio $\hat{X}/\hat{Y}$ of optimal holdings is a reflected diffusion process in a fixed interval; more precisely, one trades only when this ratio hits the endpoints of the interval, and in such a way as to keep the ratio inside the interval. In our case, under the assumptions of Example 17.5, and with $U(x) = \log x$, one obtains from (17.13) that

$$(\hat{X}(t) + \hat{R}(t)\hat{Y}(t))/B(t) = (\hat{\zeta}\hat{Z}_0(t))^{-1}, \quad 0 \leq t \leq T.$$

Comparing the stochastic integral representation of $(\hat{\zeta}\hat{Z}_0(\cdot))^{-1}$ with the equation (15.8), one obtains

$$\frac{\hat{R}(t)\hat{Y}(t)}{B(t)}(\hat{\theta}_1(t) - \hat{\theta}_0(t)) = -\frac{\hat{\theta}_0(t)}{\hat{\zeta}\hat{Z}_0(t)}, \quad 0 \leq t \leq T.$$

The last two equations imply

$$\hat{X}(t)/\hat{Y}(t) = -\hat{R}(t)\left(\frac{\hat{\theta}_1(t)}{\hat{\theta}_0(t)}\right), \quad 0 \leq t \leq T,$$

provided $\hat{Y}(t)\hat{\theta}_0(t) \neq 0, \forall\, t \in [0,T]$. While $\hat{R}(\cdot)$ is a reflected process in a fixed interval, it is not clear what happens to the second factor, either for fixed T or as $T \to \infty$.

17.6 Remark: Assuming $y = 0$, it can be argued, using the arguments as in the case of constraints, that the utility based price $V(0)$ of a claim $C = (C_0, C_1)$ in our setting should be the expected value of the discounted claim evaluated under the optimal shadow state-price densities of the dual problem, i.e., by

$$V(0) = E\left[\hat{Z}_0(T)\frac{C_0}{B(T)} + \hat{Z}_1(T)\frac{C_1}{S(T)}\right],$$

provided that the dual optimization problem of (17.3) has a *unique* solution $(\hat{Z}_0(\cdot), \hat{Z}_1(\cdot))$. Notice that this price does not depend on the initial holdings x in stock, if the solution to the dual problem does not depend on ζ, as in Assumption 17.1. However, $V(0)$ *does* depend in general on the return rate of the stock $b(\cdot)$.

18. ON PORTFOLIO OPTIMIZATION UNDER "DRAWDOWN" CONSTRAINTS.

Let us consider again the standard model of a financial market $\mathcal{M}$, as in Section 2. We give a different representation for the wealth process of a financial agent with initial capital $x > 0$:

$$dX^\pi(t) = r(t)X^\pi(t)dt + \left(X^\pi(t) - \frac{\alpha M^\pi(t)}{\gamma_0(t)}\right)\pi^*(t)\left[(b(t) - r(t)\mathbf{1})dt + \sigma(t)dW(t)\right]$$
$$\tag{18.1}$$

$$X^\pi(0) = x,$$

and we require that it satisfies the "drawdown constraint"

$$P\left[\gamma_0(t)X^\pi(t) > \alpha M^\pi(t), \quad \forall 0 \leq t < \infty\right] = 1. \tag{18.2}$$

Here $\alpha \in (0,1)$ is a given constant, and

$$M^\pi(t) := \max_{0 \le s \le t} (\gamma_0(s) X^\pi(s)). \tag{18.3}$$

The interpretation is this: the agent does not tolerate the "drawdown $1 - \frac{\gamma_0(t) X^\pi(t)}{M^\pi(t)}$ of his discounted wealth, from its maximum-to-date", to be greater than or equal to the constant $1 - \alpha$, *at any time $t \ge 0$*; thus, he imposes the (almost sure) constraint (18.2).

More precisely, we say that a portfolio $\pi = (\pi_1, \ldots, \pi_d)$ is admissible if $\pi_i(t)$ is the proportion of the difference $X^\pi(t) - \alpha \frac{M^\pi(t)}{\gamma_0(t)} > 0$ invested in the ith stock, $i = 1, \ldots, d$, and the remainder is invested in the bank account; we also require that $\pi(\cdot)$ is measurable and adapted process, and that, for any $T > 0$, $\int_0^T \|\pi(t)\|^2 dt < \infty$. We denote by $\mathcal{A}_\alpha(x)$ the class of admissible portfolios.

18.1 Lemma: *If $\pi(\cdot) \in \mathcal{A}_\alpha(x)$, then (18.1) has a unique solution and (18.2) is satisfied.*

Proof: Without loss of generality we set $r(\cdot) \equiv 0$ and $\sigma(\cdot)$ equal to identity matrix. We therefore have to show that there is a unique solution to

$$dX(t) = (X(t) - \alpha M(t)) \pi^*(t) dW_0(t), \quad M(t) = \max_{0 \le s \le t} X(s); \quad X(0) = x, \tag{18.4}$$

that satisfies a.s.

$$X(t) > \alpha M(t), \quad \forall 0 \le t < \infty. \tag{18.5}$$

Suppose that $X(\cdot)$ is an adapted process that satisfies (18.4), (18.5). Observe that

$$d\left(\frac{X(t)}{M(t)} - \alpha\right) = \left(\frac{X(t)}{M(t)} - \alpha\right) \pi^*(t) dW_0(t) - \frac{dM(t)}{M(t)},$$

whence

$$d\left(\log\left(\frac{X(t)}{M(t)} - \alpha\right)\right) = d\xi(t) - \frac{1}{1-\alpha} \frac{dM(t)}{M(t)},$$

where

$$\xi(t) := \int_0^t \pi^*(s) dW_0(s) - \frac{1}{2} \int_0^t \|\pi(s)\|^2 ds.$$

Therefore,

$$0 \le R(t) := \log(1 - \alpha) - \log\left(\frac{X(t)}{M(t)} - \alpha\right) = -\xi(t) + \log\left(\frac{M(t)}{x}\right)^{\frac{1}{1-\alpha}}.$$

Clearly, the continuous increasing process $K(t) := \log\left(\frac{M(t)}{x}\right)^{\frac{1}{1-\alpha}}$ is flat away from the set $\{t \ge 0 / X(t) = M(t)\}$, i.e., away from the zero-set of the continuous nonnegative process $R(\cdot)$. From the theory of the Skorohod equation (e.g. Karatzas & Shreve (1991), §3.6) we have then $K(t) = \max_{0 \le s \le t} \xi(s)$, and from this:

$$M(t) \equiv \tilde{M}(t) := x \exp\left\{(1 - \alpha) \max_{0 \le s \le t} \xi(s)\right\},$$

$$X_t \equiv \tilde{X}_t := x \exp\left\{(1-\alpha)\max_{0 \le s \le t}\xi_s\right\}\left[\alpha + (1-\alpha)\exp\left\{\xi_t - \max_{0 \le s \le t}\xi_s\right\}\right].$$

It is straightforward to check that $\tilde{X}(\cdot)$ satisfies (18.4), (18.5).

18.2 Problem *(Grossman & Zhou (1993))*: For some given $0 < \delta < 1$, maximize the long-term rate of growth

$$\mathcal{R}(\pi) := \varlimsup_{T \to \infty}\frac{1}{T}\log E(X^\pi(T))^\delta \qquad (18.6)$$

of expected power-utility, over $\pi \in \mathcal{A}_\alpha(x)$. In particular, compute

$$v(\alpha) := \sup_{\pi \in \mathcal{A}_\alpha(x)}\mathcal{R}(\pi) \qquad (18.7)$$

and find $\pi \in \mathcal{A}_\alpha(x)$, for which the limit $\lim_{T \to \infty}\frac{1}{T}\log E(X^{\hat{\pi}}(T))^\delta = \mathcal{R}(\hat{\pi})$ exists and achieves the supremum in (18.7).

In order to solve Problem 18.2, we introduce an auxiliary process and problem, as follows: For any portfolio process $\pi \in \mathcal{A}_\alpha(x)$, consider the auxiliary process

$$N_\alpha^\pi(t) := \left(X^\pi(t) - \alpha\frac{M^\pi(t)}{\gamma_0(t)}\right)(M^\pi(t))^{\frac{\alpha}{1-\alpha}}, \quad 0 \le t < \infty. \qquad (18.8)$$

Because the increasing process $M^\pi(\cdot)$ is flat off the set $\{t \ge 0/\gamma_0(t)X^\pi(t) = M^\pi(t)\}$, we have:

$$d(\gamma_0(t)N_\alpha^\pi(t)) = (\gamma_0(t)N_\alpha^\pi(t))\pi^*(t)\sigma(t)dW_0(t), \qquad (18.9)$$

Notice also, in the notation of the proof of Lemma 18.1, that

$$\gamma_0(t)N_\alpha^{\hat{\pi}}(t) = (1-\alpha)x^{\frac{1}{1-\alpha}}e^{\xi(t)}. \qquad (18.10)$$

Recall also the notation

$$Z_0(t) := \exp\left\{-\int_0^t\theta^*(s)dW(s) - \frac{1}{2}\int_0^t\|\theta(s)\|^2ds\right\}, \quad H_0(t) := \gamma_0(t)Z_0(t).$$

¿From the product rule we obtain: $d(H_0(t)N_\alpha^\pi(t)) = H_0(t)N_\alpha^\pi(t)(\pi^*(t)\sigma(t) - \theta^*(t))dW(t)$. In other words, for any $\pi \in \mathcal{A}_\alpha(x)$ the process

$$H_0(t)N_\alpha^\pi(t)$$

$$= (1-\alpha)x^{\frac{1}{1-\alpha}}\exp\left\{\int_0^t(\pi^*\sigma - \theta^*)(s)dW(s) - \frac{1}{2}\int_0^t\|\pi^*\sigma - \theta^*\|^2(s)ds\right\}$$

is a positive local martingale, hence supermartingale, which thus satisfies

$$E[H_0(T)N_\alpha^\pi(T)] \le (1-\alpha)x^{\frac{1}{1-\alpha}}, \quad \forall T \in (0,\infty). \qquad (18.11)$$

We now pose an auxiliary stochastic control problem, involving the process $N_\alpha^\pi(\cdot)$.

18.3 An Auxiliary, Finite-Horizon, Control Problem: *For a given $T \in (0,\infty)$ and utility function $U : (0,\infty) \to \mathbb{R}$, denote by $\mathcal{A}_\alpha(x,T)$ the class of* *admissible portfolios $\pi(\cdot)$ on the finite horizon $[0,T]$, and find $\hat{\pi}(\cdot) \in \mathcal{A}_\alpha(x,T)$ which achieves*

$$V(\alpha;T,x) := \sup_{\pi \in \mathcal{A}_\alpha(x,T)} EU(N_\alpha^\pi(T)). \qquad (18.12)$$

As we know from earlier sections, the solution is found as follows: Choose $\hat{y}$ such that

$$E[H_0(T)I(\hat{y}H_0(T))] = (1-\alpha)x^{1/1-\alpha}.$$

Choose portfolio $\hat{\pi} \in \mathcal{A}_\alpha(x)$ by introducing the positive martingale

$$\mathcal{Q}(t) := E[H_0(T)I(\hat{y}H_0(T))\|\mathcal{F}(t)] = (1-\alpha)x^{\frac{1}{1-\alpha}} + \int_0^t \mathcal{Q}(s)\varphi^*(s)dW(s),$$

for $0 \le t \le T$, and setting

$$\hat{\pi}(\cdot) = ((\theta^* + \varphi^*)\sigma^{-1})^*(\cdot) \in \mathcal{A}_\alpha(x,T).$$

18.4 Remark: Let us consider now Problem 18.2 with utility function

$$U(x) = \frac{1}{\gamma}x^\gamma \quad \text{for } \gamma := \delta(1-\alpha), \quad 0 < \delta < 1.$$

Then, with $\mu := \frac{\gamma}{1-\gamma}$, we get

$$\hat{y}^{-\frac{1}{1-\gamma}} = \frac{(1-\alpha)x^{\frac{1}{1-\alpha}}}{E[(H_0(T))^{-\mu}]}, \quad V(\alpha;T,x) = \frac{1}{\gamma}\left((1-\alpha)x^{\frac{1}{1-\alpha}}(E(H_0(T))^{-\mu})^{1/\mu}\right)^\gamma.$$

$$(18.13)$$

If, in addition, the coefficients $r(\cdot)$, $b(\cdot)$, $\sigma(\cdot)$ are deterministic, then

$$(H_0(t))^{-\mu} = \exp\left[\mu \int_0^t \theta^*(s)dW(s) - \frac{\mu^2}{2}\int_0^t \|\theta(s)\|^2 ds\right]$$

$$\times \exp\left\{\mu \int_0^t \left(r(s) + \frac{1+\mu}{2}\|\theta(s)\|^2\right) ds\right\}$$

and we obtain

$$\mathcal{Q}(t) = (1-\alpha)x^{\frac{1}{1-\alpha}}\exp\left\{\mu\int_0^t \theta^*(s)dW(s) - \frac{\mu^2}{2}\int_0^t \|\theta(s)\|^2 ds\right\}, \quad \varphi(t) = \mu\theta(t)$$

$$(18.14)$$

$$\hat{\pi}^*(t)\sigma(t) = (1+\mu)\theta^*(t) = \frac{1}{1-\delta(1-\alpha)}\theta^*(t), \quad \text{independent of } T, \quad (18.15)$$

$$V(\alpha;T,x) = \frac{1}{\gamma}\left((1-\alpha)x^{\frac{1}{1-\alpha}}\exp\left\{\int_0^T \left(r(t) + \frac{1+\mu}{2}\|\theta(t)\|^2\right) dt\right\}\right)^\gamma.$$

$$(18.16)$$

In order to solve the original, Grossman-Zhou problem, we shall assume that the coefficients $r(\cdot)$, $b(\cdot)$, $\sigma(\cdot)$ are deterministic, and $r_* := \lim_{T\to\infty} \frac{1}{T}\int_0^T r(s)ds$, $\|\theta_*\|^2 := \lim_{T\to\infty} \frac{1}{T}\int_0^T \|\theta(s)\|^2 ds$ exist and are finite.

18.5 Theorem: (Grossman & Zhou (1993), Cvitanić & Karatzas (1995)) *Under the above assumptions, the portfolio $\hat{\pi}(\cdot)$ of (18.15) is optimal for the Problem 18.2. In fact, we have*

$$\lim_{T\to\infty} \frac{1}{T} \log E(X^{\hat{\pi}}(T))^{\delta} = \mathcal{R}(\hat{\pi}) = v(\alpha) = V(\alpha) + \alpha\delta r_*, \qquad (18.17)$$

where

$$V(\alpha) := \lim_{T\to\infty} \frac{1}{T} \log V(\alpha; T, x) = \gamma r_* + \frac{\gamma}{2}(1+\mu)\|\theta_*\|^2$$
$$= \delta(1-\alpha)\left[r_* + \frac{\|\theta_*\|^2}{2}\frac{1}{1-\delta(1-\alpha)}\right]. \qquad (18.18)$$

In order to establish this result, it will be helpful to consider the auxiliary problem

$$\bar{v}(\alpha) := \sup_{\pi\in\mathcal{A}_\alpha(x)} \bar{\mathcal{R}}_\alpha(\pi), \quad \bar{\mathcal{R}}_\alpha(\pi) := \varlimsup_{T\to\infty} \frac{1}{T}\log E(N^\pi(T))^{\delta(1-\alpha)}. \qquad (18.19)$$

¿From the fact that the portfolio $\hat{\pi}(\cdot)$ of (18.15) does not depend on the horizon $T \in (0, \infty)$, it is clear that

$$\lim_{T\to\infty}\frac{1}{T}\log E(N^{\hat{\pi}}_\alpha(T))^{\delta(1-\alpha)} = \bar{\mathcal{R}}_\alpha(\hat{\pi}) = \bar{v}(\alpha) = V(\alpha). \qquad (18.20)$$

It will also be helpful to note that

$$(N^\pi_\alpha(t))^{\delta(1-\alpha)} = (\gamma_0(t))^{\alpha\delta}(X^\pi(t))^\delta \left(f_\alpha\left(\frac{\alpha M^\pi(t)}{\gamma_0(t)X^\pi(t)}\right)\right)^\delta, \qquad (18.21)$$

where the function $f_\alpha(x) := \left(\frac{x}{\alpha}\right)^\alpha (1-x)^{1-\alpha}$, $0 \le x \le 1$ is strictly increasing on $(0, \alpha)$ and strictly decreasing on $(\alpha, 1)$.

Proof of Theorem 18.5: From (18.21) we obtain

$$E(N^\pi_\alpha(T))^{\delta(1-\alpha)} \le (\gamma_0(T))^{\alpha\delta}(1-\alpha)^{\delta(1-\alpha)}E(X^\pi(T))^\delta,$$

whence

$$\bar{\mathcal{R}}_\alpha(\pi) \le \mathcal{R}(\pi) - \alpha\delta r_* \le v(\alpha) - \alpha\delta r_*, \quad \forall \pi \in \mathcal{A}_\alpha(x)$$

and therefore $V(\alpha) \le v(\alpha) - \alpha\delta r_*$. In order to establish the reverse inequality, take $\eta \in (0, \alpha)$ close enough to α so that $f_\eta(\eta) \ge f_\eta(\eta/\alpha)$, and observe from (18.21) that for an arbitrary $\pi \in \mathcal{A}_\alpha(x)$ ($\subseteq \mathcal{A}_\eta(x)$) we have

$$E(N^\pi_\eta(T))^{\delta(1-\eta)} \ge (\gamma_0(T))^{\eta\delta} (f_\eta(\eta/\alpha))^\delta E(X^\pi(T))^\delta$$
$$= (\gamma_0(T))^{\eta\delta} \left(\alpha^{-\eta}(1-\eta/\alpha)^{1-\eta}\right)^\delta E(X^\pi(T))^\delta.$$

Consequently

$$V(\eta) \ge \bar{\mathcal{R}}_\eta(\pi) \ge \mathcal{R}(\pi) - \eta\delta r_*, \quad \forall \pi \in \mathcal{A}_\alpha(x),$$

whence $V(\eta) \geq v(\alpha) - \eta \delta r_*$; letting $\eta \uparrow \alpha$ and invoking the continuity of the function $V(\cdot)$, we obtain $V(\alpha) \geq v(\alpha) - \alpha \delta r_*$ and thus the third equality of (18.17):

$$v(\alpha) = V(\alpha) + \alpha \delta r_* = \delta r_* + \frac{\|\theta_*\|^2}{2} \frac{\delta(1-\alpha)}{1 - \delta(1-\alpha)}.$$

To obtain the second equality, it suffices to observe that

$$v(\alpha) \geq \mathcal{R}(\hat{\pi}) \geq \bar{\mathcal{R}}_\alpha(\hat{\pi}) + \alpha \delta r_* = V(\alpha) + \alpha \delta r_* = v(\alpha).$$

Finally, the first equality, i.e., the existence of the indicated limit, follows from the double inequality

$$-\frac{\delta(1-\alpha)}{T} \log(1-\alpha) + \frac{\alpha\delta}{T} \int_0^T r(s)ds + \frac{1}{T} \log E(N^{\hat{\pi}}(T))^{\delta(1-\alpha)}$$

$$\leq \frac{1}{T} \log E(X^{\hat{\pi}}(T))^\delta$$

$$\leq -\frac{\delta}{T} \log\left(\alpha^{-\eta}(1 - \frac{\eta}{\alpha})^{1-\eta}\right) + \frac{\eta\delta}{T} \int_0^T r(s)ds + \frac{1}{T} \log E(N^{\hat{\pi}}(T))^{\delta(1-\eta)}$$

by passing to the limit as $T \to \infty$ and then letting $\eta \uparrow \alpha$.

The methods above can also be used to show that the portfolio

$$\pi_*(t) = (\theta^*(t)\sigma^{-1}(t))^*, \quad 0 \leq t < \infty$$

is optimal for the problem of *maximizing the long-term rate of expected logarithmic utility* under the drawdown constraint:

$$\varlimsup_{T \to \infty} \frac{1}{T} E(\log X^\pi(T)) \leq \lim_{T \to \infty} \frac{1}{T} E(\log X^{\pi_*}(T)) = (1-\alpha)\left(\bar{r} + \frac{\|\bar{\theta}\|^2}{2}\right) + \alpha\bar{r},$$
$$(18.22)$$

for all $\pi \in \mathcal{A}_\alpha(x)$ (with $\bar{r}, \|\bar{\theta}\|^2$ defined below). It turns out that this holds for *general random, adapted coefficients* $r(\cdot)$, $b(\cdot)$, $\sigma(\cdot)$, for which the conditions of the model are satisfied and the limits

$$\bar{r} := \lim_{T \to \infty} \frac{1}{T} \int_0^T Er(t)dt, \quad \|\bar{\theta}\|^2 := \lim_{T \to \infty} \frac{1}{T} \int_0^T E\|\theta(t)\|^2 dt$$

exist and are finite.

More important than the optimality property (18.22), however, is the fact that the portfolio $\pi_*(\cdot)$ *maximizes the long-term growth rate from investment*

$$\mathcal{S}(\pi) := \varlimsup_{T \to \infty} \frac{1}{T} \log X^\pi(T) \leq \lim_{T \to \infty} \frac{1}{T} \log X^{\pi_*}(T) = (1-\alpha)\left(r^* + \frac{\|\theta^*\|^2}{2}\right) + \alpha r^*,$$
$$(18.23)$$

over all $\pi \in \mathcal{A}_\alpha(x)$. Again, this comparison is valid for *general random, adapted coefficients* in the model, under the proviso that the limits

$$r^* := \lim_{T \to \infty} \frac{1}{T} \int_0^T r(t)dt, \quad \|\theta^*\|^2 := \lim_{T \to \infty} \frac{1}{T} \int_0^T \|\theta(t)\|^2 dt$$

exist and are finite, almost surely.

In order to prove this, let us start by noticing that $\Lambda(t) := N_\alpha^\pi(t)/N_\alpha^{\pi*}(t)$, $0 \leq t < \infty$ satisfies the stochastic equation

$$d\Lambda(t) = \Lambda(t)(\pi^*(t)\sigma(t) - \theta^*(t))dW(t), \quad \Lambda(0) = 1$$

and is thus a positive supermartingale, for any $\pi \in \mathcal{A}_\alpha(x)$. It follows readily from this (see Karatzas (1989), p.1243)) that $\overline{\lim}_{t\to\infty} \frac{1}{t} \log \Lambda(t) \leq 0$, or equivalently

$$\bar{S}_\alpha(\pi) := \varliminf_{T\to\infty} \frac{1}{T} \log(N_\alpha^\pi(T))^{1-\alpha} \leq \lim_{T\to\infty} \frac{1}{T} \log(N^{\pi*}(T))^{1-\alpha} =$$

$$= (1-\alpha)\left(r^* + \frac{\|\theta^*\|^2}{2}\right) =: \bar{s}(\alpha), \text{ a.s.}$$

$$(18.24)$$

The existence of this last limit, and its value follows from (18.10). On the other hand, similarly as above, one gets

$$\frac{1}{T} \log(N_\alpha^\pi(T))^{1-\alpha} + \frac{\alpha}{T} \int_0^T r(s)ds - \frac{1-\alpha}{T} \log(1-\alpha) \leq \frac{1}{T} \log X^\pi(T)$$

$$(18.25)$$

$$\leq \frac{1}{T} \log(N_\eta^\pi(T))^{1-\eta} + \frac{\eta}{T} \int_0^T r(s)ds - \frac{1}{T} \log\left(\alpha^{-\eta}(1-\eta/\alpha)^{1-\eta}\right)$$

almost surely, for any $\pi \in \mathcal{A}_\alpha(x) \subseteq \mathcal{A}_\eta(x)$ and any $\eta \in (0,\alpha)$ sufficiently close to α. In particular,

$$\bar{S}_\alpha(\pi) + \alpha r^* \leq s(\alpha) := \operatorname*{esssup}_{\pi \in \mathcal{A}_\alpha(x)} \lim S(\pi), \text{ a.s.}$$

whence $\bar{s}(\alpha) + \alpha r^* \leq s(\alpha)$, a.s.; similarly,

$$S(\pi) - \eta r^* \leq \bar{S}_\eta(\pi) \leq \bar{s}(\eta), \text{ whence } s(\alpha) - \eta r^* \leq \bar{s}(\eta)$$

and in the limit as $\eta \uparrow \alpha$: $s(\alpha) - \alpha r^* \leq \bar{s}(\alpha)$, a.s. It develops that $s(\alpha) = \bar{s}(\alpha) + \alpha r^* = (1-\alpha)\left(r^* + \frac{1}{2}\|\theta^*\|^2\right) + \alpha r^*$, and it remains to show the existence of the limit and the equality in (18.23). But both of these follow by writing the double inequality (18.25) with $\pi \equiv \pi_*$, letting $T \to \infty$ to obtain in conjunction with (18.24)

$$s(\alpha) = \bar{s}(\alpha) + \alpha r^* \leq \varliminf_{T\to\infty} \frac{1}{T} \log X^{\pi*}(T) \leq \varlimsup_{T\to\infty} \frac{1}{T} \log X^{\pi*}(T) \leq s(\eta),$$

and then letting $\eta \uparrow \alpha$ to conclude $\lim_{T\to\infty} \frac{1}{T} \log X^{\pi*}(T) = s(\alpha)$, almost surely.

A. APPENDIX.

Most of the results in this section are taken from Karatzas & Shreve (1991) and Protter (1990).

A.1 Definition: A real valued continuous process $W^{(1)}(\cdot)$ is called a *standard Brownian motion* if $W^{(1)}(0) = 0$, if it has independent increments, and, for all $u, t \geq 0$, and the law of the increment $W^{(1)}(t) - W^{(1)}(u)$ is normal with mean zero and variance $t - u$. A vector process W of d independent Brownian motions, $W = (W^{(1)}, \ldots, W^{(d)})^*$, where a^* is vector a transposed, is a *standard Brownian motion* in $\mathbb{R}^d$. These are defined on a complete probability space $(\Omega, \mathcal{F}, \mathbf{P})$, and we shall denote by $\{\mathcal{F}_t\}$ the **P**-augmentation of the filtration $\mathcal{F}_t^W = \sigma(W(s); \ 0 \leq s \leq t)$ generated by W. All our processes $X(\cdot)$ (unless otherwise mentioned) will be *adapted* to the filtration $\{\mathcal{F}_t\}$, i.e., for all $t \geq 0$, $X(t)$ is $\mathcal{F}_t$−measurable random variable. They will also always be *right-continuous* (sometimes *left-continuous*).

Brownian motion in $\mathbb{R}$ is a *martingale*, namely

$$E[W^{(1)}(t)|\mathcal{F}_s] = W(s), \quad s \leq t. \qquad (A.1)$$

Processes X for which we have $\leq$ (resp., $\geq$) instead of equality in (A.1) are called *supermartingales* (resp., *submartingales*) with respect to the filtration $\{\mathcal{F}_t\}_{0 \leq t \leq \infty}$. If the property holds only for processes $X^{(n)}(t) = X(t \wedge \tau_n)$, for each n, where τ_n is a nondecreasing sequence of *stopping times* converging to infinity, then we say that $X(\cdot)$ is a *local martingale* (local super/sub-martingale). A stopping time τ is a nonnegative random variable for which $\{\tau \leq t\} \in \mathcal{F}_t$, for all $t \geq 0$.

A.2 Lemma: *A local martingale bounded from below is a supermartingale.*
Proof: Fatou's lemma for conditional expectations.
A.3 Theorem (Doob-Meyer decomposition): *A supermartingale* X *can be uniquely decomposed as*

$$X = X(0) + M - A \qquad (A.2)$$

where M *is a local martingale,* A *is nondecreasing (locally natural) process,* $EA(T) < \infty$, $M(0) = A(0) = 0$. *If* X *is also positive and the family* $\{X(\tau) : \tau$ a *stopping time bounded by* $\alpha\}$ *is uniformly integrable for all* $\alpha > 0$ *(we say that* $X(\cdot)$ *is of class DL), then* M *is a martingale.*

¿From now on (unless otherwise mentioned) we restrict ourselves to the finite time-horizon $[0, T]$, $T < \infty$. For a given $\mathbb{R}^d$−valued adapted process $\pi(\cdot)$ with $\int_0^T |\pi(s)|^2 ds < \infty$, it is possible to define *Ito stochastic integral* $\int_0^t \pi(s) dW(s)$, which is a continuous local martingale process. If, moreover, $E \int_0^T |\pi(s)|^2 ds < \infty$, then $E[\int_0^t \pi(s) dW(s)]^2 = E \int_0^t \pi^2(s) ds$. Integrals with respect to finite variation processes are defined ω by ω in the usual way.
A.4 Theorem (martingale representation theorem): *Let* $M(\cdot)$ *be an* $\mathcal{F}_t$− *adapted local martingale with RCLL (right-continuous with left limits) paths and* $M(0) = 0$. *Then there exist an* $\mathbb{R}^d$-*valued process* $\varphi(\cdot)$ *such that*

$$M(t) = \int_0^t \varphi(s) dW(s), \quad \int_0^T |\varphi(s)|^2 ds < \infty.$$

In particular, M has to be continuous. Moreover, if $EM^2(T) < \infty$, then

$$E \int_0^T |\varphi(s)|^2 ds < \infty,$$

and $M(\cdot)$ is a martingale. If $\tilde{\varphi}(\cdot)$ is another such a process, then

$$\int_0^T |\varphi(s) - \tilde{\varphi}(s)|^2 ds = 0.$$

1.5 Definition: A *semimartingale* $X(\cdot)$ is a process of the form

$$X = X(0) + M + A,$$

where $M(t) = M(0) + \int_0^t \varphi(s)dW(s)$ is a local martingale and A is a process of finite total variation. Its *quadratic variation process* is given by

$$< X, X > (t) = < M, M > (t) := \int_0^t \varphi^2(s)ds.$$

A *cross-variation* of semimartingales X and $\tilde{X}$ (with the corresponding representation) is given by

$$< X, \tilde{X} > (t) := \int_0^t \varphi^*(s)\tilde{\varphi}(s)ds.$$

In particular, cross-variation of process A of finite variation and any other semimartingale is zero.

A.6 Theorem (Ito's rule): Let $X = (X_1, \ldots, X_k)$ be a vector of continuous semimartingales of the form

$$X_i(t) = X_i(0) + \int_0^t \varphi_i(s)dW(s) + A_i(t)$$

and let $g : \mathbb{R}^k \mapsto \mathbb{R}$ be twice continuously differentiable function. Then

$$g(X(t)) = g(X(0)) + \sum_{i=1}^k \int_0^t \frac{\partial g}{\partial x_i}(X(u))dX_i(u)$$

$$+ \frac{1}{2} \sum_{i,j=1}^k \int_0^t \frac{\partial^2 g}{\partial x_i \partial x_j}(X(u))\varphi_i^*(u)\varphi_j(u)du.$$

In particular,

$$X_1(t)X_2(t) = X_1(0)X_2(0) + \int_0^t X_1(u)dX_2(u) + \int_0^t X_2(u)dX_1(u)$$

$$+ \int_0^t \varphi_1^*(u)\varphi_2(u)du.$$

A.7 Proposition: *Consider the following one-dimensional linear Stochastic Differential Equation with possibly random, locally bounded coefficients* A, a, S_j, σ_j:

$$dX(t) = [A(t)X(t) + a(t)]dt + \sum_{j=1}^{d}[S_j(t)X(t) + \sigma_j(t)]dW^{(j)}(t),$$

which is understood as $X(t) = X(0) + \int_0^t \ldots ds + \int_0^t \ldots dW^{(j)}(s)$. *Its unique solution is given by*

$$X_t = Y_t \left[X_0 + \int_0^t \frac{1}{Y_u}\{a(u) - \sum_{j=1}^{d} S_j(u)\sigma_j(u)\}du + \sum_{j=1}^{d} \int_0^t \frac{\sigma_j(u)}{Y_u}dW_u^{(j)} \right]$$

where

$$Y(t) = \exp\{\int_0^t A(u)du\}Z(t),$$

and

$$Z(t) = \exp\{\sum_{j=1}^{d} \int_0^t S_j(u)dW^{(j)}(u) - \frac{1}{2}\sum_{j=1}^{d}\int_0^t S_j^2(u)du\}. \qquad (A.3)$$

Notice then, denoting $S = (S_1, \ldots, S_d)$, that Z itself is a solution to

$$dZ(t) = S(t)Z(t)dW(t), \quad Z(0) = 1. \qquad (A.4)$$

Therefore, $Z(\cdot)$ is a positive local martingale, hence a supermartingale. If it is also a martingale, then one can define a new probability measure $\tilde{\mathbf{P}}$ by $\frac{d\tilde{\mathbf{P}}}{d\mathbf{P}} = Z(T)$. We have the following *Bayes rule*:

A.8 Lemma: *If* $0 \leq s \leq t \leq T$ *and* Y *is an* $\mathcal{F}_t$-*measurable random variable satisfying* $\tilde{E}|Y| < \infty$ *(where* $\tilde{E}$ *is the expectation under the* $\tilde{\mathbf{P}}$ *measure), then*

$$\tilde{E}[Y|\mathcal{F}_s] = \frac{1}{Z(s)}E[YZ(t)|\mathcal{F}_s].$$

A.9 Theorem (Girsanov-Cameron-Martin): *Define a process* $\tilde{W}$ *by*

$$\tilde{W}^{(i)}(t) := W^{(i)}(t) - \int_0^t S_i(s)ds.$$

Then the process $\{\tilde{W}(t), \mathcal{F}_t; 0 \leq t \leq T\}$ *is a* d−*dimensional Brownian motion on* $(\Omega, \mathcal{F}_T, \tilde{\mathbf{P}})$.

A.10 Remark: In general, the filtration $\{\tilde{\mathcal{F}}_t\}$ corresponding to $\tilde{W}$ will not be equal to the filtration $\{\mathcal{F}_t\}$. Nevertheless, one can show, using the Bayes rule, that the martingale representation theorem still holds, namely that the $\tilde{\mathbf{P}}$-(local) martingales can be represented as stochastic integrals with respect to $\tilde{W}$, with $\{\mathcal{F}_t\}$-adapted integrands.

We finish this section by some results on connections with PDE's (Partial Differential Equations). Consider the SDE in $\mathbb{R}^k$

$$dX(t) = b(t, X(t))dt + \sigma(t, X(t))dW(t), \qquad (A.5)$$

where the coefficients $b_i(t, x), \sigma_{ij}(t, x) : [0, T] \times \mathbb{R}^k \mapsto \mathbb{R}$ are Lipshitz in x (uniformly in t) and of linear growth in x. Define the "infinitesimal generator" of $X(\cdot)$ as the second-order differential operator

$$\mathcal{A}_t V(x) := \frac{1}{2} \sum_{i=1}^{k} \sum_{j=1}^{k} a_{ij}(t, x) \frac{\partial^2 V(x)}{\partial x_i \partial x_j} + \sum_{i=1}^{k} b_i(t, x) \frac{\partial V(x)}{\partial x_i}; \quad V \in C^2(\mathbb{R}^k), \quad (A.6)$$

where $a(t, x)$ is a $k \times k$ matrix $a(t, x) := \sigma(t, x)\sigma^*(t, x)$.

A.11 Theorem (Feynman-Kac): *Let $f : \mathbb{R}^k \mapsto [0, \infty)$, $g, h : [0, T] \times \mathbb{R}^k \mapsto [0, \infty)$ be given continuous functions and let $V(t, x) : [0, T] \times \mathbb{R}^k \mapsto \mathbb{R}$ be a function which is continuous, twice continuously differentiable on $[0, T) \times \mathbb{R}^k$ and of polynomial growth in x, uniformly in t, satisfying the PDE*

$$\frac{\partial V}{\partial t} + \mathcal{A}_t V - hV + g = 0, \text{ in } [0, T) \times \mathbb{R}^k,$$

with the terminal condition

$$V(T, x) = f(x); \quad x \in \mathbb{R}^k.$$

Then, $V(t, x)$ admits the stochastic representation

$$V(t, x) = E^{t,x} \left[f(X(T)) \exp \left\{ - \int_t^T h(s, X_s)ds \right\} \right.$$
$$\left. + \int_t^T g(u, X_u) \exp \left\{ - \int_t^u h(s, X_s)ds \right\} du \right].$$

In particular, such a solution is unique.

Here, $E^{t,x}$ means that we are taking expectations of the functional of the process $X(\cdot)$ which is assumed to start at x at time t, $X(t) = x$.

Suppose now that we can control our SDE (A.5) by choosing $\{\mathcal{F}_t\}$–adapted processes $\pi(\cdot)$, which are now arguments of the coefficients, i.e., (A.5) becomes

$$dX(t) = b(t, X(t), \pi(t))dt + \sigma(t, X(t), \pi(t))dW(t), \qquad (A.5)'$$

Suppose also that we are considering the following *stochastic control problem*:

$$V(t, x) = \sup_{\pi(\cdot)} E^{t,x} \left[\int_t^T g(s, X_s, \pi_s)ds + f(X(T)) \right],$$

for appropriate functions f, g. Then, under suitable conditions, the *value function* $V(t, x)$ solves the *Hamilton-Jacobi-Bellman* equation

$$\frac{\partial V}{\partial t} + \sup_{\pi}[\mathcal{A}_t V + g] = 0, \text{ in } [0, T) \times \mathbb{R}^k, \qquad (A.7)$$

with the terminal condition

$$V(T, x) = f(x); \quad x \in \mathbb{R}^k.$$

Moreover, if the supremum in (A.7) is attained for some $\pi = \pi(t, x)$, then, again under suitable conditions, the *feedback control* process $\pi(t, X(t))$ is the optimal control.

19. REFERENCES.

BLACK, F. & SCHOLES, M. (1973) The pricing of options and corporate liabilities. *J. Polit. Economy* **81**, 637-659.

BROADIE, M., CVITANIĆ, J. & SONER, H. M. (1996) On the cost of super-replication under portfolio constraints. Preprint.

COX, J. & HUANG, C. F. (1989) Optimal consumption and portfolio policies when asset prices follow a diffusion process. *J. Econ. Theory* **49**, 33-83.

CVITANIĆ, J. & KARATZAS, I. (1992) Convex duality in constrained portfolio optimization. *Ann. Appl. Probab.* **2**, 767-818.

CVITANIĆ, J. & KARATZAS, I. (1995) On portfolio optimization under "drawdown constraints". *The IMA volumes in mathematics and its applications* **65**, 35-46.

CVITANIĆ, J. & KARATZAS, I. (1993) Hedging contingent claims with constrained portfolios. *Ann. Appl. Probab.* **3**, 652-681.

CVITANIĆ, J. & KARATZAS, I. (1996) Hedging and portfolio optimization under transaction costs: a martingale approach. *Mathematical Finance* **6**, 133-165.

CVITANIĆ, J. & MA, J. (1996) Hedging options for a large investor and Forward -Backward SDE's. *Annals of Applied Probability* **6**, 370-398.

DAVIS, M.H.A. (1994) A general option pricing formula. Preprint, Imperial College, London.

DAVIS, M. H .A. & CLARK, J. M. C. (1994) A note on super-replicating strategies. *Phil. Trans. Royal Soc. London A* **347**, 485-494.

DAVIS, M. H. A. & NORMAN, A. (1990) Portfolio selection with transaction costs. *Math. Operations Research* **15**, 676-713.

DUFFIE, D. (1992) *Dynamic Asset Pricing Theory*. Princeton University Press.

EKELAND, I. & TEMAM, R. (1976) *Convex Analysis and Variational Problems*. North-Holland, Amsterdam and Elsevier, New York.

EL KAROUI, N., PENG, S. & QUENEZ, M.C. (1993) Backwards Stochastic Differential Equations in Finance and Optimization. Preprint.

EL KAROUI, N. & QUENEZ, M.C. (1995) Dynamic programming and pricing of contingent claims in an incomplete market. *SIAM J. Control & Optimization*, **33**, 29-66.

FLEMING, W.H. & SONER, H.M. (1993) *Controlled Markov Processes and Viscosity Solutions*. Springer-Verlag, New York.

GROSSMAN, S. J. & ZHOU, Z. (1993) Optimal investment strategies for controlling drawdowns. *Math. Finance* **3** (3), (1993), pp. 241–276

HARRISON, J. M. & KREPS, D. M. (1979) Martingales and arbitrage in multiperiod security markets. *J. Econom. Theory* **20**, 381–408.

HARRISON, J. M. & PLISKA, S. R. (1981) Martingales and stochastic integrals in the theory of continuous trading. *Stochastic Processes and Appl.* **11**, 215–260.

HARRISON, J. M. & PLISKA, S. R. (1983) A stochastic calculus model of continuous time trading: complete markets. *Stochastic Processes and Appl.* **15**, 313–316.

HE, H. & PEARSON, N. (1991) Consumption and portfolio policies with incomplete markets and short-sale constraints: The infinite-dimensional case. *J. Econ. Theory* **54**, 259-304.

JOUINI, E. & KALLAL, H. (1995) Martingales and arbitrage in securities markets with transaction costs. *J. Econ. Theory* **66**, 178-197.

KARATZAS, I. (1989) Optimization problems in the theory of continuous trading. *SIAM J. Control & Optimization* **27**, 1221-1259.

KARATZAS, I. & KOU, S-G. (1994) On the pricing of contingent claims under constraints. To appear in *The Annals of Applied Probability.*

KARATZAS, I., LEHOCZKY, J. P. & SHREVE, S. E. (1987) Optimal portfolio and consumption decisions for a "small investor" on a finite horizon. *SIAM J. Control Optimization* **25**, 1557-1586.

KARATZAS, I., LEHOCZKY, J. P., SHREVE, S. E. & XU, G. L. (1991) Martingale and duality methods for utility maximization in an incomplete market. *SIAM J. Control Optimization* **29**, 702,730.

KARATZAS, I. & SHREVE, S.E. (1991) *Brownian Motion and Stochastic Calculus* (2nd edition), Springer-Verlag, New York.

KUSUOKA, S. (1995) Limit theorem on option replication with transaction costs. *Ann. Appl. Probab.* **5**, 198-221.

LADYŽENSKAJA, O.A., SOLONNIKOV, V.A. & URAL'TSEVA, N.N. (1968) *Linear and Quasilinear Equations of Parabolic Type.* Translations of Mathematical Monographs, Vol. 23, American Math. Society, Providence, R.I.

LEVENTAL, S. & SKOROHOD, A. V. (1995) On the possibility of hedging options in the presence of transactions costs. Preprint, Michigan State University.

MA, J., PROTTER, P. & YONG, J. (1994) Solving Forward-Backward Stochastic Differential Equations Explicitly—A Four Step Scheme. *Probability Theory and Related Fields*, **98**, 339-359

MERTON, R.C. (1973) Theory of rational option pricing. *Bell J. Econ. Manag. Sci.* **4**, 141-183.

NEVEU, J. (1975) *Discrete-Parameter Martingales.* North-Holland, Amsterdam.

PLATEN, E. & SCHWEIZER, M. (1994) On smile and skewness. Preprint.

PROTTER,P.(1990) *Stochastic Integration and Differential Equations.*Springer-Verlag, New York.

ROCKAFELLAR, R.T. (1970) *Convex Analysis.* Princeton University Press, Princeton.

SHREVE, S. E. & SONER, H. M. (1994) Optimal investment and consumption with transaction costs, *Ann. Appl. Probab.* **4**, 609-692.

SONER, H. M., SHREVE, S. E. & CVITANIĆ, J. (1995) There is no nontrivial hedging portfolio for option pricing with transaction costs, *Ann. Appl. Probab.* **5**, 327-355.

XU, G. & SHREVE, S. E. (1992) A duality method for optimal consumption and investment under short-selling prohibition. I. General market coefficients. II. Constant market coeficients. *Ann. Appl. Probab.* **2**, 87-112, 314-328.

NON-LINEAR PRICING THEORY

and

BACKWARD STOCHASTIC DIFFERENTIAL

EQUATIONS

El Karoui.N.[1], Quenez.M.C.[2]

Keywords: backward stochastic equation, mathematical finance, pricing, hedging portfolios, incomplete market, constrained portfolio, recursive utility, american contingent claim, viscosity solution of PDE.

[1] Laboratoire de Probabilités, CNRS-URA 224, Université de Paris VI
4 Place Jussieu, 75232 Paris Cedex 05, E-Mail : ne@ccr.jussieu.fr
[2] Equipe de Mathématiques, Université de Marne la Vallée, 2 rue de la Butte Verte, 93.166 Noisy-Le-Grand. E-Mail : quenez@math.univ-mlv.fr

Contents

Introduction

The main purpose of these lectures is to show that the Theory of Backward Stochastic Differential Equations (BSDE) is a useful tool to study the problem of pricing contingent claims in Finance.

Recall that a contingent claim is a contract which pays the amount ξ at time T. The basic idea of valuing a contingent claim $\xi \geq 0$ with maturity T is to construct a hedging portfolio, i.e., a portfolio holding some number of shares ϕ of traded stocks and some number of shares of a money market, so that the hedging portfolio replicates the claim. At any time, the price (X_t) of the contingent claim has the same value as the hedging portfolio if the market is arbitrage free and satisfies the dynamics

$$dX_t = r_t X_t dt + \phi_t^* \sigma_t [dW_t + \theta_t dt] \, ; \, X_T = \xi.$$

where r, θ are the interest rate and the risk premium processes. Actually, there exist an infinite number of pairs (X, ϕ) satisfying this equation. Using the BSDEs theory, we will show that the problem is well-posed, that is there exist an unique price X and a unique hedging portfolio ϕ, by restricting to square integrable solutions.

These equations were first introduced by Bismut (1973) for the linear case and in the general case by Pardoux and Peng (1990). The solution of a BSDE consists of a pair of adapted processes (Y, Z) satisfying:

$$- dY_t = f(t, Y_t, Z_t) \, dt - Z_t^* dW_t \, ; \quad Y_T = \xi, \qquad (0.1)$$

where f is called the driver and ξ the terminal condition.

In a complete market, the price process is solution of a BSDE with a linear driver. However, if the market is imperfect as, for example, in the case of a higher interest rate for borrowing, or different risk-premia for the seller and the buyer, the dynamics of the wealth or price process are given by a BSDE with a non-linear driver.

Recall that Duffie and Epstein (1992) presented a stochastic differential formulation of recursive utility for the consumption, as the objective function of a problem of investment-consumption. Recursive utility is an extension of the standard additive utility with the instantaneous utility depending not only on the instantaneous consumption rate c_t but also on the future utility. In fact, standard utilities correspond to BSDEs with linear driver whereas generalized utilities correspond to a non-linear one. In their paper, Duffie and Epstein (1992) showed that, under Lipschitz conditions, the recursive utility exists and satisfies the usual properties of standard utilities (concavity with respect to consumption if the BSDE is concave, consistency...). The BSDE point of view gives a simple formulation of more general recursive utilities and their properties.

In this work, using the BSDE's theory, we develop a non-linear arbitrage pricing theory in an imperfect market. First, we recall the pricing theory in the linear classical framework. Then, we present some typical situations of imperfect markets where the price system is no longer linear. We show the existence of such non-linear prices, and, using BSDE's technics such as the comparison theorem, we state some properties quite similar to the properties of utility functions: consistency but also absence of arbitrage, admissibility for sellers, that is sublinearity of the associated price system (in the spirit of Harrison and Kreps (1979) and Müller (1987)... The non-linear prices which are admissible for sellers correspond to BSDE's with a sublinear and hence convex driver; some examples are given by a higher interest rate for borrowing (Bergman 1991, Korn 1992, Cvitanic and Karatzas 1993), the case of short sales constraints (Jouiny Kallal

1992, He and Pearson 1991, Jouiny 1996). Using the comparison theorem, we show that such a price process corresponds to the upper price with respect to fictitious linear markets. Hence, as the utility function, the price can be written as the maximum of "ex-post" prices, taken over all feasible changes of "numeraire". Also, the maximum is attained for an optimal change of numeraire (which corresponds to an optimal fictitious market). Furthermore, in a Markovian framework (first studied by Peng (1991) and Pardoux and Peng (1992)), the price of the European option corresponds to the unique viscosity solution of a non-linear parabolic partial differential equation.

Then, we study the problem of pricing an American contingent claim in an imperfect market. We will see that the price process corresponds to the solution of a new type of backward equations called reflected BSDEs. The "reflection" keeps the solution of such an equation to stay above a given stochastic process called the obstacle. An increasing process is introduced which pushes the solution to stay above the obstacle.

In the case of an American option, the obstacle is given by the payoff of the option. Furthemore, the price corresponds to the maximum, over all stopping times $\tau \leq T$, of the price processes of European options associated with exercise time τ. Actually, this property is well known in the classical case of a complete market (Karatzas, 1988) and the driver of the associated reflected BSDE is linear. In the case of an imperfect market, the driver of the reflected BSDE is no longer linear but still convex. From this fact, we extend to the pricing of American options the interpretation of the price as the upper price over a family of fictitious markets which had been given for the European options. Furthermore, we state a separation theorem: that is the optimal exercise time for the American option corresponds, as in the classical case, to the first time the price process attains the payoff. Then, the optimal fictitious market or optimal deflator is the one which corresponds to the price of the European option associated with the optimal exercise time. Moreover, in a Markovian framework, the price of the American option corresponds to the unique viscosity solution of an obstacle problem for a parabolic partial differential equation.

1   Classical Linear Pricing Theory

1.1   The model

The probabilistic setting

We begin[2] with the typical set-up for continuous-time asset pricing: the basic securities consist of $n + 1$ assets. One of them is a non-risky asset (the money market instrument or bond), with price per unit S^0 governed by the equation,

$$dS_t^0 = S_t^0 \, r_t \, dt, \tag{1.1}$$

---

[2] We adopt the same framework and same notation as Cvitanic. The only difference is in the definition of portfolios

where r_t is the short rate. In addition to the bond, n risky securities (the stocks) are continuously traded. The price process S^i for one share of ith stock is modeled by the linear stochastic differential equation,

$$dS_t^i = S_t^i[b_t^i\, dt + \sum_{j=1}^{n} \sigma_t^{i,j}\, dW_t^j], \qquad (1.2)$$

driven by a standard n-dimensional Wiener process $W = (W^1, \ldots, W^n)^*$, defined on a filtered probability space $(\Omega, (\mathcal{F}_t)_{t\in[0,T]}, \mathbb{P})$. In general, we assume the filtration $(\mathcal{F}_t)$ generated by the Wiener process W and complete. $\mathbb{P}$ is said to be the "objective" probability measure.

Generally speaking, the coefficients $\sigma^i = (\sigma^{i,j})_{j=1}^{n}$, b^i and r are assumed to be bounded and progressively measurable processes, with values in $\mathbb{R}^n$, $\mathbb{R}$ and $\mathbb{R}$, respectively. For notational convenience, we will sometimes write $\sigma = [\sigma^{i,j}]$ to denote the random $n \times n$ volatility matrix, and $b = (b^i)$ to denote the stock appreciation rates vector. The assumption that the number of risky securities is equal to the dimension of the underlying noise is introduced for sake of simplicity.

To ensure the absence of arbitrage opportunities in the market, we assume that there exists a n-dimensional bounded progressively measurable vector process θ, such that

$$b_t - r_t\mathbf{1} = \sigma_t\, \theta_t, \qquad dt \otimes \mathbb{P}\ a.s,$$

where $\mathbf{1}$ is the vector whose every component is 1. θ is said to be a risk premium vector or "relative risk" process. We will assume more generally that the square matrix σ_t has a full rank for any fixed $t \in [0, T]$. Under these assumptions, the market is dynamically complet.

Portfolios and Wealth process

Let us consider a small investor, whose actions cannot affect market prices, and who can decide at time $t \in [0, T]$ what amount[3] ϕ_t^i of the wealth V_t to invest in the ith stock, $i = 1, \ldots, n$. Of course, his decisions can only be based on the current information $(\mathcal{F}_t)$, i.e., $\phi = (\phi^1, \phi^2, \ldots, \phi^n)^*$ [4] and $\phi^0 = V - \sum_{i=1}^{n}\phi^i$ are progressively measurable processes. Following Harrison and Pliska (1981), a strategy is self-financing if the wealth process $V = \sum_{i=0}^{n}\phi^i$ satisfies the equality,

$$V_t = V_0 + \int_0^t \sum_{i=0}^{n} \phi_s^i\, \frac{dS_s^i}{S_s^i},$$

or equivalently, if the wealth process is governed by the linear stochastic differential equation (LSDE),

$$dV_t = r_tV_t dt + \phi_t^*(b_t - r_t\mathbf{1})dt + \phi_t^*\sigma_t dW_t = r_tV_t dt + \phi_t^*\sigma_t[dW_t + \theta_t dt]. \quad (1.3)$$

---

[3]Unlike Cvitanic (1996 this volume), our portfolios φ are defined by the amounts invested in the risky assets, whereas Cvitanic defines the portfolios π by the proportion of the wealth invested in the risky assets. With this definition, we have some trouble with the null wealth.

[4]We denote by σ^* the transpose of the matrix σ

Generally, the initial wealth $x = V_0$ is taken as a primitive, and for an initial endowment and portfolio process (x, ϕ), there exists a unique wealth process V that is solution of the linear equation (1.3) with initial condition $V_0 = x$. Therefore, there exists a one-to-one correspondence between pairs (x, ϕ) and trading strategies (V, ϕ) which it will be sometimes convenient to use.

1.2   Pricing and hedging contingent claims in complete markets

A European contingent claim ξ settled at time T is a $\mathcal{F}_T$-measurable random variable. It can be thought as a contract which pays ξ at maturity T. The pricing of a contingent claim is based on the following principle: if we start with the price of the claim as initial endowment and invest it in the $n+1$ assets, the value of the portfolio at time T must be enough to finance ξ. A contingent claim ξ is said to be attainable, if it can be replicated by means of a self-financing trading strategy. In general, an attainable contingent claim may be replicated by an infinite number of self-financing strategies [5]. Using the theory of backward stochastic differential equations, we will show that the pricing problem is well-posed, that is the replicating strategy is unique, if we consider only square-integrable replicating strategies. In this framework, the notion of admissible strategy is defined in reference to the objective probability measure $I\!\!P$, unlike most of the of classical presentations of Arbitrage Pricing Theory, where the integrability constraint is related to a risk-adjusted probability measure.

Definition and Theorem 1.1 *A hedging strategy against a contingent claim ξ is a self-financing strategy (V, ϕ) such that $V_T = \xi$ and $I\!\!E \int_0^T |\sigma_t^* \phi_t|^2 dt < +\infty$. An attainable square integrable contingent claim ξ is replicated by a unique hedging strategy (X, ϕ).*

PROOF: Let us give a short proof of the uniqueness property. By linearity, it is clearly equivalent to show that the portfolio equal to 0 is the only hedging portfolio with terminal value 0. Let (X, ϕ) be such a hedging square integrable strategy. Ito's formula applied to $e^{\beta s}|X_s|^2$ between t and T leads to

$$
\begin{aligned}
e^{\beta t}|X_t|^2 &= 0 - \int_t^T e^{\beta s}|X_s|^2(\beta + 2r_s)ds - \int_t^T e^{\beta s}|\sigma_s^* \phi_s|^2 ds. \\
&- \int_t^T e^{\beta s} 2X_s \phi_s^* \sigma_s[dW_s + \theta_s ds]
\end{aligned}
\tag{1.4}
$$

For large β, the quadratic form

$$
x^2(\beta + 2r_s) + 2xz^* \sigma_s \theta_s + |\sigma_s^* z|^2 = |\sigma_s^* z + x\theta|^2 + x^2(\beta + 2r_s - |\theta_s|^2)
$$

is positive and $e^{\beta t}|X_t|^2$ is a positive local submartingale with terminal value 0. By semimartingale inequalities, the square integrable semimartingale X is

---
[5] They can be constructed from the suicide strategy (such that $V_0 = 1$, $V_T = 0$) given by Harrison and Pliska (1981)

uniformly bounded in $L^2(\mathbb{R})$ (i.e.,$\sup_{t\in[0,T]}|X_t|\in L^2(\mathbb{R})$), so by Fatou's lemma $\mathbb{E}[e^{\beta t}|X_t|^2]\leq 0$, that is ended the proof.   □

According to the result of Pardoux and Peng (1990) on the existence and uniqueness of the solutions of backward stochastic differential equations, that we state in Section 2, we prove that all the square integrable claims are attainable by square integrable replicating strategies.

Theorem 1.2 *Any square integrable contingent claim is attainable; the market is said to be complete. In other words, for any square integrable ξ, there exists a unique pair (X,ϕ) such that $\mathbb{E}\int_0^T |\sigma_t^*\phi_t|^2 dt < +\infty$ and*

$$dX_t = r_t X_t dt + \phi_t^*\sigma_t\theta_t dt + \phi_t^*\sigma_t dW_t; \qquad X_T = \xi. \qquad (1.5)$$

X_t *is the price of the claim at time t, given by the closed formula,*

$$X_t = \mathbb{E}[H_T^t\xi|\mathcal{F}_t], \qquad\qquad (1.6)$$

where H_s^t is the deflator process, started at time t, such that

$$dH_s^t = -H_s^t[r_s ds + \theta_s^* dW_s]; \qquad H_t^t = 1. \qquad (1.7)$$

PROOF: By Ito's calculus, the process [6] $\{H_tX_t, t\in[0,T]\}$ is a stochastic integral such that $dH_tX_t = H_t[\phi_t^*\sigma_t - X_t\theta_t^*]dW_t$. Classical results about the solution of the forward linear stochastic differential equation (1.7) with bounded coefficients give that the semimartingale H is uniformly bounded in L^2, and that the process $(H_tX_t, t\in[0,T])$ is a uniformly integrable martingale. So ,

$$H_tX_t = \mathbb{E}[H_T\xi|\mathcal{F}_t], \text{ or equivalently } X_t = \mathbb{E}[H_T^t\xi|\mathcal{F}_t].$$

The closed form of the deflator process $H_s^t = \exp -[\int_t^s r_u du + \int_t^s \theta_s^* dW_s + 1/2\int_t^s |\theta_u|^2 du]$ leads to the more classical formulation of the price of contingent claim as

$$X_t = \mathbb{E}[e^{-[\int_t^T \theta_u^* dW_u + 1/2\int_t^T |\theta_u|^2 du]}e^{-\int_t^T r_u du}\xi|\mathcal{F}_t] = \mathbb{E}_\mathbb{Q}[e^{-\int_t^T r_u du}\xi|\mathcal{F}_t]$$

where $\gamma_0(t) = e^{-\int_t^T r_u du}$ is the discounted factor over $[t,T]$ and $\mathbb{Q}$ the risk-adjusted probability measure defined by its Radon-Nikodym derivative with respect to $\mathbb{P}$,

$$\frac{d\mathbb{Q}}{d\mathbb{P}} = \exp -[\int_0^T \theta_s^* dW_s + \frac{1}{2}\int_0^T |\theta_s|^2 ds].$$

Notice that $\mathbb{Q}$ is a martingale-measure, that is the discounted wealth processes are $\mathbb{Q}$-local martingales.   □

Corollary 1.3 *Suppose the contingent claim to be a non-negative variable different from 0. Then, the price process is strictly positive.*

In particular, there do not exist arbitrage opportunities, that is admissible (i.e. non-negative) self-financing strategies (V,ϕ), worthless at time t on a set $A\in\mathcal{F}_t$, and such that V_T is non-negative and different from 0 on A.

---

[6]For notational simplicity we denote the process H_t^0 by H_t

PROOF: Since the price of a square integrable non-negative contingent claim is given throught an integral of a positive variable with respect of a martingale -measure, the only possibility for the price to be equal to 0 on A is to be associated with a null variable on the set A. In others words, arbitrage opportunities cannot exist on this market.   □.

Classically, the price is defined by arbitrage by considering only hedging-strategies for which the discounted wealth process is a $\mathbb{Q}$-martingale (Harrison and Pliska (1981), Musiela and Rutkowski (1997), Delbaen and Schachermayer (1994)) or by considering only hedging strategies such that $\mathbb{E}_{\mathbb{Q}} \int_0^T |\sigma_t^* \phi_t|^2 dt <$ $+\infty$. The major drawback of this point of view is its dependence on the martingale measure $\mathbb{Q}$.

Another approach is proposed by Karatzas and Shreve (1987) who consider trading strategies for which the wealth is bounded from below (in fact positive), instead of satisfying some integrability conditions. These authors prove that the **fair price** X_t^f at time t of a square integrable positive contingent claim ξ, defined as the smallest endowment which allows to hedge ξ, is still given by $X_t^f = \mathbb{E}[H_T^t \xi | \mathcal{F}_t]$. With a positive hedging strategy (V, ϕ) is associated a positive local martingale $\{H_t V_t, t \in [0, T]\}$. So, $\{H_t V_t, t \in [0, T]\}$ is a supermartingale with terminal value $H_T \xi$, which dominates the martingale $\{H_t X_t^f, \in [0, T]\}$. A nice feature of this point of view is to be not depending on the risk-neutral probability measure. The drawback is that the class of trading strategies is not a linear space. Notice that it is not easy to deduce directly from the closed form of the price $X_t = \mathbb{E}[H_T^t \xi | \mathcal{F}_t]$, that X is the wealth of a square integrable strategy under the "objective" probability.

1.3   Incomplete Markets: Föllmer-Schweizer Hedging Strategy

We consider a complete market modeled as before, where only some securities, called primary securities, must be held in a hedging portfolio. The seller of a contingent claim cannot have a perfect hedge because of the constraints on the portfolio. Föllmer and Schweizer (1990) introduce the notion of self-financing in mean strategy, which minimize the variance of the tracking error. As we will see above, this point of view yields to a linear price system of contingent claim. El Karoui and Quenez (1991) propose to introduce a seller's price, as the smallest investment which is enough to finance the claim with constrained portfolio. The associated price system is sublinear.

For sake of simplicity, we introduce the following conventional notation: the primary securities are the first j ones, the $j \times n$ volatility matrix of the primary securities is denoted by $\sigma_t^1 = (\sigma_t^{i,k})_{i=1..j, k=1..n}$, and the volatility matrix of the others is denoted by σ_t^2. Let us remark that the matrix σ_t^1 has a full rank, since the global matrix σ_t has a full rank, and so the matrix $(\sigma_t^1)(\sigma_t^1)^*$ is invertible. The amount of a general portfolio ϕ_t invested in the primary securities is denoted by $^1\phi_t$ and the amount in the others is denoted by $^2\phi_t$, such that

$$\phi_t^* \sigma_t = (^1\phi_t)^* \sigma_t^1 + (^2\phi_t)^* \sigma_t^2$$

So, an admissible hedging portfolio $(^2\phi_t = 0)$ is to be constrained to $\phi_t^* \sigma_t = (^1\phi_t)^* \sigma_t^1$, and the admissible wealth is modeled by

$$dV_t = r_t V_t \, dt + (^1\phi_t)^* \sigma_t^1 (dW_t + \theta_t dt).$$

This equation is unchanged if θ_t is replaced by the "minimal risk premium" θ_t^1 defined as the orthogonal projection of θ_t onto the range of $(\sigma_t^1)^*$ since $\theta_t - \theta_t^1$ belongs to the kernel of σ_t^1. Classical results from linear algebra allow us to give a closed formula for θ_t^1,

$$\theta_t^1 = (\sigma_t^1)^* [(\sigma_t^1)(\sigma_t^1)^*]^{-1} \sigma_t^1 \theta_t. \tag{1.8}$$

In what follows, the process θ^1 is supposed to be bounded. Given a square integrable contingent claim ξ, there does not still exist an admissible hedging portfolio which finances ξ; in other words, the BSDE

$$dX_t = r_t X_t + (^1\phi_t)^* \sigma_t^1 (dW_t + \theta_t dt) \qquad X_T = \xi,$$

can have no solution. So Föllmer and Schweitzer (1990) introduced the notion of self-financing in mean strategies (in short FS-strategies), $(V, ^1\phi, \Phi)$

$$dV_t = r_t V_t + (^1\phi_t)^* \sigma_t^1 (dW_t + \theta_t^1 dt) - d\Phi_t$$

where, by definition, the tracking error Φ is a square integrable martingale, orthogonal to the martingale $\{\int_0^t \sigma_s^1 dW_s, t \in [0,T]\}$, and prove the existence of a unique FS-strategy which finances a square integrable contingent claim ξ. Actually, the FS-strategy is simply given by the solution of a linear BSDE.

Theorem 1.4 *The FS-strategy is the hedging strategy (X, ψ) in a market with the "minimal risk premium" θ^1, that is*

$$dX_t = r_t X_t dt + \psi_t^* \sigma_t \theta_t^1 dt + \psi_t^* \sigma_t dW_t ; \quad X_T = \xi. \tag{1.9}$$

More precisely, let us denote by (q_t^1, q_t^2) the orthogonal decomposition of $(\sigma_t)^* \psi_t$ onto the range of $(\sigma_t^1)^*$ and denote by $^1\phi_t$ the vector process such that $(\sigma_t^1)^* (^1\phi_t) = q_t^1$. Then $\{X_t, ^1\phi_t, \Phi_t = \int_0^t (q_s^2)^* dW_s, t \in [0,T]\}$ is the unique FS-strategy associated with ξ.

PROOF: Let (X, ψ) be the no constrained hedging strategy, solution of (1.9). Project the vector $q_t = \sigma_t^* \psi_t$ orthogonally onto the range of $(\sigma_t^1)^*$, so

$$q_t = q_t^1 + q_t^2 \qquad \text{where } q_t^1 \in \text{Range}[(\sigma_t^1)^*] \qquad \text{and } q_t^2 \in Ker(\sigma_t^1).$$

We have an explicit formula for q^1 ,

$$\begin{aligned} q_t^1 &= (\sigma_t^1)^* [(\sigma_t^1)(\sigma_t^1)^*]^{-1} \sigma_t^1 [(\sigma_t^1)^* (\psi_t^1) + (\sigma_t^2)^* \psi_t^2] \\ &= (\sigma_t^1)^* (\psi_t^1 + [(\sigma_t^1)(\sigma_t^1)^*]^{-1} (\sigma_t^2)^* \psi_t^2) = (\sigma_t^1)^* (^1\phi_t) \end{aligned}$$

Since θ_t^1 is the orthogonal projection of θ_t onto the range of $(\sigma_t^1)^*$,

$$\psi_t^* \sigma_t \theta_t^1 = (q_t^1)^* \theta_t^1 = (q_t^1)^* \theta_t.$$

The martingale $\{\Phi_t = \int_0^t (q_t^2)^* dW_s, t \in [0, T]\}$ is orthogonal to $\int_0^t \sigma_s^1 dW_s$ and $(X, {}^1\phi, \Phi)$ is a FS-strategy.

Conversely, let $(X, {}^1\phi, \Phi)$ be a FS-strategy. Let q^2 be such that $\Phi_t = \int_0^t (q_t^2)^* dW_s$ and put $(\sigma_t)^* \psi_t = (\sigma_t^1)^* ({}^1\phi_t) + q_t^2$. Then (X, ψ) is solution of BSDE (1.9) and it follows from the uniqueness that the FS-strategy is unique.
□

Remark 1: The admissible portfolio is given by ${}^1\phi_t = ({}^1\psi_t) + [(\sigma_t^1)(\sigma_t^1)^*]^{-1} \cdot \sigma_t^1 (\sigma_t^2)^* \psi_t$. In particular, if the matrix $\sigma_t^1 (\sigma_t^2)^*$ is the null matrix, or in more financial point of view, if the no primary securities do not introduce supplementary risk on the admissible portfolios, then the FS-strategy consists to hold the amount ${}^1\phi_t = {}^1\psi_t$ in the primary securities. In general the FS-strategy does not depend on the matrix σ^2. In particular the simplest way to calculate the FS-strategy is to complete the " primary market "by introducing other securities whose volatility matrix satisfies $\sigma_t^1 (\sigma_t^2)^* = 0$.

Remark 2: Notice that θ^1 is the risk premium associated with the probability measure $\mathbb{Q}^1$ which is the minimal martingale-measure introduced by Föllmer and Schweitzer as a martingale measure such that any $\mathbb{P}$-local martingale, orthogonal to the martingale $\int_0^t \sigma_s^1 dW_s$ is a $\mathbb{Q}^1$-local martingale. So the price X_t of the claim associated with the FS-strategy is the conditional expectation of the discounted contingent claim computed under the minimal martingale measure.

2 Non-linear Arbitrage Pricing Theory and Backward Stochastic Differential Equations

We first give some examples of non-linear problems in Finance before stating and proving the main results on BSDEs. A more exhaustive exposition can be found in El Karoui, Peng and Quenez (1997).

2.1 Hedging in imperfect markets

Bid-ask spread for interest rates

We present an example of imperfect market, which takes into account a higher interest rate for borrowing than for lending. As Bergman (1991), Korn (1992) and Cvitanic and Karatzas (1993) we consider the following problem: the investor is allowed to borrow money at an interest rate $R_t > r_t$, the bond rate. Both processes R and r are supposed to be adapted and bounded. It is not reasonable to borrow money and to invest money in the bond at the same time. Therefore, we restrict to policies for which the amount borrowed at time t is equal to $(V_t - \sum_{i=1}^n \phi_t^i)^-$. Then the square integrable strategy (wealth, portfolio)

(V, ϕ) satisfies

$$dV_t = r_t V_t \, dt + \phi_t^* \sigma_t \theta_t dt + \phi_t^* \sigma_t dW_t - (R_t - r_t)(V_t - \sum_{i=1}^{n} \phi_t^i)^- \, dt. \qquad (2.1)$$

Given an inital investment $V_0 = x$ and a risky portfolio ϕ, there exists a unique solution to this forward stochastic differential equation with Lipschitz coefficients.

The **fair price** of a claim defined as the minimal endowment V which allows to finance an admissible strategy ϕ which guarantees ξ at time T, $(\xi = V_T)$ appears as the solution of the non-linear backward stochastic differential equation, where the non-linear term is depending on both processes, wealth and portfolio. The driver of this BSDE is given by

$$b(t, x, z) = -r_t x - z^* \theta_t + (R_t - r_t)(x - \mathbf{1}^*(\sigma_t^*)^{-1}z)^-. \qquad (2.2)$$

Short sales constraints

Similar equations appear in continuous trading with short sales constraints or transaction costs (Jouiny and Kallal (1992) and He and Pearson (1991), Jouiny (1996) this volume). Let $\theta^l - \theta^s$ be the difference in excess return between long and short positions in the stock. Then the present value corresponding to the strategy ϕ satisfies

$$dV_t = r_t V_t \, dt + [\phi_t^*]^- \sigma_t [\theta_t^l - \theta_t^s] dt + \phi_t^* \sigma_t (dW_t + \theta_t^l dt), \qquad (2.3)$$

In such constrained market, the fair price for a contingent claim ξ, is solution of the BSDE with driver

$$b(t, x, z) = -r_t x - z^* \theta_t^l - (z^-)^* [\theta_t^l - \theta_t^s].$$

We develop below (Subsection 2.3) a general pricing theory for contingent claims with respect to convex constrained portfolios.

Price pressure

We come back to the example of incomplete markets, where only the primary securities may be traded, and assume for simplicity that $\sigma_t^1 (\sigma_t^2)^*$ is the null matrix.

We penalize the portfolios which hold some shares of the no basic securities, i.e., such that $(^2\phi_t \neq 0)$. Recall that the assumption $(\sigma_t^1)^* \sigma_t^2 = 0$ implies that the projection $(Proj_t)$ of the vector $\sigma_t^* \phi_t$ on the kernel of σ_t^1 is $(\sigma_t^2)^* (^2\phi_t)$. Let us introduce the processes (X^k, ϕ^k) solutions of the following BDSEs,

$$
\begin{aligned}
dX_t^k &= r_t X_t^k \, dt - k||Proj_t(\sigma_t^* \phi_t^k)|| dt + (\phi_t^k)^* \sigma_t [dW_t + \theta_t^l dt] \\
&= r_t X_t^k \, dt + (^1\phi_t^k)^* \sigma_t^1 [dW_t + \theta_t^1 dt] - k||(\sigma_t^2)^* (^2\phi_t^k)|| dt + (^2\phi_t^k)^* \sigma_t^2 dW_t \\
X_T^k &= \xi
\end{aligned}
$$

These processes can be viewed as the wealth processes of superstrategies, that is strategies whose tracking error $(\Phi_t^k = \int_0^t k||(\sigma_s^2)^*(^2\phi_s^k)||ds - \int_0^t (^2\phi_s^k)^*\sigma_s^2 dW_s)$ is a submartingale. The penalizing process given by $k\int_0^t ||Proj_t(\sigma_t^*\phi_t^k)||dt$ has an intensity porportional to the lenght of the non admissible part of $(\sigma_t^*\phi_t)$. The more this lenght is large, the more the local "variance" of the non admissible martingale part $\int_0^t (^2\phi_s^k)^*\sigma_s^2 tdW_s$ is large and the more the penalty is expansive.

By the comparison theorem for BSDEs stated in Subsection 2.4 below, the sequence of the penalized processes X^k is increasing, and by using their interpretation as value functions of control problem, its limit is a super-strategy. We shall come back to this problem in Section 3 and show that the interpretation as value function of a control problem only results from the convexity of the function $Proj_t$.

2.2   An Existence and Uniqueness Result for BSDE's

We now establish the main properties of non-linear BSDE's. The first paper concerned with such equations is to our knowledge the paper of Bismut (1978), where he introduced a non-linear Riccati BSDE for which he showed existence and uniqueness of bounded solutions. Pardoux and Peng (1990) were the first who considered general BSDE's. They introduced the notion of an adapted solution of a backward stochastic differential equation on the probability space of Brownian motion as a pair of adapted processes (Y, Z) satisfying the BDSE and the terminal condition $Y_T = \xi$.

In this subsection, we first state some a priori estimates for the spread between the solutions of two BSDE's, from which we derive the results of existence and uniqueness. Secondly we give different properties concerning BSDE's, in particular a useful comparison theorem.

Notation and Definition

First fix some notation. On a probability space $(\Omega, \mathcal{F}, I\!\!P)$ equipped with the filtration $(\mathcal{F}_t)$ of an $I\!\!R^n$-valued Brownian motion W, we consider all kind of spaces of variable or processes:

- For $x \in I\!\!R^d$, $|x|$ will denote its Euclidian norm, and $\langle x, y \rangle$ the inner product. An element $y \in I\!\!R^{n \times d}$ will be considered as a $n \times d$ matrix; note that its Euclidean norm is given by: $|y| = \sqrt{trace(y\,y^*)}$; $\langle y, z \rangle = trace(yz^*)$.

- $I\!\!L_T^{2,d} = \{X \in I\!\!R^d;\ X \in \mathcal{F}_T\text{-measurable} ;\ ||X||^2 = I\!\!E(|X|^2) < +\infty\}$,

- $I\!\!H_T^{2,d} = \{\varphi;\ \varphi_t \in I\!\!R^d\ ,\ \text{progressively measurable},\ ||\varphi||^2 = I\!\!E \int_0^T |\varphi_t|^2 dt < +\infty\}$. Such processes are said to be square integrable (in short).

- For $\beta > 0$ and $\varphi \in I\!\!H_T^{2,d}$, $||\varphi||_\beta^2$ will denote $I\!\!E \int_0^T e^{\beta t}|\varphi_t|^2 dt$. $I\!\!H_\beta^{2,d}$ denotes the space $I\!\!H_T^2$ endowed with the norm $||.||_\beta$.

- $I\!\!H_T^{1,d} = \{\varphi;\ \varphi_t \in I\!\!R^d;\ \text{progressively measurable};\ I\!\!E\sqrt{\int_0^T |\varphi_t|^2 dt} < +\infty\}$.

Consider the BSDE

$$- dY_t = f(t, Y_t, Z_t)\, dt - Z_t^* dW_t \; ; \quad Y_T = \xi, \tag{2.4}$$

or equivalently

$$Y_t = \xi + \int_t^T f(s, Y_s, Z_s)\, ds - \int_t^T Z_s^* dW_s, \tag{2.5}$$

where:

- W is a n-dimensional Brownian motion

- $f : \Omega \times I\!R^+ \times I\!R^d \times I\!R^{n \times d} \longrightarrow I\!R^d$ is $\mathcal{P} \otimes \mathcal{B}^d \otimes \mathcal{B}^{n \times d}$-measurable.

- $\xi \in I\!L_T^{2,d}$ and $f(.,0,0) \in I\!H_T^{2,d}$.

- f is uniformly lipschitz, i.e., there exists $C > 0$ such that $dt \otimes dI\!Pa.s.$, for all (y_1, z_1, y_2, z_2),

$$|f(t, y_1, z_1) - f(t, y_2, z_2)| \leq C\,(|y_1 - y_2| + |z_1 - z_2|)$$

- Y and Z are $I\!R^d$ and $I\!R^{n \times d}$-valued progressively measurable processes and the process Y is continuous.

- f is called **the driver** of the BSDE and ξ the terminal value. If (f, ξ) satisfies the above assumptions, the pair (f, ξ) is said to be **standard** data for the BSDE.

Theorem 2.1 (Pardoux-Peng 1990) *Given standard data (f, ξ), there exists a unique pair $(Y, Z) \in I\!H_T^{2,d} \times I\!H_T^{2,n \times d}$ which solves equation(2.4)*

A proof can be found in Pardoux and Peng (1990). We give here a shorter direct proof, using useful a priori estimates for the solutions.

A priori estimates

Proposition 2.2 *Let $((f^i, \xi^i); i=1,2)$ be two BSDE's standard data and $((Y^i, Z^i); i=1,2)$ be the square integrable associated solutions. Let C be a Lipschitz constant for f^1, and put $\delta Y_t = Y_t^1 - Y_t^2$, $\delta Z_t = Z_t^1 - Z_t^2$ and $\delta_2 f_t = f^1(t, Y_t^2, Z_t^2) - f^2(t, Y_t^2, Z_t^2)$.*
For any $\beta > C(2 + C)$, the a priori estimates hold:

$$\| \delta Y \|_\beta^2 \leq T[e^{\beta T} I\!E(|\delta Y_T|^2) + \frac{1}{\beta - 2C - C^2} \| \delta_2 f \|_\beta^2] \tag{2.6}$$

$$\| \delta Z \|_\beta^2 \leq (2 + 2C^2 T)e^{\beta T} I\!E(|\delta Y_T|^2) + \frac{2 + 2C^2 T}{\beta - 2C - C^2} \| \delta_2 f \|_\beta^2. \tag{2.7}$$

Remark: By classical results on the norm of semimartingales, we prove similarly that

$$E[\sup_{t \le T} |\delta Y_t|^2] \le K_T[E(|\delta Y_T|^2 + \frac{1}{\beta - 2C - C^2} \parallel \delta_2 f \parallel_\beta^2],$$

where K_T is a positive constant only depending on T.

PROOF : Let $(Y, Z) \in I\!H_T^{2,d} \times I\!H_T^{2,n \times d}$ be a solution of (2.4). The equation (2.5) gives that

$$|Y_t| \le E[|\xi| + \int_0^T |f(s, Y_s, Z_s)| \, ds + \sup_t |\int_t^T Z_s^* dW_s| \, |\mathcal{F}_t].$$

It follows from Burkholder–Davis–Gundy inequalities (Karatzas and Shreve,1987, Th.3.28) that

$$E[\sup_t |\int_t^T Z_s^* dW_s|^2] \le 2E[|\int_0^T Z_s^* dW_s|^2]$$

$$+ 2E[\sup_t |\int_0^t Z_s^* dW_s|^2] \le 4E[\int_0^T |Z_s|^2 ds].$$

Now since (f, ξ) are standard data, $|\xi| + \int_0^T |f(s, Y_s, Z_s)| \, ds$ belongs to $I\!L_T^2(I\!R^+)$ and $\sup_{s \le T} |Y_s| \in I\!L_T^2(I\!R^+)$.

Now consider (Y^1, Z^1) and (Y^2, Z^2), two solutions associated with (f^1, ξ^1) and (f^2, ξ^2), respectively. From Itô's formula applied from $s = t$ to $s = T$ to the semimartingale $e^{\beta t} |\delta Y_t|^2$, it follows that:

$$e^{\beta t} |\delta Y_t|^2 = e^{\beta T} |\delta Y_T|^2 + 2 \int_t^T e^{\beta s} < \delta Y_s, f^1(s, Y_s^1, Z_s^1) - f^2(s, Y_s^2, Z_s^2) > ds$$

$$- \beta \int_t^T e^{\beta s} |\delta Y_s|^2 ds - \int_t^T e^{\beta s} |\delta Z_s|^2 ds - 2 \int_t^T e^{\beta s} < \delta Y_s, \delta Z_s^* dW_s > .$$

Since $\sup_{s \le T} |\delta Y_s|$ belongs to $L_T^{2,1}$, $\delta Z \, \delta Y$ belongs to $I\!H_T^{1,n}$ and the stochastic integral $\int_t^T e^{\beta s} < \delta Y_s, \delta Z_s^* dW_s >$ is $I\!P$-integrable, with zero expectation. Moreover, from the Lipschitz property of the driver f^1, it follows that,

$$|f^1(s, Y_s^1, Z_s^1) - f^2(s, Y_s^2, Z_s^2)| \le |f^1(s, Y_s^1, Z_s^1) - f^1(s, Y_s^2, Z_s^2)| + |\delta_2 f_s|$$

$$\le C[|\delta Y_s| + |\delta Z_s|] + |\delta_2 f_s|.$$

The quadratic form $Q(y, z) = -\beta |y|^2 + 2C|y|^2 + 2C|y||z| + 2|\delta_2 f_s||y| - |z|^2$, which appears in the right hand side of the above equality can be reduced to

$$Q(y, z) = -\beta |y|^2 + 2C|y|^2 + C^2|y|^2 + 2|\delta_2 f_s||y| - (|z| - C|y|)^2$$

$$= -\beta_C ((|y| - \beta_C^{-1} |\delta_2 f_s|)^2 - (|z| - C|y|)^2 + \beta_C^{-1} |\delta_2 f_s|^2$$

where $\beta_C := \beta - 2C - C^2$ is assumed to be strictly positive.

$$
\mathbb{E}[e^{\beta t}|\delta Y_t|^2] \;+\; \beta_C \int_t^T e^{\beta s}(|\delta Y_s| - \beta_C^{-1}|\delta_2 f_s|)^2 ds + \int_t^T e^{\beta s}|\delta Z_s - C\delta Y_s|^2\, ds]
$$

$$
\leq\; \mathbb{E}[e^{\beta T}|\delta Y_T|^2] + \mathbb{E}\int_t^T e^{\beta s}\frac{|\delta_2 f_s|^2}{\beta_C} ds.
$$

$$(2.8)$$

This inequality leads to the estimates of the β-norm of Y by two ways: first by integrating the right side between 0 and T, secondly by using the inequality $|\delta Y_s|^2 \leq 2[(|\delta Y_s| - \beta_C^{-1}|\delta_2 f_s|)^2 + \beta_C^{-2}|\delta_2 f_s|^2]$. The first way gives better estimates for short time T, the second for large T, and large β of course. The control of the norm of the process δZ follows by the above inequality (2.8). □

PROOF of Theorem 2.1: We use a fixed point theorem for the mapping from $\mathbb{H}_T^{2,d} \times \mathbb{H}_T^{2,n\times d}$ into itself, which maps (y, z) onto the solution (Y, Z) of the BSDE with driver $f(t, y_t, z_t)$, i.e.,

$$
Y_t = \xi + \int_t^T f(s, y_s, z_s)ds - \int_t^T Z_s^* dW_s\,.
$$

Let us remark that the assumption that (f, ξ) are standard parameters implies that $\{f(t, y_t, z_t); t \in [0, T]\}$ belongs to $\mathbb{H}_T^{2,d}$. The solution (Y, Z) is defined by considering the continuous version M of the square integrable martingale $\mathbb{E}[\int_0^T f(s, y_s, z_s)ds + \xi|\mathcal{F}_t]$. By the martingale representation theorem for the Brownian motion, (Karatzas and Shreve (1987) Th.4.15), there exists a unique integrable process $Z \in \mathbb{H}_T^{2,n\times d}$ such that $M_t = M_0 + \int_0^t Z_s^* dW_s$. Define the adapted and continuous process Y by $Y_t = M_t - \int_0^t f(s, y_s, z_s)ds$. Notice that Y is also given by

$$
Y_t = \mathbb{E}[\int_t^T f(s, y_s, z_s)ds + \xi|\mathcal{F}_t].
$$

The square integrability of Y follows from the above assumptions.

Let (y^1, z^1), (y^2, z^2) be two elements of $\mathbb{H}_T^{2,d} \times \mathbb{H}_T^{2,n\times d}$ and let (Y^1, Z^1) and (Y^2, Z^2) be the associated solutions. Proposition 2.2 applied with $C = 0$ gives

$$
\| \delta Y \|_\beta^2 \leq \frac{T}{\beta}\mathbb{E}\int_0^T e^{\beta s}|f(s, y_s^1, z_s^1) - f(s, y_s^2, z_s^2)|^2 ds
$$

and

$$
\| \delta Z \|_\beta^2 \leq \frac{2}{\beta}\mathbb{E}\int_0^T e^{\beta s}|f(s, y_s^1, z_s^1) - f(s, y_s^2, z_s^2)|^2 ds\,.
$$

Now since f is Lipschitz with constant C, we have

$$
\| \delta Y \|_\beta^2 + \| \delta Z \|_\beta^2 \leq \frac{2(2 + T)C^2}{\beta}[\| \delta y \|_\beta^2 + \| \delta z \|_\beta^2].
$$

$$(2.9)$$

Choosing β such that $2(2 + T)C^2 < \beta$, we see that this mapping is contracting from $(\mathbb{H}_T^{2,d} \otimes \mathbb{H}_T^{2,n\times d}, \| \cdot \|_\beta + \| \cdot \|_\beta)$ onto itself, and that there exists a fixed point, which is the unique continuous solution of the BSDE. □

From the proof of Proposition 2.2 (and more precisely from the estimate (2.9)), we derive the following useful corollary on the almost convergence of the Picard approximates.

Corollary 2.3 *Let β be such that $2(2+T)C^2 < \beta$. Let (Y^k, Z^k) be the Picard sequence defined recursively by $(Y_0 = 0; Z_0 = 0)$ and*

$$-dY_t^{k+1} = f(t, Y_t^k, Z_t^k)dt - (Z_t^{k+1})^* dW_t \; ; \; Y_T^{k+1} = \xi.$$

Then

$$\|Y^{k+1} - Y^k\|_\beta^2 + \|Z^{k+1} - Z^k\|_\beta^2 \le \epsilon^k K.$$

where $K = \|Y^1 - Y^0\|_\beta^2 + \|Z^1 - Z^0\|_\beta^2$ and $\epsilon = \dfrac{2(2+T)C^2}{\beta} < 1$.

Also, the sequence (Y^k, Z^k) converges to (Y, Z), $d\mathbb{P} \otimes dt$ a.s. and in $\mathbb{H}_T^{2,d} \otimes \mathbb{H}_T^{2,n \times d}$ as k goes to $+\infty$.

Remark: Again for Y it is possible to consider the norm $\|\sup_{s \in [0,T]} |Y_s^k - Y_s|\|_2^2$ instead of $\|Y\|_\beta^2$; consequently we also have that $\sup_{s \in [0,T]} |Y_s^k - Y_s|$ converges $\mathbb{P}$ a.s. to 0.

Extensions

A lot of recent working papers extend this result of existence and uniqueness under milder assumptions.

By introducing generalized BSDEs, it is possible to relax the assumption of Brownian filtration and continuous solution (Buckdahn 1993, El Karoui and Huang 1996...).

In the vector case, El Karoui and Huang (1996) show that the assumption of a driver with uniformly Lipschitz constant bounds may be relaxed but the existence and uniqueness result is stated under stronger integrability conditions for the solutions. Another main extension is obtained under the assumption that the driver is monotone with respect to y instead of being uniformly Lipschitz. More precisely, the driver is assumed to be Lipschitz with respect to z , with linear growth with respect to y; moreover, it is supposed that there exists C such that $dt \times d\mathbb{P}$, for any (z, y^1, y^2)

$$\langle f(t, y^1, z) - f(t, y^2, z), y^1 - y^2 \rangle \le C|y^1 - y^2|^2$$

Pardoux (1996) states that existence and uniqueness hold in the set of square integrable solutions.

In the one-dimensional case, the existence of a square integrable solution is stated in Lepeltier and San Martin (1996) under the only assumption of a continuous driver with linear growth. Uniqueness does not hold in general.

Extending the result of Bismut (1978) on the existence of a bounded solution for a Riccati BSDE, Barles and Kobylansky (1996) show that there exists a $\mathbb{H}_T^\infty \times \mathbb{H}_T^{2,d}$ solution for a continuous driver bounded with respect to y, and with quadratic growth with respect to z, and that uniqueness holds under the stronger assumption,

$$\partial_y f(t, y, z) \leq \beta + C|z|^2, \qquad |\partial_z f(t, y, z)| \leq \gamma + C|z|.$$

The results only hold in the one-dimensional case, because the authors need the comparison theorem that we state in the following subsubsection.

Linear BSDE

In the linear and one dimensional case, Theorem 2.1 specifies the integrability properties of the solution of the standard pricing problem.

Proposition 2.4 *Let (β, γ) be a bounded $(\mathbb{R}, \mathbb{R}^n)$-valued progressively measurable process, φ be an element of $\mathbb{H}_T^{2,1}$, and ξ be an element of $\mathbb{L}_T^{2,1}$. Then the LBSDE*

$$dY_t = [\varphi_t + Y_t \beta_t + Z_t^* \gamma_t]\,dt - Z_t^*\,dW_t\,; \qquad Y_T = \xi \qquad (2.10)$$

has a unique solution (Y, Z) in $\mathbb{H}_T^{2,1} \times \mathbb{H}_T^{2,n}$ given explicitly by:

$$\Gamma_t Y_t = E\left[\xi \Gamma_T + \int_t^T \Gamma_s \varphi_s ds\,|\mathcal{F}_t\right], \qquad (2.11)$$

where Γ_t is the adjoint process defined by the forward LSDE,

$$d\Gamma_t = \Gamma_t[\beta_t\,dt + \gamma_t^*\,dW_t]\,; \qquad \Gamma_0 = 1. \qquad (2.12)$$

In particular if ξ and φ are non-negative, the process (Y_t) is non-negative. If in addition $Y_0 = 0$, then for any t, $Y_t = 0$ a.s., $\xi = 0$ a.s. and $\varphi_t = 0$ $dt \otimes d\mathbb{P}$-a.s .

PROOF: By Theorem 2.1 there is a unique solution (Y, Z) of the BSDE. By standard calculations we see that $\Gamma_t Y_t + \int_0^t \Gamma_s \varphi_s ds$ is a local martingale. Now, $\sup_{s \leq T} |Y_s|$ and $\sup_{s \leq T} |\Gamma_s|$ belong to $\mathbb{L}_T^{2,1}$ and $\sup_{s \leq T} |Y_s| \times sup_{s \leq T}|\Gamma_s|$ belongs to $\mathbb{L}^{1,1}$. Therefore the local martingale $\{(\Gamma_t Y_t + \int_0^t \Gamma_s \varphi_s ds), t \in [0, T]\}$ is a uniformly integrable martingale, whose t-time value is the $\mathcal{F}_t$-conditional expectation of its terminal value. In particular, if ξ and φ are non-negative, Y_t is also non-negative. If in addition $Y_0 = 0$, the expectation of the non-negative variable $\xi \Gamma_T + \int_0^T \Gamma_s \varphi_s ds$ is equal to 0. So, $\xi = 0$, $\varphi_t = 0$ and $Y_t = 0$ a.s.$dt \otimes d\mathbb{P}$.
$\square$

Comparison theorem

As an immediate consequence of Proposition 2.4, we state in the one dimensional case a comparison theorem first obtained by Peng (1992a).

Theorem 2.5 (Comparison theorem) *Let (f^1, ξ^1) and (f^2, ξ^2) be two standard data of BSDE's, and let (Y^1, Z^1) and (Y^2, Z^2) be the associated solutions. We suppose that $\xi^1 \geq \xi^2$ $I\!\!P$ a.s., and $\delta_2 f_t = f^1(t, Y_t^2, Z_t^2) - f^2(t, Y_t^2, Z_t^2) \geq 0$ $dt \times dI\!\!P$ a.s.. Then we have $Y^1 \geq Y^2$ $I\!\!P$ a.s.. Moreover the comparison is strict, that is, on the event $\{Y_t^1 = Y_t^2\}$, we have $\xi^1 = \xi^2$, $f^1(s, Y_s^2, Z_s^2) = f^2(s, Y_s^2, Z_s^2)$, $ds \times dI\!\!P$ a.s. and $Y_s^1 = Y_s^2$, $t \leq s \leq T$ a.s.*

By applying the comparison theorem to $Y^2 = 0$, we deduce a sufficient condition for non-negativity :

Corollary 2.6 *If $\xi \geq 0$ a.s. and $f(t, 0, 0) \geq 0$ $dt \times dI\!\!P$ a.s., then $Y \geq 0$ a.s.*

PROOF: We use notation of Proposition 2.2. For sake of simplicity, we suppose that $n = 1$. The pair $(\delta Y, \delta Z)$ is the solution of the following LBSDE :

$$\begin{cases} -d\delta Y_t = [\Delta_y f^1(t)\delta Y_t + \Delta_z f^1(t)\delta Z_t + \delta_2 f_t]dt - \delta Z_t^* dW_t \\ \delta Y_T = \xi^1 - \xi^2, \end{cases} \qquad (2.13)$$

where $\Delta_y f^1(t) = \frac{f^1(t, Y_t^1, Z_t^1) - f^1(t, Y_t^2, Z_t^1)}{Y_t^1 - Y_t^2}$ if $Y_t^1 - Y_t^2 \neq 0$, whereas $\Delta_y f^1(t) = 0$ otherwise. Also, $\Delta_z f^1(t)$ is defined in the same manner.

Now, since by assumption the driver f^1 is uniformly Lipschitz with respect to (y, z), it follows that $\Delta_y f^1$ and $\Delta_z f^1$ are bounded processes. Also by assumption $\delta_2 f_t$ and δY_T are non-negative. Proposition 2.4 gives that the unique solution $(\delta Y, \delta Z)$ of the LBSDE (2.13) is non-negative and satisfies

$$\Gamma_t \delta Y_t = I\!\!E[(\xi^1 - \xi^2)\Gamma_T + \int_t^T \Gamma_s \delta_2 f_s ds | \mathcal{F}_t], \qquad (2.14)$$

where Γ is the positive adjoint process of the above LBSDE. $\square$

2.3   Non-linear European Option Pricing

A general setting of the wealth equation (which extends the examples of Subection 2.1) is

$$- dX_t = b(t, X_t, \sigma_t^* \phi_t)dt - \phi_t^* \sigma_t dW_t . \qquad (2.15)$$

Here, b is a real-valued process defined on $[0, T] \times \Omega \times I\!\!R \times I\!\!R^n$ satisfying the standard hypotheses of a driver. The classical case (Section 1) corresponds to a linear functional

$$b(t, x, z) = -r_t x - z^* \theta_t ,$$

where θ is the bounded risk premium vector, and r the bounded spot rate process. Notice that, since b is Lipschitz, given an initial investment x and a risky portfolio ϕ, there exists a unique wealth process of the forward equation 2.15 with initial value x.

Let us extend the definition of price systems given in Harrison and Kreps (1979) and Müller (1987),

Definition 2.7 *A non-linear price system* Ψ *maps a square integrable contingent claim* ξ *onto its continuous adapted price process* $\{\Psi_t(\xi), t \in [0, T]\}$ *such that,*

- *the price process* $\Psi(0)$ *for a null contingent claim is the null process,*

- *the price system is increasing, that is* $\{\xi^1 \geq \xi^2 \Rightarrow \Psi(\xi^1) \geq \Psi(\xi^2)\}$,

- *no arbitrage holds, i.e. if* $\xi^1 \geq \xi^2$ *and* $\Psi_t(\xi^1) = \Psi_t(\xi^2)$ *on a event* $A \in \mathcal{F}_t$, *then* $\xi^1 = \xi^2$ *on A a.s.*

- *Time-consistency : for any stopping time* $S \leq T$, $\Psi_t(\xi, T) = \Psi_t(\Psi_S(\xi, T), S)$

Furthemore, a price system Ψ *is admissible for the sellers if the price* $\Psi(\xi)$ *is sublinear with respect to* ξ.
A seller price system Ψ *is consistent with the basic securities prices, if* $\Psi_t(S_T^i) = S_t^i$ *a.s. at any time t.*

Consider an imperfect market where the admissible portfolios are constrained by the dynamics (2.15) and introduce the map $\Psi^b : \xi \mapsto X(\xi)$, where $X(\xi)$ is the solution of the BSDE (2.15).

Theorem 2.8 *Let b be a standard driver. The map* Ψ^b *defines a price system as soon as* $b(t, 0, 0) \equiv 0$ $dt \otimes d\mathbb{P}a.s.$ *The price system is admissible for sellers if* $b(t, x, z)$ *is sublinear, and consistent with the basic securities if for each* $i = 0, ..., n,$

$$-b(t, S_t^i, \sigma_t^i) = r_t S_t^i + \theta_t^* \sigma_t^i, \qquad dt \otimes d\mathbb{P}a.s.$$

PROOF: The comparison theorem gives the main properties, since b is a standard driver. The assumption $b(t, 0, 0) \geq 0$ implies that the price of a non-negative contingent claim is non-negative. Moreover, when the terminal condition is equal to 0, the null process is the unique solution of this equation if $b(t, 0, 0) \equiv 0$.
How to understand the arbitrage condition? if all the agents in the market agree that the portfolio dynamics are non-linear, no arbitrage opportunity is equivalent to the property $\xi^1 \geq \xi^2$, $X_t(\xi^1) = X_t(\xi^2) \Rightarrow \xi^1 = \xi^2$. But, by the comparison theorem, this property holds.
The time-consistency property is known in the vocabulary of the BSDEs as the flow property. By conventional notation, we define the price process, and the portfolio for $t \geq T$ by $(X_t = \xi, Z_t = 0)$. So if $T' \geq T$, then $(X_t, Z_t ; t \leq T')$ is the non-linear price portfolio strategy associated with the driver $f(t, y, z) 1_{\{t \leq T\}}$ and the claim ξ. Now let $S \leq T$ be a stopping time, and denote by $X(S, \xi_S)$ the price strategy associated with $(T, \xi_S, f(t, y, z) 1_{\{t \leq S\}})$. Both of the processes $(X_t(S, X_S), Z_t(S, X_S); t \in [0, T])$ and $(Y_{t \wedge S}(T, \xi), Z_t(T, \xi) 1_{[0,S]}(t); t \in [0, T])$ are solutions of the BSDE with parameters $(T, X_S, f(t, y, z) 1_{\{t \leq S\}})$. The uniqueness property for BSDE's gives that these processes are the same $d\mathbb{P} \otimes dt$ a.s.
A price system admissible for the sellers must be sublinear. Let us show that it is the case for the system Ψ^b if b is sublinear. Let (X^1, Z^1), (X^2, Z^2) and

(X, Z) be the price strategies respectively associated with contingent claims ξ^1, ξ^2, $\alpha\xi^1 + \beta\xi^2$, $(\alpha, \beta \geq 0)$. We have

$$-dX_t = b(t, X_t, Z_t)dt - Z_t^* dW_t \; ; \; \xi = \alpha\xi^1 + \beta\xi^2.$$

We have also that

$$
\begin{aligned}
-d(\alpha X_t^1 + \beta X_t^2) &= (b(t, \alpha X_t^1 + \beta X_t^2, \alpha Z_t^1 + \beta Z_t^2) + \Psi_t)dt - (\alpha Z_t^1 + \beta Z_t^2)^* dW_t \; ; \\
\xi &= \alpha\xi^1 + \beta\xi^2,
\end{aligned}
$$

where Ψ is a positive square integrable process. The comparison theorem gives that a.s.

$$\alpha X_t^1 + \beta X_t^2 \geq X_t, \; 0 \leq t \leq T$$

The consistency property with respect to the basic securities is a simple consequence of the uniqueness property.   □

What are the relations between such a price system and the linear price system in a classical market? The next section is devoted to solve this question

3   Convex price systems and upper-pricing with respect to fictitious markets

The aim of this section is to give a financial interpretation of a convex price system in terms of the classical valuation in fictitious markets, first introduced by Cvitanic and Karatzas (1993). More generally, under some mild conditions, there exists an one to one correspondence between the supremum of price systems and the supremum of the associated drivers. In other words,

$$X_t(\sup f^\alpha) = \operatorname{ess\,sup} X_t(f^\alpha).$$

From the description of a convex driver as supremum of a family of affine functions it follows that a price system is the supremum of affine price systems, that is as we have shown in Section 1, the supremum of prices over a family of standard markets.

3.1   Price systems and Optimization

A general result on the maximum or maxmin of price systems

Proposition 3.1 *Let* $\{(b, b^\alpha); \alpha \in \mathcal{A}\}$ *be a family of standard drivers and* $\{(\xi, \xi^\alpha); \alpha \in \mathcal{A}\}$, *be a family of square integrable contingent claims. Let* (X, Z) *and* (X^α, Z^α) *be the associated (price, portfolio)-strategies. Suppose that there exists a parameter* $\overline{\alpha} \in \mathcal{A}$ *such that*

$$
\begin{aligned}
b(t, X_t, Z_t) &= \operatorname*{ess\,sup}_\alpha b^\alpha(t, X_t, Z_t) = b^{\overline{\alpha}}(t, X_t, Z_t) && d\mathbb{P} \otimes dt \, a.s. \\
\xi &= \operatorname*{ess\,sup}_\alpha \xi^\alpha = \xi^{\overline{\alpha}} && \mathbb{P} \, a.s. && (3.1)
\end{aligned}
$$

Then the [7] price processes X and X^α satisfy:

$$X_t = \operatorname{ess\,sup}_\alpha X_t^\alpha = X_t^{\overline{\alpha}}, \quad \forall t \in [0, T], \ P \ a.s. \tag{3.2}$$

PROOF: (X, Z) and (X^α, Z^α) are solutions of two BDSE's whose drivers and terminal conditions satisfy the assumptions of the comparison theorem (2.5). Hence, for any α, $X_t \geq X_t^\alpha$ and consequently $X_t \geq \operatorname{ess\,sup} X_t^\alpha$ for any time t P a.s.

We now prove the equality using the uniqueness theorem for BSDE's and the existence of a parameter $\overline{\alpha}$ such that $b(t, X_t, Z_t) = b^{\overline{\alpha}}(t, X_t, Z_t)$ and $\xi = \xi^{\overline{\alpha}}$ a.s.. Hence (X, Z) and $(X^{\overline{\alpha}}, Z^{\overline{\alpha}})$ are both solutions of the same BSDE with parameters $(b^{\overline{\alpha}}, \xi^{\overline{\alpha}})$, the uniqueness property gives that they are the same. So,

$$\operatorname{ess\,sup} X_t^\alpha \geq X_t = X_t^{\overline{\alpha}} \geq \operatorname{ess\,sup} X_t^\alpha \quad \forall t \in [0, T], \ P \ a.s.$$

□

Corollary 3.2 *The same result holds if the drivers only satisfy:*
- *the drivers b^α are equi-Lipschitz with the same constant C,*
- *for each $\epsilon > 0$, there exists a parameter α^ϵ such that*

$$
\begin{aligned}
b(t, X_t, Z_t) &= \operatorname{ess\,sup} b^\alpha(t, X_t, Z_t) \leq f^{\alpha^\epsilon}(t, X_t, Z_t) + \epsilon, && dP \otimes dt \ a.s. \\
\xi &= \operatorname{ess\,sup}_\alpha \xi^\alpha \leq \xi^{\alpha^\epsilon} + \epsilon, && P \ a.s.. \tag{3.3}
\end{aligned}
$$

Similar results were extended to minimax problems in Hamadene and Lepeltier (1995) in connection with stochastic differential games. These techniques are also useful for solving optimization problems with recursive utilities and constrained portfolios (Quenez (1993), El Karoui, Peng and Quenez (1994)).

Corollary 3.3 *Let $(f, f^{\alpha, \beta})$, be a family of standard drivers, and $(\xi, \xi^{\alpha, \beta})$ a family of square-integrable contingent claims. We denote by (X, Z) and $(X^{\alpha, \beta}, Z^{\alpha, \beta})$ the associated (price, portfolios) strategies. Suppose that f and $f^{\alpha, \beta}$ (resp. ξ and $\xi^{\alpha, \beta}$) are linked by a minimax relation and that there exists a pair of parameters $(\overline{\alpha}, \overline{\beta})$ such that the following formulation of the Isaac condition holds:*

$$
\begin{aligned}
f(t, X_t, Z_t) &= \operatorname*{ess\,inf}_\alpha \sup_\beta f^{\alpha, \beta}(t, X_t, Z_t) = f^{\overline{\alpha}, \overline{\beta}}(t, X_t, Z_t), && dP \otimes dt \ a.s. \\
\xi &= \operatorname*{ess\,inf}_\alpha \sup_\beta \xi^{\alpha, \beta} = \xi^{\overline{\alpha}, \overline{\beta}}, && P \ a.s. \tag{3.4}
\end{aligned}
$$

---

[7]Dellacherie (1977) introduced the notion of esssup of processes in the following manner:
- a process U is said dominating the process U^α if $\{\omega; \exists t \in [0, T] U_t(\omega) > U_t^\alpha(\omega)\}$ is a P-null set.
- a process U is said to be the esssup U^α if for any α, U is greater than U^α, and if a process V lower than U^α for each α dominates U. Moreover, for right-continuous left-limited processes U^α, the esssup U^α exists and there exists a denumberable family (α_n) such that $U = \sup_n U^{\alpha_n}$.

Then the price systems X and $X^{\alpha,\beta}$ are also linked by a minimax relation with saddle point $(\overline{\alpha}, \overline{\beta})$, that is the Isaac condition is satisfied:

$$X_t = \operatorname*{ess\,inf}_{\alpha} \sup_{\beta} X_t^{\alpha,\beta} = X_t^{\overline{\alpha},\overline{\beta}} = \operatorname*{ess\,sup}_{\beta} \inf_{\alpha} X_t^{\alpha,\beta} \quad \forall t \in [0,T], \ I\!\!P \ a.s. \qquad (3.5)$$

PROOF: Use the fact that $(\overline{\alpha}, \overline{\beta})$ is a saddle point, i.e.,

$$\operatorname*{ess\,sup}_{\beta} f^{\overline{\alpha},\beta}(X_t, Z_t) \geq f(t, X_t, Z_t) = f^{\overline{\alpha},\overline{\beta}}(t, X_t, Z_t) \geq \operatorname*{ess\,inf}_{\alpha} f^{\alpha,\overline{\beta}}(t, X_t, Z_t).$$

The sames inequalities hold for the terminal conditions. As a consequence of the previous proposition, the same inequalities hold for the price processes:

$$\operatorname*{ess\,sup}_{\beta} X_t^{\overline{\alpha},\beta} \geq X_t = X_t^{\overline{\alpha},\overline{\beta}} \geq \operatorname*{ess\,inf}_{\alpha} X_t^{\alpha,\overline{\beta}} \quad \forall t \in [0,T], \ I\!\!P \ a.s.$$

These inequalities imply that the Isaac condition is satisfied for these processes.□

3.2 Upper-price with respect to fictitious markets and Convex price system

Upper-price with respect to a family of controlled markets

A lot of pricing problems in imperfect markets, such as those studied by Cvitanic and Karatzas (1993), may be restated as follows: the market parameters, short rate and risk premium processes belong to a family of controlled processes $d(t, u_t)$ and $n(t, u_t)$ such that the deflator processes satisfy to

$$dH_t^u = H_t^u[-d(t, u_t)dt - n(t, u_t)^* dW_t]; \quad H_0^u = 1, \qquad (3.6)$$

where $d(t, u)$ and $n(t, u)$ are progressively measurable processes uniformly bounded by δ and ν respectively. A feasible control $\{u_t, t \in [0, T]\}$ is a progressively measurable process valued in a (polish) space U. The set of feasible controls is denoted by $\mathcal{U}$. The market with parameters $d(t, u_t)$ and $n(t, u_t)$ is referred by $\mathcal{M}^u$. A cost process $k(t, u_t)$ is devoted to the market $\mathcal{M}^u$. We suppose that these progressively measurable cost processes are bounded by a progressively measurable square integrable process k.

The $\mathcal{M}^u$- price process $X^u(\xi)$ of a contingent claim ξ can be viewed as associated with the linear BSDE with driver

$$b^u(t, y, z) = k(t, u_t) - d(t, u_t).y - n(t, u_t)^* z.$$

Definition 3.4 *Given the markets family $\{\mathcal{M}^u, u \in \mathcal{U}\}$, and the cost process $\{k(t, u_t), 0 \leq t \leq T, u \in \mathcal{U}\}$, the upper-price at time t of the contingent claim ξ is defined as*

$$\overline{X}_t(\xi) = \operatorname*{ess\,sup}\{X_t^u(\xi) \mid u \in \mathcal{U}\}$$

The lower-price process is defined as

$$\underline{X}_t(\xi) = \operatorname*{ess\,inf}\{X_t^u(\xi) \mid u \in \mathcal{U}\}$$

The interval $[\underline{X}_t(\xi), \overline{X}_t(\xi)]$ is the arbitrage prices interval for the claim ξ w.r. to the markets family $\{\mathcal{M}^u, u \in \mathcal{U}\}$.

Proposition 3.5 *Under the previous assumptions, the drivers $(\overline{b}, \underline{b})$ defined by*

$$\overline{b}(t, y, z) = \text{esssup}\{b^u(t, y, z) \mid u \in \mathcal{U}\}, \qquad \underline{b}(t, y, z) = \text{essinf}\{b^u(t, y, z) \mid u \in \mathcal{U}\}$$

are standard drivers.

Let $X^{\overline{b}}$ be the $\overline{b}$-price process of ξ. Then, $X^{\overline{b}}$ is the upper-price of ξ with respect to the $\mathcal{M}^u$ markets. Similary, the $\underline{b}$-price process $X^{\underline{b}}$ is the lower-price of ξ with respect to the $\mathcal{M}^u$-markets.

PROOF: In order to show that $(\overline{b}, \underline{b})$ are standard drivers, we have to overcome some measurability questions. In the following, we only consider the driver $\overline{b}$. For given (ω, t), $\overline{b}(\omega, t, y, z) = \sup\{k(\omega, t, u) - d(\omega, t, u)y - n(\omega, t, u)^*z \mid u \in \mathcal{U}\}$ is a convex function with respect to (y, z), with bounded derivatives. By taking the supremum only over a denumberable dense family $\{(y_n, z_n)\}$ we define, for each n, a measurable process $\overline{b}(t, y_n, z_n)$ and a $d\mathbb{P} \otimes dt$-null set N such that for $(\omega, t) \in N^c, \overline{b}(\omega, t, y_n, z_n) = \sup\{k(\omega, t, u) - d(\omega, t, u)y_n - n(\omega, t, u)^*z_n \mid u \in \mathcal{U}\}$. For $(\omega, t) \in N^c$, $\overline{b}(\omega, t, y, z)$ is defined as the limit of the Cauchy sequence $b(\omega, t, y_n, z_n)$ as (y_n, z_n) goes to (y, z). So, the supremum of these linear drivers defines a standard driver $\overline{b}$.

In order to apply the previous results on the supremum of standard drivers, we will use the following lemma:

Lemma 3.6 *For each $\epsilon > 0$, there exists a feasible control u^ϵ such that*

$$\overline{b}(t, X_t^{\overline{b}}, Z_t^{\overline{b}}) = \text{ess sup}\, b^u(t, X_t^{\overline{b}}, Z_t^{\overline{b}}) \le b^{u^\epsilon}(t, X_t^{\overline{b}}, Z_t^{\overline{b}}) + \epsilon, \text{ a.s. } d\mathbb{P} \otimes dt \quad (3.7)$$

PROOF of Lemma 3.6: For each $(\omega, t) \in \Omega \times [0, T[$, the sets given by

$$\{u \in U, \overline{b}(t, X_t^{\overline{b}}(\omega), Z_t^{\overline{b}}(\omega)) \le k(t, \omega, u) - d(t, \omega, u).X_t^{\overline{b}}(\omega) - n(t, \omega, u)^*Z_t^{\overline{b}}(\omega) + \epsilon\}$$

are non-empty. Hence, by a measurable selection theorem (see for example, Dellacherie and Meyer(1980) or Benes (1971)), there exists a U-valued predictable processes u^ϵ such that

$$\overline{b}(t, X_t^{\overline{b}}, Z_t^{\overline{b}}) = \text{ess sup}\, b^u(t, X_t^{\overline{b}}, Z_t^{\overline{b}}) \le b^{u^\epsilon}(t, X_t^{\overline{b}}, Z_t^{\overline{b}}) + \epsilon, \text{ a.s. } d\mathbb{P} \otimes dt. \quad (3.8)$$

PROOF of the Theorem 3.5: Corollary 3.2 and the above lemma give directly the desired result. □

Corollary 3.7 (Optimality criterion) *Define a 0-optimal market $\mathcal{M}^{u^*}$ for the claim ξ, as the market such that the upper-price at time 0 is equal to the price at time 0 in the $\mathcal{M}^{u^*}$-market.*
A market $\mathcal{M}^{u^*}$ is 0-optimal if and only if

$$b(s, X_s^{\overline{b}}, Z_s^{\overline{b}}) = b^{u^*}(s, X_s^{\overline{b}}, Z_s^{\overline{b}}) \quad d\mathbb{P} \otimes ds \text{ a.s.} \quad (3.9)$$

In this case, u^* is also optimal for the problem starting at time t, that is, $X_t^{\overline{b}} = X_t^{u^*}$.

PROOF: It is an easy consequence of the second part of the comparison theorem
2.5.   □

Remark: From Proposition 3.5, for each feasible control u, $H_t^u X_t^{\overline{b}} +$
$\int_0^t H_s^u k(s, u_s) ds$ is a uniformly integrable supermartingale with increasing process given by $\int_0^t H_s^u K_s^u ds$ where

$$K_t^u = \overline{b}(t, X_t^{\overline{b}}, Z_t^{\overline{b}}) - b^u(t, X_t^{\overline{b}}, Z_t^{\overline{b}}).$$

Furthermore, the optimality criterion yields that u^0 is 0-optimal if and only if
$\xi = \xi^{u^0}$ and $K_t^{u^0} = 0$; in other words, $H_t^{u^0} X_t^{\overline{b}} + \int_0^t H_s^{u^0} k(s, u_s^0) ds$ is a martingale.
Consequently, these properties of the upper-price correspond to the classical
properties of the value function of a stochastic control problem (El Karoui (1981)
Theorem 3.2).

Notice that, in this example the driver $\overline{b}$ is convex. We shall see in the next
subsection that conversely, a convex price system is always associated with an
upper-price system given a family of fictitious markets, and a family of running
costs.

Convex Price system as upper-price

Here we fix some notation and recall a few properties of convex analysis (whose
proofs are, for example,in Ekeland and Turnbull, 1979) in order to show that a
convex driver is a supremum of linear drivers. Let $f(t, y, z)$ be a standard driver
of a BSDE, convex with respect to y, z and let $F(t, \beta, \gamma)$ be the polar process
associated with f,

$$F(\omega, t, \beta, \gamma) = \inf_{(y,z) \in \mathbb{R} \times \mathbb{R}^n} [f(\omega, t, y, z) + \beta y + \gamma^* z]. \qquad (3.10)$$

The *effective domain of F* is by definition

$$\mathcal{D}_F = \{(\omega, t, \beta, \gamma) \in \Omega \times [0, T] \times \mathbb{R} \times \mathbb{R}^n | \ F(\omega, t, \beta, \gamma) > -\infty\}.$$

Notice [8] that, since by assumption f is uniformly Lipschitz with Lipschitz cons-
tant C, the (ω, t)-section of $\mathcal{D}_F$ denoted by $\mathcal{D}_F^{(\omega, t)}$ is included in the bounded
domain $K = [-C, C]^{n+1}$ of $\mathbb{R} \times \mathbb{R}^n$. Since f is convex, f is continuous with
respect to (y, z), and (f, F) satisfies the conjugacy relation

$$f(\omega, t, y, z) = \sup\{F(\omega, t, \beta, \gamma) - \beta y - \gamma^* z \ | \ (\beta, \gamma) \in \mathcal{D}_F^{(\omega, t)}\}.$$

For every (ω, t, y, z) the maximum is achieved in this relation by a pair (β, γ)
which depends on (ω, t). [9]

---

[8] Indeed, if for example β satisfies $|\beta| > C$, then

$$f(\omega, t, y, z) - \beta y - \gamma^* z \leq +C|y| + f(\omega, t, 0, z) - \beta y - \gamma^* z.$$

Now, $\inf_{y \in \mathbb{R}}[C|y| - \beta y] = -\infty$. Hence, $(\beta, \gamma) \notin \mathcal{D}_F^{(\omega, t)}$.

[9] Indeed, for fixed (ω, t) there exists a sequence $(\beta^k, \gamma^k)_{k \in \mathbb{N}} \in \mathcal{D}_F^{(\omega, t)}$ such that

$$f(\omega, t, y, z) = \lim_{k \to +\infty} \{F(\omega, t, \beta^k, \gamma^k) - \beta^k y - (\gamma^k)^* z\}.$$

We want to associate with the polar process F a wide enough family of linear *standard* drivers $f^{\beta,\gamma}$ such that the assumptions of Theorem 5.10 hold. Let us denote

$$f^{\beta,\gamma}(t,y,z) = F(t,\beta_t,\gamma_t) - \beta_t\, y - \gamma_t^*\, z\,,$$

where (β,γ) are progressively measurable processes, called *control parameters*. Recall that by the conjugacy relation, f is also the supremum of $f^{\beta,\gamma}$. To ensure that $f^{\beta,\gamma}$ is a standard driver, it is sufficient to suppose that (β,γ) belongs to $\mathcal{A}$ defined by

$$\mathcal{A} = \{(\beta,\gamma) \in \mathcal{P}, \mathrm{K} - \text{valued} \mid I\!E \int_0^T F(t,\beta_t,\gamma_t)^2\, dt < +\infty\,\}.$$

$\mathcal{A}$ is said to be the set of *admissible control parameters*. Let (X,Z) be the unique solution of the BSDE with convex standard driver f and terminal value ξ. In order to apply Theorem 5.10, it remains to show the following Lemma (which is similar to Lemma 3.6):

Lemma: There exists an optimal control $(\overline{\beta},\overline{\gamma}) \in \mathcal{A}$, such that

$$f(t,X_t,Z_t) = f^{\overline{\beta},\overline{\gamma}}(t,X_t,Z_t) \qquad dI\!P \otimes dt \ a.s.$$

PROOF: Recall that for each (t,ω,y,z), the supremum in the conjugacy relation is achieved, since f is convex uniformly Lipschitz. Also, by a measurable selection theorem and since $f(.,X,Z)$, X and Z are progressively measurable processes, there exists a pair of progressively measurable (bounded) processes $(\overline{\beta},\overline{\gamma})$ such that

$$f(t,X_t,Z_t) = f^{\overline{\beta},\overline{\gamma}}(t,X_t,Z_t) \qquad dI\!P \otimes dt \ a.s.$$

Since by assumption $f(.,X,Z)$, Z and X are square integrable and $\overline{\beta},\overline{\gamma}$ are bounded, $F(.,\overline{\beta},\overline{\gamma})$ belongs also to $I\!H_T^{2,1}$. Hence, the pair $(\overline{\beta},\overline{\gamma})$ which achieves the infimum in the conjugacy relation belongs to $\mathcal{A}$. $\square$

Let us interpret a convex price system with driver f as associated with the upper-price with respect to the family of the dual markets $\mathcal{M}^{\beta,\gamma}$ where $(\beta,\gamma) \in \mathcal{A}$ are respectively the interest rate process and the risk premium process in this market. Given the running cost $F(t,\beta_t,\gamma_t)$, and a contingent claim ξ, the dual (β,γ) strategy $(X^{\beta,\gamma}, Z^{\beta,\gamma})$ is the unique solution of the LBSDE with data $(f^{\beta,\gamma},\xi)$. Thus, Theorem 5.10 gives directly the following result.

Proposition 3.8 *Let f be a convex standard driver and $\{\mathcal{M}^{\beta,\gamma}; (\beta,\gamma) \in \mathcal{A}\}$ be the dual markets associated with the running cost F. Then the f-price system is the upper-price system with respect to the $\mathcal{M}^{\beta,\gamma}$ family and to the running cost function F. Moreover, the f-price of a contingent claim ξ is the standard price*

Now since the sequence (β^k,γ^k) is bounded, there exists a subsequence still denoted by (β^k,γ^k) which converges in K to (β,γ). Also, (β,γ) achieves the maximum since F is continuous and

$$F(\omega,t,\beta,\gamma) - \beta y - \gamma^* z = \lim_{k\to+\infty}[F(\omega,t,\beta^k,\gamma^k) - \beta^k y - (\gamma^k)^* z] = f(\omega,t,y,z).$$

in an optimal fictitious market, associated with $(\overline{\beta}, \overline{\gamma})$, *where* $(\overline{\beta}, \overline{\gamma})$ *achieves the supremum in the conjugacy relation*

$$f(t, X_t, Z_t) = F(t, \overline{\beta}_t, \overline{\gamma}_t) - \overline{\beta}_t X_t - (\overline{\gamma})_t^* Z_t \qquad d\mathbb{P} \otimes dt \, a.s. \qquad (3.11)$$

The only difference with the non-constrained case is the fact that the optimal fictitious market depends on the claim to be priced (and also the introduction of a cost function).

Remark:
Suppose that the driver f has an affine part, that is f can be written

$$f(t, y, z) = \phi_t + b_t y + c_t^* z + g(t, y, z),$$

where ϕ is a square integrable adapted process and b, c are bounded adapted processes and where g is a convex driver. In this case, it is more simple to use the polar function of g and the associated conjugacy relation.

Example 1:
We come back to Example 1 seen in Subsection 2.1 (Bid-ask spread with interest rate) and solved by Cvitanic and Karatzas (1993) under slightly different assumptions. Here the matrix $(\sigma^*)^{-1}$ is suppsed to be bounded . The hedging strategy (wealth, portfolio) (X, ϕ) satisfies

$$dX_t = r_t X_t \, dt + \phi_t^* \sigma_t \theta_t \, dt + \phi_t^* \sigma_t dW_t - (R_t - r_t)(X_t - \sum_{i=1}^n \phi_t^i)^- \, dt \,; \quad X_T = \xi.$$
$$(3.12)$$

Like the other coefficients, the process R $(R_t \geq r_t)$ is supposed to be bounded. The driver b of this LBSDE is given by the convex process:

$$
\begin{aligned}
b(t, x, \sigma_t^* \phi) \qquad &= -r_t x - \phi^* \sigma_t \theta_t + (R_t - r_t)(x - \phi_t^* 1)^- \\
&= -r_t x - \phi^* \sigma_t \theta_t + (R_t - r_t) \sup_{u \in [0,1]}[(-x + \phi_t^* 1)u] \\
&= \sup_{\beta \in [r_t, R_t]}[-\beta x - [\sigma_t \theta_t + (r_t - \beta)1]^* \phi] \qquad (3.13)
\end{aligned}
$$

By Proposition 3.8, it follows that the unique solution $(X, \sigma^* \phi)$ of the BSDE (3.12) satisfies

$$X_t = \operatorname{ess\,sup}\{X_t^\beta \mid r_t \leq \beta_t \leq R_t\}$$

where

$$-dX_t^\beta = -\beta_t X_t^\beta - [\sigma_t \theta_t + (r_t - \beta_t)1]^* \phi_t^\beta \, dt - (\phi_t^\beta)^* \sigma_t dW_t \,; \quad X_T = \xi.$$

Example 2 : the large investor
In our formulation, the example of the non-linear portfolio dynamics-large investor given in Cvitanic's lectures gives an example of classical convex BSDE's. Recall that contrary to Cvitanic's modelisation, the amounts of the wealth invested in the different assets of the market are not supposed to be proportional to the wealth. More precisely, using our notation, for each i, ψ^i denotes the

amount of the wealth invested in asset i (and not the proportion). The prices of the assets are driven by

$$dS_t^0 = S_t^0 \left[r_t + f_0(X_t, \psi_t) \right] dt,$$

$$dS_t^i = S_t^i \left[(b_t^i + f_i(X_t, \psi_t)) \, dt + \sum_{j=1}^n \sigma_t^{i,j} \, dW_t^j \right].$$

Here, $(f_i : I\!\!R^+ \times I\!\!R^n \to I\!\!R, \ 0 \le i \le n)$ are some given functions which traduce the effect of the strategies chosen by the investor on the prices. The self-financing assumption gives that the dynamics of the total wealth X are given by

$$-dX_t = b(t, X_t, \sigma_t^* \psi_t) dt - \psi_t^* \sigma_t dW_t,$$

with $f = (f^1, f^2, ..., f^n)^*$, and

$$b(t, x, z) = -r_t x - (x - (\sigma_t^*)^{-1} z) f_0(x, (\sigma_t^*)^{-1} z) - z^* \sigma_t^{-1} [(b_t - r_t 1 + f(x, (\sigma_t^*)^{-1} z)].$$

Let ξ be a square integrable contingent claim. The problem is to price ξ in this context. Suppose that b is a standard driver, convex with respect to x, z. Then, the price process X for ξ corresponds to the solution of the BSDE associated with driver b and terminal condition ξ. Let $B(t, \beta, \gamma)$ be the polar function of $b(t, ., .)$. The previous results give directly that for each t,

$$X_t = \mathrm{esssup}\{X_t^{\beta,\gamma} \mid (\beta, \gamma) \in \mathcal{A}\},$$

where

$$\mathcal{A} = \{(\beta, \gamma) \in \mathcal{P} \mid I\!\!E \int_0^T B(t, \beta_t, \gamma_t)^2 \, dt < +\infty \}$$

and where, for each control parameter $(\beta, \gamma) \in \mathcal{A}$, the ex-post strategy $(X^{\beta,\gamma}, \psi^{\beta,\gamma})$ corresponds to the unique solution of the LBSDE,

$$\begin{cases} -dX_t^{\beta,\gamma} = (B(t, \beta_t, \gamma_t) - \beta_t X_t^{\beta,\gamma} - (\gamma_t)^* \sigma_t^* \psi_t^{\beta,\gamma}) dt - (\psi^{\beta,\gamma})_t^* \sigma_t dW_t \\ X_T^{\beta,\gamma} = \xi. \end{cases}$$

(3.14)

Note that Cvitanic's formulation does not correspond exactly to this case since the driver of the BSDE satisfied by X is not a standard driver.

Suppose now that ξ is a function of the underlying asset S. Then, the problem is more complex; it is considered below in the Markovian case in the subsection on Forward Backward equations for a large investor (Section 4).

Extension

The problems with constraints on the wealth (El Karoui, Kapoudjan, Pardoux and Peng (1995)) or on the portfolio (El Karoui and Quenez (1991,1995) and Cvitanic and Karatzas (1993)) may be formally formulated in the same way but with a driver which can be infinite, with non bounded effective domain. The constraints on the wealth yield to adopt the point of view of American options detailed in Section 5. Also, the case of an incomplete market and more generally, the case of the portfolio process ψ being constrained to take values in a convex set K (Cvitanic and Karatzas (1993)) corresponds formally to an upper-price $(X_t, 0 \leq t \leq T)$ solution of BSDE (2.4) with driver

$$b(t, x, \sigma_t^* \psi) = -r_t x - \psi^* \sigma_t \theta_t + \chi_K(\psi),$$

where $\chi_K(\psi)$ is the indicator function of K in the sense of convex analysis, namely equal to 0 if $\psi \in K$ and equal to ∞ otherwise. Notice that the example of an incomplete market corresponds to $K = \{\psi \in \mathbb{R}^n / \psi_k = 0, j \leq k \leq n\}$.

The variational formulation of the price remains almost the same as the one described above. However, the effective domain is not bounded; generally, the supremum is not attained and does not correspond to the solution of a classical BSDE. More precisely, the polar function B of χ_K is given explicitly by,

$$\begin{cases} B(\nu) & = -\infty \qquad \text{if } \nu \notin \tilde{K} \\ & = -\delta(\nu) \qquad \text{if } \nu \in \tilde{K}. \end{cases} \tag{3.15}$$

where $\delta(x) = \sup_{\psi \in K}(-\psi^* x)$ is the support function of the convex $-K$ and $\tilde{K} = \{x / \delta(x) < \infty\}$ is the effective domain. For each (t, ω), the domain of the concave function $B(\omega, .)$ is equal to $\tilde{K}$ and the conjuacy relation can be written

$$\chi_K(\psi) = \sup_{\nu \in \tilde{K}} \{-\delta(\nu) - \psi^* \nu\}.$$

Define now the right continuous left limited (RCLL) adapted process

$$\overline{X}_t = \operatorname*{ess\,sup}_{\nu \in \mathcal{D}} X_t^\nu, \tag{3.16}$$

where X^ν is the price of the claim ξ associated with ν, that is the solution of the BSDE

$$\begin{cases} -dX_t^\nu & = \left(-r_t X_t^\nu - (\psi_t^\nu)^* \sigma_t \theta_t - \delta(\nu_t) - (\psi_t^\nu)^* \nu_t\right) dt - (\psi_t^\nu)^* \sigma_t dW_t \\ X_T^\nu & = \xi. \end{cases} \tag{3.17}$$

and where $\mathcal{D}$ denotes the subset consisting of $\tilde{K}$-valued bounded processes ν such that $\mathbb{E} \int_0^T \delta(\nu(t))^2 dt < +\infty$.

Note that, in general, the domain $\mathcal{D}$ is not bounded and consequently that the supremum in (3.16) is not attained. The process $\overline{X}$ is no longer solution of a classical BSDE. Recall (see El Karoui and Quenez (1991,1995) for the incomplete case, and Cvitanic and Karatzas (1993)) that there exists a RCLL process (C_t) and a square integrable K-valued process $\hat{\psi}$ such that

$$d\overline{X}_t = r_t \overline{X}_t dt + \hat{\psi}_t^* \sigma_t \theta_t dt + \hat{\psi}_t^* \sigma_t dW_t - dC_t \, ; \, \overline{X}_T = \xi.$$

In other words, the strategy $(\overline{X}, \hat{\psi})$ is a superhedging strategy.

Furthermore, the process $\overline{X}$ is the smallest of the processes satisfying these properties; using the terminology of Cvitanic, $\overline{X}_0$ is equal to the hedging price at time 0 with K-constrained portfolios of the contingent claim ξ.

Nevertherless, in the case where this supremum is attained, then the hedging price $\overline{X}$ is solution of a classical BSDE; in the terminology of Cvitanic, the contingent claim ξ is K-hedgeable. More precisely, recall the following property (El Karoui and Quenez (1991,1995) and Cvitanic and Karatzas (1993)) expressed in terms of BSDE's

Proposition 3.9 *The following properties are equivalent:*
- λ *achieves the supremum in* $\overline{X}_0 = \sup_{\nu \in \mathcal{D}} X_0^\nu$.
- *There exists* $\lambda \in \mathcal{D}$ *such that*

$$0 = \sup_{\nu \in \mathcal{D}} \{-\delta(\nu_t) - \hat{\psi}_t^* \nu\} = -\delta(\lambda_t) - \hat{\psi}_t^* \lambda_t \qquad \mathbb{P} \times dt\, a.s.$$

- $(\overline{X}, \hat{\psi})$ *is solution of the following BSDE*

$$-d\overline{X}_t = -r_t \overline{X}_t dt - \hat{\psi}_t^* \sigma_t \theta_t dt - \hat{\psi}_t^* \sigma_t dW_t; \quad \overline{X}_T = \xi.$$

PROOF : The proof follows from a mild generalization of the comparison theorem.□

If the supremum is not attained, the hedging price $\overline{X}$ is no longer solution of a BSDE but can be written as the increasing limit of a sequence of "penalized" price processes $\overline{X}^n$ which are solutions of classical BSDEs. More precisely, $\overline{X}$ can be written

$$\overline{X}_t = \lim \uparrow \overline{X}_t^n,$$

where $\overline{X}^n$ is the RCLL process such that for each t,

$$\overline{X}_t^n = \operatorname{ess} \sup_{\nu \in \mathcal{D}/|\nu| \le n} X_t^\nu.$$

Let us introduce

$$d_n(\psi) = \sup_{\nu \in \tilde{K}/|\nu| \le n} \{-\delta(\nu) - \psi^* \nu\}.$$

With this notation, the penalized price $\overline{X}^n$ is solution of the BSDE

$$-d\overline{X}_t^n = -r_t \overline{X}_t^n dt - \psi_t^{n,*} v_t \theta_t dt + d_n(\psi_t^n) dt - \psi_t^{n,*} \sigma_t dW_t; \quad \overline{X}_T^n = \xi.$$

Notice that if the convex K is a cone, then $d_n(\psi) = nd(\psi)$ where

$$d(\psi) = \sup_{\nu \in \tilde{K}/|\nu| \le 1} \{-\delta(\nu) - \psi^* \nu\}.$$

Notice that this property has already been seen in the case of an incomplete market in Subsection 2.1 (Price pressure).

3.3 Recursive utility

Properties of recursive utilities

In the continuous-time, deterministic case, recursive utilities were first (to our knowledge) introduced by Epstein and Zin (1989). Let us consider a small agent who can consume between time 0 and time T. Let c_t be the (positive) consumption rate at time t. We suppose that there exists a terminal reward Y at time T. The utility at time t is a function of the instantaneous consumption rate c_t and of the future utility (corresponding to the future consumption). In fact, the recursive utility Y is assumed to satisfy the following differential equation,

$$- dY_t = f(c_t, Y_t)\, dt\,; \qquad Y_T = Y. \tag{3.18}$$

f is called **the driver**.

Thus at time 0 the utility of the consumption path $(c_t,\ 0 \le t \le T)$ is given by

$$Y_0 = Y + \int_0^T f(c_s, Y_s) ds.$$

Under uncertainty, Duffie and Epstein (1992a, 1992b) introduce the following class of recursive utilities:

$$- dY_t = \left[f(c_t, Y_t) - A(Y_t) \frac{1}{2} Z^* Z_t \right] dt - Z_t^* dW_t\,; \qquad Y_T = Y, \tag{3.19}$$

where A is the "variance multiplier". We can give another representation of the utility at time t of the future consumption $(c_s;\ t \le s \le T)$:

$$Y_t = E \left[Y + \int_t^T [f(c_s, Y_s) - A(Y_s) \frac{1}{2} Z_s^* \, Z_s] ds \, | \mathcal{F}_t \right],$$

Because of their economic motivations, we provide the following examples of recursive utilities, following Duffie and Epstein (1992a).

Examples

• Standard additive utility
The driver of the standard utility is given by:

$$f(c, y) = u(c) - \beta\, y,$$

The recursive utility is given by:

$$Y_t = E \left[Y\, e^{-\beta(T-t)} + \int_t^T u(c_s)\, e^{-\beta(s-t)} ds \, | \mathcal{F}_t \right].$$

• Uzawa utility
The driver has the same form as additive utility, but the discounting rate β is depending on the consumption rate c_t:

$$f(c, y) = u(c) - \beta(c)\, y.$$

• <u>Kreps-Porteus utility</u> Let $0 \neq \rho \leq 1, 0 \leq \beta$. The driver is defined by

$$f(c, y) = \frac{\beta}{\rho} \frac{c^\rho - y^\rho}{y^{\rho-1}}.$$

In general utilities must satisfy the following classical properties:
  • monotonicity with respect to the terminal value and to the consumption,
  • concavity with respect to the consumption,
  • time consistency: it means that, for any two consumption processes c^1 and c^2 and any time t, if c^1 and c^2 are identical up to time t, and if the continuation of c^1 is preferred to the continuation of c^2 at time t, then c^1 is preferred to c^2 at time 0.

Duffie and Epstein (1992a) show that if f is Lipschitz with respect to y, the following equation,

$$-dY_t = f(t, c_t, Y_t) \, dt - Z_t^* dW_t; \qquad Y_T = Y,$$

has a unique solution. Also, they state that if f is concave with respect to (c, y) and increasing with respect to c, the above properties are satisfied.

In fact, we may consider a more general class of recursive utilities, defined as associated with the solution of a general BSDE given by :

$$- dY_t = f(t, c_t, Y_t, Z_t) \, dt - Z_t^* dW_t; \qquad Y_T = Y. \qquad (3.20)$$

with concave driver f. In general, we suppose that the consumption process c belongs to $I\!H_T^{2,1}$ and that for such a c, $(f(., c, ., .), Y)$ are standard parameters of BSDE (in particular [10] $f(., c, 0, 0) \in I\!H_T^{2,1}$. The existence and uniqueness of solutions of (3.20) follow from Theorem 2.1. Also, as the properties of the price systems, the above properties are obtained as direct consequences of a comparison theorem. The main result is the interpretation of recursive utility as the value function of a control problem.

Variational formulation of the recursive utility

A natural assumption for a recursive utility is the concavity of the driver f with respect to (c, y, z). Consequently, by the results on concave BSDE's, the recursive utility can be written as the value function of a control problem.

Fix ξ a terminal reward which belongs to $I\!L_T^2(I\!R)$ and c a consumption-process in $I\!H_T^{2,1}$. Let $Y^{c,\xi}$ be the associated recursive utility (that is the solution of the BSDE associated with driver $f(t, c_t(.), ., .)$ and terminal value ξ. For a consumption rate c_t, let $F(t, c_t, ., .)$ be the polar function of $f(t, c_t, ., .)$,i.e.,

$$F(t, c_t, \beta, \gamma) = \sup_{(y,z) \in I\!R \times I\!R^n} [f(t, c_t, y, z) - \beta.y - \gamma.z].$$

---

[10] For example, it suffices that $|f(t, c, 0, 0)| \leq k_1 + k_2|c|$ *P.a.s.*

Let $\mathcal{A}(c)$ be the set of admissible processes (β, γ) such that $\mathbb{E} \int_0^T F(t, c_t, \beta_t, \gamma_t)^2 \, dt <$ $+\infty$. Then by the results stated in the previous subsections, the recursive utility can be written as

$$Y_t^{c,\xi} = \text{ess} \inf_{(\beta,\gamma) \in \mathcal{A}} \mathbb{E} \left[\int_t^T \Gamma_{t,s}^{\beta,\gamma} F(s, c_s, \beta_s, \gamma_s) ds + \Gamma_{t,T}^{\beta,\gamma} \xi \mid \mathcal{F}_t \right].$$

Hence, the recursive utility $Y^{c,\xi}$ can be defined through a felicity function F firstly introduced by Geoffard (1995) in the deterministic case. The felicity function $F(t, c, \beta, \gamma)$ at some current time t, expressed in terms of current time unit, is a function of current consumption c, current rate $-\beta$, and risk premium $-\gamma$. This function can be thought as an ex-post felicity when the agent knows the current rate and the risk-premium.

Notice that the adjoint processes $\Gamma^{\beta,\gamma}$ can be interpreted as a deflator (Duffie (1992), or Duffie and Skiadias (1991)). Also, the process $X_t^{\beta,\gamma}$ can be interpreted as an ex-post utility, when the deflator is given by $\Gamma_t^{\beta,\gamma}$. Hence, the utility is equal to the minimum of ex-post utilities over all price deflators. Ex-ante, the optimal deflator is the one that minimizes the agent's ex-post utility.

4   Forward-Backward Stochastic Differential Equations and Partial Differential Equations

4.1   Forward-Backward Stochastic Differential Equations

Notation and assumptions

In this subsection we consider the non-linear pricing problem in a market where the dynamics of some basic securities are modeled by some forward classical stochastic differential equations. For any given $(r, y), \in [0, T] \times \mathbb{R}^p$, consider the following classical Itô stochastic differential equation,

$$\begin{cases} dS_u = b(u, S_u)du + \sigma(u, S_u)dW_u , \ r \le u \le T, \\ S_u = y , \ 0 \le u \le r. \end{cases} \tag{4.1}$$

We then consider the associated BSDE,

$$- dX_u = f(u, S_u^{r,y}, X_u, Z_u)du - Z_u^* dW_u , \qquad X_T = \Psi(S_T^{r,y}). \tag{4.2}$$

The solutions will be denoted $(S_u^{r,y}, X_u^{r,y}, Z_u^{r,y}, 0 \le u \le T)$. The system (4.1,4.2) is called a forward-backward stochastic differential equation (FBSDE). Here, f (resp. Ψ) is a $\mathbb{R}^d$-valued Borel function defined on $[0, T] \times \mathbb{R}^p \times \mathbb{R}^d \times \mathbb{R}^{n \times d}$ (resp. on $\mathbb{R}^p$), and b (resp. σ) is a $\mathbb{R}^p$-valued (resp. $\mathbb{R}^{p \times n}$-valued) function defined on $[0, T] \times \mathbb{R}^p$. The function f is supposed to be uniformly Lipschitz with respect to (y, z) with Lipschitz constant C, that is, for any $(t, x, y_1, y_2, z_1, z_2)$

$$|f(t, x, y_1, z_1) - f(t, x, y_2, z_2)| \le C[|y_1 - y_2| + |z_1 - z_2|]. \tag{4.3}$$

Moreover, b, σ are supposed to be uniformly Lipschitz with respect to x. Finally, we will suppose that there exists a constant C such that for a real $p \geq 1/2$ and for each (u, x, y, z),

$$|\sigma(u, x)| + |b(u, x)| \leq C(1 + |x|); \qquad (4.4)$$

$$|f(u, x, y, z)| + |\Psi(x)| \leq C(1 + |x|^p). \qquad (4.5)$$

When these assumptions hold, it is well-known that the forward equation has a unique solution, which satisfies the Markov property,

$$S_t^{r,y} = S_t^{u, S_u^{r,y}}, \qquad t \geq u \geq r. \qquad (4.6)$$

Efficient Markets

In an efficient market, the knowledge of the t-prices of the basic securities is sufficient to give the price at time t of an European contingent claim. In a standard market, this well-known property is at the basis of the Black-Scholes formula. In terms of FBSDE's, that means that the solution $X_u^{r,y}$ of a linear BSDE is a deterministic function of $S_u^{r,y}$. We can see this from the Markov property of the underlying securities prices and the risk-neutral pricing rule. The price process is said to be **Markovian**. Furthermore, the portfolio process ϕ_u is also a deterministic function of $S_u^{r,y}$. This result can be obtained from Cinlar, Jacod, Protter and Sharpe (1980)'s study on the functional representation of the additive martingale of a diffusion process.

Lemma 4.1 *Let $\mathcal{B}_e$ be the filtration on $\mathbb{R}^p$ generated by the functions $\mathbb{E} \int_0^T \phi(u, S_u^{r,y}) du$ where ϕ is a continuous bounded $\mathbb{R}^p$-valued function. Then, for any f and Ψ, $\mathcal{B}_e$-measurable such that $\mathbb{E} \int_0^T f^2(u, S_u^{r,y}) du < +\infty$; $\mathbb{E}[\Psi^2(S_T^{r,y})] < +\infty$, the semimartingale $X_v^{r,y} = \mathbb{E}[\Psi(S_T^{r,y}) + \int_v^T f(u, (S_u^{r,y})) dr | \mathcal{F}_v]$ admits a continuous version given by $u(v, (S_v^{r,y}))$ where $u(r, y) = \mathbb{E}[\Psi(S_T^{r,y}) + \int_r^T f(v, S_v^{r,y}) dr]$ is $\mathcal{B}_e$-measurable.*

Moreover, $\int_r^s f(v, S_v^{r,y}) dv + X_v^{r,y}$ is an additive martingale which admits the following representation:

$$\int_r^s f(v, S_v^{r,y}) dv + X_s^{r,y} = u(r, y) + \int_r^s d(v, S_v^{r,y})^* \sigma(v, S_v^{r,y}) dW_v \; ; \; r \leq s \leq T$$

where $d(r, y)$ is $\mathcal{B}_e$-measurable.

By using the almost sure convergence of the Picard approximates of the solution of the BSDE, this result may be extended to general BSDEs.

Theorem 4.2 *There exist two deterministic functions $u(r, y)$ and $d(r, y)$ such that the non-linear hedging strategies of a contingent claim $\Psi(S_T^{r,y})$, $(X_v^{r,y}, \sigma^*(v, S_v^{r,y}) \phi_v^{r,y})$ as solution of the FBSDE (4.2) is given by*

$$X_v^{r,y} = u(v, S_v^{r,y}); \quad \phi_v^{r,y} = d(v, S_v^{r,y}), 0 \leq v \leq T \; dv \otimes d\mathbb{P} \; a.s.$$

Moreover, the function $u(t, y)$ is continuous.

PROOF: This result is shown by the iterative procedure used in the proof of the existence of the solution of a Backward Equation. Let $(X^{(r,y),k}, Z^{(r,y),k})$ be the sequence defined recursively by $(X^{(r,y),0} = 0,, Z^{(r,y),0} = 0)$ and

$$- dX_s^{k+1} = f(s, X_s^{r,y}, X_s^k, Z_s^k)ds - (Z_s^{k+1})^* dW_s \,; \ X_T^{k+1} = \Psi(S_T^{r,y}) \,. \qquad (4.7)$$

Corollary 2.3 gives that the sequence $(X_t^{(r,y),k}, r \leq t \leq T)$ converges uniformly with respect to t to $(X_t^{(r,y)}, r \leq t \leq T)$ and $Z^{(r,y),k}$ converges $dt \otimes d\mathbb{P}$ a.s. to $Z^{r,y}$ as $k \to +\infty$, where $(X^{r,y}, Z^{r,y})$ is the unique solution of the BSDE. We have also that $\sup_{s \in [t,T]} |X_s^{(r,y),k} - X_s^{(r,y)}|$ converges $\mathbb{P}$ a.s to 0. By applying Lemma 4.1, we derive by recursion the existence of some $\mathcal{B}_e$-measurable functions (u_k, d_k) such that $(X_v^{(r,y),k} = u_k(v, S_v^{r,y}) \,; \ Z_v^{(r,y),k} = \sigma(v, S_v^{r,y})^* d_k(v, S_v^{r,y}))$ converge to $(X_v^{(r,y)}, Z_v^{(r,y)})$.
Put

$$u^i(r, y) = \limsup_{k \to +\infty} u_k^i(r, y) \,; \ d^{i,j}(r, y) = \limsup_{k \to +\infty} d_k^{i,j}(r, y);$$

where $u = (u^i)_{1 \leq i \leq d}$ and $d = (d^{i,j})_{1 \leq i \leq p, 1 \leq j \leq d}$. Notice that from the a.s.convergence of the sequence $(X^{(r,y),k}, Z^{(r,y),k})$ to $(X^{r,y}, Z^{r,y})$, it follows that $\mathbb{P}$ a.s., $\forall v \in [t, T]$,

$$u^i(v, S_v^{r,y}) = (\limsup_{k \to \infty} u_k^i)(v, S_v^{r,y}) = \limsup_{k \to \infty}(u_k^i(v, S_v^{r,y})) = \lim_{k \to \infty} X_v^{i,(r,y),k} = X_v^{i,(r,y)}.$$

The same properties hold for d and we derive that $d(v, S_v^{r,y}) = Z_v^{(r,y)} \ d\mathbb{P} \otimes dv$ a.s.. This completes the proof. $\square$

Non-linear Partial Differential Equations

From the Black-Scholes formula, the link between prices and partial differential equations (PDE's) is useful for the aim of effective calculation. A similar relation holds for imperfect markets.

First we give a generalization of the Feynman Kac formula first stated by Pardoux and Peng (1992). Then we recall that, conversely, under some smoothness assumptions, the function $u(t, x) = Y_t^{t,x}$ is solution in some sense of a PDE.

Proposition 4.3 (Generalization of the Feynman Kac formula) *Let w be a d-valued function of class $\mathcal{C}^{1,2}$ (or smooth enough to be able to apply Itô's formula to $w(v, S_v^{r,y})$) and suppose that there exists a constant C such that, for each (r, y),*

$$|w(r, y)| + |\partial_y w(r, y)\sigma(r, y)| \leq C(1 + |y|)$$

Also, w is supposed to be the solution of the following system of quasilinear parabolic partial differential equation:

$$\left\{ \begin{array}{r} \partial_r w(r, y) + \mathcal{L}w(r, y) + f(r, y, w(r, y), \partial_y w(r, y)\sigma(r, y)) = 0 \\ w(T, y) = \Psi(y) \,. \end{array} \right. \qquad (4.8)$$

where $\partial_y w$ is the gradiant of w and $\mathcal{L}_{(r,y)}$ denotes the following second order differential operator:

$$\mathcal{L}_{(r,y)} = \sum_{i,j} a_{ij}(r,y)\partial^2_{y_i y_j} + \sum_i b_i(r,y)\partial_{y_i}; \quad a_{ij} = \frac{1}{2}[\sigma\sigma^*]_{ij}. \quad (4.9)$$

Then, $w(r,y) = X_r^{(r,y)}$, where $\{(X_s^{r,y}, Z_s^{r,y}), r \le s \le T)\}$ is the price process and the hedging portfolio of the claim. Also,

$$\{(X_u^{r,y}, Z_u^{r,y}) = (w(u, S_u), \partial_y v(u, S_u)\sigma(u, S_u)), r \le u \le T\}.$$

PROOF : By applying Itô's formula to $w(u, S_u^{r,y})$, we obtain that,

$$d\,w(u, S_u) = (\partial_t w(u, S_u) + \mathcal{L}w(u, S_u))\,du + \partial_x w(u, S_u)\sigma(u, S_u)dW_u.$$

Since w solves (4.8), it follows that

$$-d\,w(u, S_u) = f(u, S_u, w(u, S_u), \partial_y w(u, S_u)\sigma(u, S_u))\,du - \partial_y w(u, S_u)\sigma(u, S_u)\,dW_u$$
$$w(T, S_T) = \Psi(S_T).$$

So $\{(w(u, S_u), \partial_y w(u, S_u)\sigma(u, S_u)); u \in [0,T]\}$ is the unique solution of BSDE (4.2), and the proof is ended. □

Now we show that conversely, under some additional assumptions, the solution of the BSDE (4.2) corresponds to the solution of the PDE (4.8). If $d = 1$, we can use the comparison theorem to show that if b, σ, f, Ψ satisfy the assumptions at the beginning of the section and if f, Ψ are supposed to be uniformly continuous with respect to x, then $u(t,x)$ is a viscosity solution of the above PDE (4.8) (see Peng (1992b) and Pardoux and Peng (1992)).

Theorem 4.4 *We suppose that $d = 1$ and that f, Ψ are uniformly continuous with respect to (r,y). Then, the function u defined by $u(r,y) = X_r^{r,y}$ is a viscosity solution[11] of PDE (4.8).*

Before giving the proof, recall the definition of a viscosity solution (Fleming and Soner (1993)).

Definition 4.5 *Suppose $u \in \mathcal{C}([0,T] \times \mathbb{R}^p)$ satisfies $u(T,x) = \Psi(x)$, $x \in \mathbb{R}^p$. Then u is called a viscosity subsolution (resp. supersolution) of PDE (4.8) if for each $(r,y) \in [0,T] \times \mathbb{R}^p$ and $\phi \in \mathcal{C}^{1,2}([0,T] \times \mathbb{R}^p)$ such that $\phi(r,y) = u(r,y)$ and (r,y) is a minimum (resp. maximum) of $\phi - u$,*

$$\partial_r \phi(r,y) + \mathcal{L}\phi(r,y) + f(r,y,\phi(r,y),\sigma(r,y)^*\partial_y\phi(r,y)) \ge 0.$$

---

[11]Furthermore, if we suppose that, in addition, for each $R > 0$, there exists a continuous function $m_R : \mathbf{R}_+ \to \mathbf{R}_+$ such that $m_R(0) = 0$ and

$$\left| f(t,x,y,z) - f(t,x',y,z) \right| \le m_R(|x - x'|(1 + |z|)), \quad (4.10)$$

for all $t \in [0,T]$, $|x|, |x'| \le R$, $|z| \le R$, $z \in \mathbf{R}^n$, then u is the unique viscosity solution of PDE (4.8).

(resp. $\partial_t \phi(r,y) + \mathcal{L}\phi(r,y) + f(r,y,\phi(r,y),\sigma(r,y)^*\partial_y\phi(r,y)) \leq 0$).

Moreover, u is called a viscosity solution of PDE (4.8) if it is both a viscosity sub-and-super-solution of PDE (4.8).

PROOF: Recall that u denotes the function introduced in Theorem 4.2, $u(S_t^{r,y},t) = X_t^{r,y}$. Fix $(r,y) \in [0,T] \times \mathbb{R}^+$ and for $\delta > 0$, introduce the stopping time $\tau = (r+\delta) \wedge \inf\{u \geq T; |S_u^{r,y} - y| \leq \varepsilon\}$. Put $\eta = S_T^{(r,y)}$. Then by the time consistency property of the price system $u(r,y) = X_r^{\tau,u(\tau,\eta)}$.

Let ϕ be a regular function such that $\phi(t,x) \geq u(t,x)$, $\forall(t,x) \in [r,T] \times \mathbb{R}^p$, and $\phi(r,y) = u(r,y)$. From the comparison theorem $X^{\tau,u(\tau,\eta)} \leq X^{\tau,\phi(\tau,\eta)}$ on $[r,\tau]$. In particular,

$$u(r,y) = X_r^{\tau,u(\tau,\eta)} \leq X_r^{\tau,\phi(\tau,\eta)}$$

Moreover, $(\phi(t,S_t^{r,y}),\sigma(t,S_t^{r,y})^*\partial_y\phi(t,S_t^{r,y}))$ is the solution of a BDSE with terminal value at time τ, $\phi(\tau,\eta)$ and driver given by $-\partial_t\phi(t,S_t^{r,y}) - \mathcal{L}\phi(t,S_t^{r,y})$.

Let us suppose that, at the fixed point (r,y),

$$F(r,y) = \partial_r\phi(r,y) + \mathcal{L}\phi(r,y) + f(r,y,\phi(r,y),\sigma(r,y)^*\partial_y\phi(r,y)) < 0.$$

By the continuity of the function F with respect to (r,y), there exists δ such that for all $(t,x) \in [r,r+\delta] \times \{|x-y| \leq \delta\}$, $F(t,x) \leq -\nu_\delta < 0$.

The driver $-\partial_t\phi(t,S_t^{r,y}) - \mathcal{L}\phi(t,S_t^{r,y})$ is strictly bigger than $f(t,S_t^{r,y},\phi(t,S_t^{r,y}),\sigma(t,S_t^{r,y})^*\partial_y\phi(t,S_t^{r,y}))$, when $t \in [r,\tau]$. So, by the comparison theorem, $\phi(t,S_t^{r,y}) > X_t^{\tau,\phi(\tau,\eta)} \geq X_t^{\tau,u(\tau,\eta)}$ for any $t \in [r,r+\delta]$. In particular, $\phi(r,y) > u(r,y) = X_r^{\tau,u(\tau,\eta)}$. The contradiction is complete.

Forward-Backward Equation for a large investor

Cvitanic presents in Section 8 of his paper an example of a more general forward-backward equation for a large investor whose portfolio strategy modifies the price processes of the underlying assets. In particular, to price a contingent claim which only depends on the value of the underlying assets, contrary to the example in Section 3, it is necessary to solve the forward-backward system, introduced by Ma, Protter, and Yong (1994), that is, for $r \leq u \leq T$,

$$
\begin{aligned}
dS_u &= b(u,S_u,X_u,Z_u)du + \sigma(u,S_u,X_u)dW_u, \\
-dX_u &= f(u,S_u,X_u,Z_u)du - Z_u^*dW_u, \\
S_r &= y, \qquad X_T = \Psi(S_T^{r,y}).
\end{aligned}
\tag{4.11}
$$

Such equations are first studied by Antonelli (1993), when the coefficients do not depend on the portfolio. In particular, he gives a very illuminating counter-example about the existence of a solution. To solve this system, Ma, Protter, and Yong (1994) use a Four Step Scheme, by using a general result on the non-linear PDE's. Recently, Hu and Peng (1995) obtain existence and uniqueness results under different assumptions on the coefficients, with a possibly degenerate price processes.

The four Step Scheme is the following: by analysis results, under the assumptions that the matrix σ is nondegenerate and that the coefficients are bounded and smooth, there exists a classical solution $\theta(u, y)$ of the following non-linear PDE:

$$\begin{cases} (\partial_t \theta + \mathcal{L}_u)\theta(u,y) & + \quad f(u, y, \theta(u,y), \sigma(u, y, \theta(u,y))^* \partial_y \theta(u,y)) = 0 \\ \theta(T, y) & = \quad \Psi(y), \end{cases}$$

$$(4.12)$$

where $\partial_y \theta$ is the gradiant of θ and $\mathcal{L}_{(u,y)}$ denotes the following second order differential operator:

$$\begin{aligned} \mathcal{L}_u \theta(u, y) &= \sum_{i,j} a_{ij}(u, y, \theta(u, y)) \partial^2_{y_i y_j} \theta(u, y) \\ &+ \sum_i b_i(u, y, \theta(u, y)), \; \sigma(u, y, \theta(u, y))^* \partial_y \theta(u, y)) \partial_{y_i} \theta(u, y); \\ a_{ij} &= \tfrac{1}{2} [\sigma \sigma^*]_{ij} \, . \end{aligned}$$

Using this solution, a verification method gives the existence and the uniqueness of the system of Forward-Backward equations. Let us precise the different assumptions:

• The functions b, f, σ and Ψ are smooth (C^∞) taking values in $I\!\!R^n$, $I\!\!R^d$, $I\!\!R^{n \times n}$, $I\!\!R^d$ respectively and with first order derivatives in y, x, z being uniformly bounded.

• There exists a function μ and constants $C > 0$ and $\alpha \in]0,1[$ such that Ψ is bounded and in $C^{2+\alpha}(I\!\!R^d)$ and for all $(u, y, x, z) \in [0, T] \times I\!\!R^n \times I\!\!R^d \times I\!\!R^{n \times d}$,

$$|\sigma(u, y, x)| + |f(u, y, 0, z)| \leq C; \qquad |b(t, y, x, 0)| \leq \mu(|x|).$$

• The functions $a_{ij}(u, y, x)$, $b_i(u, y, x)$, $f(u, y, x, z)$ are smooth (C^∞) and for all (u, y, x),

$$\nu(|x|)I \leq (a_{ij}(u, y, x)) \leq \mu(|x|)I \, ; \qquad |b(u, y, x, z)| \leq \mu(|y|)(1 + |z|).$$

for some continuous functions $\mu(.)$ and $\nu(.)$ with $\nu(r) > 0$;

$$|f(u, y, x, z)| \leq [\epsilon(|x|) + P(|z|, |x|)](1 + |z|^2)$$

where $P(|z|, |x|) \to 0$ as $|z| \to \infty$ and $\epsilon(|x|)$ is small enough. Furthermore,

$$f(u, y, x, z)^* y \leq L(1 + |x|^2).$$

for some constant $L > 0$.

Proposition 4.6 *Under the above assumptions, PDE (4.12) admits a unique classical solution $\theta(u, y)$ which is bounded and $\theta_u, \theta_y, \theta_{yy}$ are bounded as well.*

SKETCH OF THE PROOF: By some analysis results, the authors know that there exists a unique classical solution of this PDE on a bounded domain with boundary conditions. The solution of PDE (4.12) on the whole space is obtained by extension of these solutions.$\square$

Theorem 4.7 *The pricing problem for the large investor (4.11) admits a unique solution* (S, X, Z), *where the price* X *and the portfolio* Z *of the claim can be computed from the price* S *of the underlying assets by by the relations:*

$$X_u = \theta(u, S_u)\,;\; Z_u = \sigma(u, S_u, \theta(u, S_u))^* \partial_y \theta(u, S_u) \qquad (4.13)$$

The modified price process S *for the underlying assets is given by the unique solution of the forward equation,*

$$\begin{cases} dS_u = b\left(u, S_u,\, \theta(u, S_u),\, \sigma(u, S_u, \theta(u, S_u))^* \partial_y \theta(u, S_u)\right) du \\ \qquad + \partial_y \theta(u, S_u)^* \sigma(u, S_u, \theta(u, S_u)) dW_u \;,\; 0 \le u \le T \\ S_0 = y \end{cases} \qquad (4.14)$$

SKETCH OF THE PROOF: Let us prove the existence. Let $(S_u, 0 \le u \le T)$ be the solution of (4.14) and $(X_u, Z_u, 0 \le u \le T)$ be the processes defined by formula (4.13). By applying Itô's formula to $\theta(u, S_u)$ and by using the fact that θ is solution of PDE (4.12), it follows that (S, X, Z) satisfies the Forward-Backward System (4.11).

It remains to show the uniqueness. Let (S, X, Z) be a solution of the system (4.11), and define

$$\tilde{X}_u = \theta(u, S_u)\,;\; \tilde{Z}_u = \sigma(u, S_u, \theta(u, S_u))^* \partial_y \theta(u, S_u).$$

By applying Itô's formula, it is possible to compute $\mathbb{E}|X_u - \tilde{X}_u|^2$ and by using the same algebraic arguments as in the classical case, to conclude that it is equal to zero. The same calculus gives that $\mathbb{E}\int_0^T |Z_u - \tilde{Z}_u|^2 du = 0$. It follows that $X_u = \tilde{X}_u$, $0 \le u \le T$, $\mathbb{P}$ a.s. and $Z_u = \tilde{Z}_u$ $d\mathbb{P} \times du$ a.s. Hence, we have showed that X and Z can be computed from S by formula (4.13), and consequently the process S satisfies the forward equation (4.14). The uniqueness follows easily. $\square$

Note that Ma, Protter and Yong consider a more general system, with a constraint on the Brownian motion coefficient of the BSDE.

$$\begin{cases} dS_u = b(u, S_u, X_u, Z_u) ds + \sigma(u, S_u, X_u) dW_u \;,\; 0 \le u \le T \\ S_0 = y \end{cases} \qquad (4.15)$$

$$\begin{cases} -dX_u = f(u, S_u, X_u, Z_u) du - \tilde{\sigma}(u, S_u, X_u, Z_u) dW_u \\ X_T = \Psi(S_T). \end{cases} \qquad (4.16)$$

In this case, the system can be solved using a supplementary step which consists to compute Z from $\tilde{\sigma}$. Under suitable assumptions on $\tilde{\sigma}$ (surjectivity of $\tilde{\sigma}$ with respect to z and invertibility of its Jacobian matrix), the following equation

$$p\sigma(u, y, x) + \tilde{\sigma}(u, y, x, z) = 0 \qquad (4.17)$$

has a unique smooth solution $z = z(u, y, x, p)$ for each u, y, x, p. Note that in the previous case where $\tilde{\sigma}z = z^*$, z is simply given by $z = \sigma(u, y, \theta(u, y))^* p$ (Note that p corresponds to $\partial_y \theta$). Solving equation (4.17) corresponds to the first step of the "four step scheme" of Ma, Protter and Yong. The three other steps remain the same except that in equations (4.12), (4.13) and (4.14), the function $\sigma(u, y, \theta(u, y))\partial_y \theta(u, y)$ has to be replaced by $z(u, y, \theta(u, y), \partial_y \theta(u, y))$. The method of Ma, Protter and Yong is used also in "Black's Consol Rate Conjecture" by Duffie, Ma and Yong (1994).

5 Non-linear Pricing of American Contingent Claims

We have seen in the previous sections that the pricing of European contingent claims, even in imperfect markets, can be formulated in terms of backward stochastic differential equations. However, the case of American options has not been considered. Recall that an American option is a contract which gives the right to the holder to exercise the option at any time before the maturity. To hedge the additional risk of early exercise, it is necessary to introduce superstrategies, with a value still greater than the payoff of the option. In a complete market, it is well-known that the price of an American option is related to the value function of an optimal stopping problem. In terms of backward equations, the price of an American option corresponds to the solution of a new type of backward equations called reflected BSDE's. The solution of such an equation is forced to stay above a given stochastic process, called the obstacle. An increasing process is introduced which pushes the solution upwards, so that it remains above the obstacle. Uniqueness and existence can both be proved by a fixed point argument, as in the classical case. We give here a complete proof; the results on the optimal stopping problem are included.

In the case of an American option, the obstacle is the payoff of the option. The solution of the RBSDE may be interpreted as the maximum price to finance the options whose maturity is a stopping time. When the driver of the equation is convex, we extend to the pricing of American options the interpretation as superprice with respect to fictitious markets given for the European price processes, and show that a separation theorem holds: first choose the optimal stopping to stop and second calculate the non-linear price of the European claim with terminal values the optimal date, and the payoff at this date. Furthermore, in a Markovian framework, the price of the American option, as for the European options, only depends on the price of the underlying asset. The associated function is the unique viscosity solution of an obstacle problem for a parabolic partial differential equation, also called solution of a variational inequality.

5.1 Pricing of American options in a complete market

Let us consider the valuation problem of an American contingent claim $\{\xi_t; t \in [0, T]\}$, where the holder has the right to exercise the option at any time between $[0, T]$. It is well known that this claim cannot be hedged by a self-financing portfolio, and that it is necessary to introduce super-strategies, with a consumption process, first introduced in Section 2, in the example of price pressure; this process can be seen sometimes as the liquidity amount necessary when some constraints are active.

Definition 5.1 *A self-financing* **super-strategy** *is a vector process* (V, π, C), *where V is the market value (or wealth process), π the portfolio process, and C the cumulative consumption process, such that*

$$dV_t = r_t V_t dt - dC_t + \pi_t^* \sigma_t [dW_t + \theta_t dt], \qquad \int_0^T |\sigma_t^* \pi_t|^2 \, dt < +\infty , \; \mathbb{P} \, a.s.$$

$$(5.1)$$

C is an increasing, right-continuous, adapted process with $C_0 = 0$.

Given a payoff $\{\xi_t; t \in [0, T]\}$, a super-strategy is called **a superhedging strategy** *if the constraint holds*

$$V_t \geq \xi_t, \quad t \in [0, T], \; \mathbb{P} \, a.s..$$

The smallest endowment to finance a superhedging strategy is the price of the American option.

The dual formulation of this problem is very natural: instead of adopting the hedging point of view, the pricing point of view leads to consider that the price could be enough to finance the payment of the payoff at any possible exercise time; in other words the price could be greater than the price of ξ_τ for any stopping time $\tau \leq T$.

According to the result on the existence and uniqueness of the solution of RB-SDE's that we state in the following subsection we prove the existence of a minimal square integrable superhedging strategy associated with a payoff bounded in $\mathbb{L}^{2,1}$, continuous on $[0, T[$.

Theorem 5.2 *Suppose the same assumptions on the market as at the beginning of this paper, and consider a square integrable payoff process ξ, continuous on $[0, T[$, with $\lim_{t \to T} \xi_t \leq \xi_T$, a.s. We suppose the payoff bounded in $\mathbb{L}^{2,1}$, that is $\sup_{t \in [0, T]} |\xi_t| \in \mathbb{L}^{2,1}$.*

The American price process Y is associated with a square integrable super-hedging strategy, that is there exists a square integrable continuous increasing process K, and a square integrable portfolio process ϕ such that for $t \in [0, T[$,

$$dY_t = r_t Y_t dt - dK_t + (\phi_t)^* \sigma_t [dW_t + \theta_t dt], \qquad \mathbb{E}[\int_0^T |\sigma_t^* \phi_t|^2 \, dt] < +\infty \quad (5.2)$$

Moreover K satisfies the minimality condition $\int_0^T (Y_s - \xi_s) dK_s = 0$.

The American price is also the maximum X^* of the European price processes associated with an exercise at the stopping time τ before T, that is

$$Y_t = X_t^* = \text{esssup}_{\tau \geq t} X_t(\tau, \xi_\tau) = \text{esssup}_{\tau \geq t} \mathbb{E}_Q[e^{-\int_t^\tau r_s \, ds} \xi_\tau | \mathcal{F}_t]. \quad (5.3)$$

The stopping time $D_t = \inf\{s \geq t, Y_s = \xi_s\}$ is optimal, that is

$$Y_t = X_t(D_t, \xi_{D_t}).$$

PROOF: We only give the proof of the equivalence between both points of view. The existence result is stated in the following subsection

Let us consider a superhedging strategy (Y, ϕ, K) and calculate the variation of Y between t and τ where τ is a stopping time taking values in $[t, T]$.

$$Y_t = \mathbb{E}\left[\int_t^T [-r_s Y_s - (\phi_s)^* \sigma_s \theta_s ds] + Y_\tau + K_\tau - K_t | \mathcal{F}_t\right]$$

$$\geq \mathbb{E}\left[\int_t^T [-r_s Y_s - (\phi_s)^* \sigma_s \theta_s ds] + \xi_\tau | \mathcal{F}_t\right].$$

By the comparison theorem, on $[t, \tau]$, Y dominates the price process for the contingent claim with exercise at time τ, $X(\tau, \xi_\tau)$. Hence

$$Y_t \geq X_t^* = \text{esssup}_{\tau \geq t} X_t(\tau, \xi_\tau).$$

We now choose an optimal elementary stopping time in order to get the reversed inequality. Let

$$D_t = \inf\{u \; ; \; t \leq u \leq T; Y_u = \xi_u\},$$

Now the condition $\int_0^T (Y_t - \xi_t) dK_t = 0$, and the continuity of K imply that $K_{D_t} - K_t = 0$. It follows that the pair $\{Y_s, \phi_s\}; s \in [t, D_t]\}$ is a (price, portfolio) strategy for the European contingent claim ξ_{D_t}. By uniqueness for BSDE's, we have $Y_s = X_s(D_t, \xi_{D_t})$, $s \in [t, D_t]$. Uniqueness for RBSDE's follows from this representation. □

5.2 Reflected BSDE, comparison theorem, existence and uniqueness

We now state the main results on the RBSDE's without using explicitly the theory of optimal stopping. The different steps are the same as for BSDE's. We give some a priori estimates and a comparison theorem and deduce existence and uniqueness from a fixed point theorem. The only difference with the non reflected case is that the existence of a solution for a given driver is now not obvious. Actually, it is the value function of an optimal stopping problem. We solve this problem as limit problem of penalized classical BSDE's. This point of view leads to a self-contained proof of the characterization of the value function of an optimal stopping problem associated with such a payoff.

First, we precisely define RBSDE's.

Put $\mathcal{S}^2 = \{\varphi$ progressively measurable; $\mathbb{E}(\sup_{0 \leq t \leq T} |\varphi_t|^2) < +\infty\}$. The processes belonging to $\mathcal{S}^2$ will be said to be bounded in $\mathbb{L}^{2,1}$.

The problem can be formulated as follows:

We are given :

– a standard $\mathbb{R}$-valued driver f ;

– an "obstacle" $\{\xi_t, 0 \leq t \leq T\}$, which is a continuous process on $[0,T[$, adapted, real valued and bounded in $\mathbb{L}^2$. We shall always assume that $\lim_{t \to T} \xi_t \leq \xi_T$ a.s.

A pair (ξ, f) satisfying the above assumptions is called a set of standard datas.

Let us now introduce the reflected BSDE. The solution is a triple $\{(Y_t, Z_t, K_t); 0 \le t \le T\}$ of processes taking values in $\mathbb{R}$, $\mathbb{R}^n$ and $\mathbb{R}_+$ respectively, and satisfying :

(i) $Z \in \mathbb{H}^{2,n}$, $Y \in \mathcal{S}^2$ and $K_T \in \mathbb{L}^{2,1}$;

(ii) $Y_t = \xi_T + \int_t^T f(s, Y_s, Z_s)\, ds + K_T - K_t - \int_t^T Z_s^*\, dW_s,$;

(iii) $Y_t \ge \xi_t, \qquad 0 \le t \le T$;

(iv) K is continuous and increasing, $K_0 = 0$, and $\int_0^T (Y_t - \xi_t)\, dK_t = 0$.

Note that from (ii) and (iv) it follows that the process Y is continuous.

Intuitively, $\dfrac{d K_t}{dt}$ represents the amount of "push upwards" that we add to $-\dfrac{dY_t}{dt}$, so that the constraint (i) is satisfied. Condition (iv) says that the push is minimal, in the sense that we push only when the constraint is saturated, i.e. when $Y_t = \xi_t$. We will see that this minimality property can be derived from a comparison theorem.

A priori estimates and Comparison theorem

Indeed, as in the classical case, we have a priori estimates for the spread of two solutions of RBSDE's with the same obstacle, and a comparison theorem even if the obstacles are not the same, from which the uniqueness for the solution of the RBSDE follows. The proof of the comparison theorem for RBSDE's differs from the proof for classical BSDE because the difference of the solutions of two RBSDE's is no longer solution of a RBSDE. In particular, we have not a strict comparison theorem. However, we shall see that the minimal property of the increasing process K allow us to derive the desired result. Moreover, small modifications of the proof give a priori estimates.

Theorem 5.3 *Let (ξ, f) and (ξ', f') be two sets of standard datas. The solutions of the associated RBSDE's are denoted by (Y, Z, K) and (Y', Z', K'). We put $\delta Y = Y - Y'$, $\delta Z = Z - Z'$ and $\partial_2 f'(t) = f(t, Y_t', Z_t') - f'(t, Y_t', Z_t')$.*
• A priori estimates,
Suppose that the obstacle is the same for the two solutions on $[0, T[$. For any $\beta > 2C + C^2$, where C is the Lipschitz constant of f, we have,

$$\mathbb{E}[\sup_{0 \le t \le T} |\delta Y_t|^2] \le K_T[e^{\beta T} \mathbb{E}(|\delta Y_T|^2) + \frac{1}{\beta - 2C - C^2} \| \delta_2 f \|_\beta^2] \qquad (5.4)$$

$$\| \delta Z \|_\beta^2 \le (2 + 2C^2 T)e^{\beta T} \mathbb{E}(|\delta Y_T|^2) + \frac{2 + 2C^2 T}{\beta - 2C - C^2} \| \delta_2 f \|_\beta^2 \qquad (5.5)$$

• *Comparison Theorem,*

Suppose that, $\forall t \in [0, T]$, $\xi_t \leq \xi_t'$ *a.s.* , *and that* $\partial_2 f'(t) = f(t, Y_t', Z_t') - f'(t, Y_t', Z_t') \leq 0$ $d\mathbb{P} \times dt$ *a.e.. Then,*

$$Y_t \leq Y_t', \qquad 0 \leq t \leq T, \quad a.s.$$

PROOF : Applying Itô's formula to $e^{\beta t}|\delta Y_t|^2$, and taking the expectation, we have since Y and Y' are bounded in $\mathbb{L}^{2,1}$,

$$\mathbb{E}[e^{\beta t}|\delta Y_t|^2] + \mathbb{E}[\int_t^T e^{\beta s}|\delta Z_s|^2 \, ds] + \mathbb{E}[\int_t^T e^{\beta s}\beta|\delta Y_s|^2 \, ds]$$

$$\leq 2\mathbb{E}\int_t^T e^{\beta s}\delta Y_s[f(s, Y_s, Z_s) - f'(s, Y_s', Z_s')] \, ds + 2\mathbb{E}\int_t^T e^{\beta s}\delta Y_s(dK_s - dK_s')$$

Since the obstacles are the same, we have

$$\int_t^T e^{\beta s}\delta Y_s(dK_s - dK_s') = \int_t^T e^{\beta s}(\xi_s - Y_s') \, dK_s + (\xi_s - Y_s) \, dK_s'. \leq 0$$

Moreover, $f(t, Y_t', Z_t') \leq f'(t, Y_t', Z_t')$, a.s. Now, since f is Lipschitz with constant C, it follows that

$$\mathbb{E}e^{\beta t}|\delta Y_t|^2 + E\int_t^T e^{\beta s}|\delta Z_s|^2 \, ds + E\int_t^T e^{\beta s}\beta|\delta Y_s|^2 \, ds$$

$$\leq 2\mathbb{E}\int_t^T e^{\beta s}\delta Y_s[f(s, Y_s, Z_s) - f(s, Y_s', Z_s') + \partial_2 f'(s)] \, ds$$

$$\leq 2C\mathbb{E}\int_t^T e^{\beta s}|\delta Y_s|(|\delta Y_s| + |\delta Z_s| + |\partial_2 f'(s)|)| \, ds.$$

As in the proof of Proposition 2.2, the quadratic form $Q(y, z) = -\beta|y|^2 + 2C|y|^2 + 2C|y||z| + 2|\delta_2 f_s'||y| - |z|^2$, can be reduced to

$$Q(y, z) = -\beta|y|^2 + 2C|y|^2 + C^2|y|^2 + 2|\delta_2 f_s'||y| - (|z| - C|y|)^2$$
$$= -\beta_C((|y| - \beta_C^{-1}|\delta_2 f_s|)^2 - (|z| - C|y|)^2 + \beta_C^{-1}|\delta_2 f_s'|^2$$

where $\beta_C := \beta - 2C - C^2$ is assumed to be strictly positive. We have exactly the same inequalities as in the non reflected case of Proposition 2.2; so the same estimates follow.

To show the comparison theorem, we now apply Itô's formula to $e^{\beta t}|(\delta Y_t)^+|^2$. The main point is now that [12] on the event $\{Y_s > Y_s'\}$, we have $Y_s > \xi_s' \geq \xi_s$; so,

$$\int_t^T e^{\beta s}(\delta Y_s)^+(dK_s - dK_s') = -\int_t^T e^{\beta s}(\delta Y_s)^+ \, dK_s' \leq 0$$

---

[12]Notice that the property $\int_0^T (Y_t' - \xi_t') \, dK_t' = 0$ is not used in the proof. Hence, the result still holds if (Y', Z', K') satisfies only (i),(ii), (iii) with K' a general continuous increasing process with $K_0' = 0$.

The same arguments as for the estimates give,

$$\mathbb{E}[e^{\beta t}|(\delta Y_t)^+|^2] + \mathbb{E}[\int_t^T e^{\beta s} 1_{\{\delta Y_s > 0\}}|\delta Z_s|^2 \, ds] + \mathbb{E}[\int_t^T e^{\beta s} \beta|(\delta Y_s)^+|^2 \, ds]$$

$$\leq 2\mathbb{E}\int_t^T e^{\beta s}(\delta Y_s)^+[f(s, Y_s, Z_s) - f'(s, Y_s', Z_s')] \, ds + 2\mathbb{E}\int_t^T e^{\beta s}(\delta Y_s)^+(dK_s - dK_s')$$

$$\leq 2\mathbb{E}\int_t^T e^{\beta s}(\delta Y_s)^+[f(s, Y_s, Z_s) - f(s, Y_s', Z_s')] \, ds$$

$$\leq 2C\mathbb{E}\int_t^T e^{\beta s}(\delta Y_s)^+(|\delta Y_s| + |\delta Z_s|) \, ds$$

The quadratic form $Q'(y, z) = -\beta|y|^2 + 2C|y|^2 + 2C|y||z| - |z|^2$ is still negative for β large enough. so

$$\mathbb{E}[e^{\beta t}|(\delta Y_t)^+|^2] \leq 0.$$

□

We deduce immediately the following uniqueness result from the comparison theorem with $\xi' = \xi$, and $f' = f$ or from the estimates. From the a priori estimates, the existence of a solution is deduced by a fixed point theorem and by the existence of a solution for a given driver that we state in the following subsection.

Theorem 5.4 *Let (ξ, f) be a set of standard data. There exists a unique square integrable solution solution $\{(Y_t, Z_t, K_t), 0 \leq t \leq T\}$ of the RBSDE associated with data (ξ, f).*

5.3   Optimal stopping problem and Existence of solutions of RBSDE's

As we have seen at the beginning of this section, the problem of finding a minimal superstrategy is classically associated with an optimal stopping time problem. The same result holds for the solution of a RBSDE. The following proposition is very similar to the one for the American options, with the same proof.

Proposition 5.5 *Let (Y, Z, K) be a square integrable solution of a standard RBSDE,*

$$Y_t = \xi_T + \int_t^T f(s, Y_s, Z_s) \, ds + K_T - K_t - \int_t^T Z_s^* \, dW_s, \, ; \, Y_t \geq \xi_t.$$

Denote $\mathcal{T}$ the set of all stopping times dominated by T, and $\mathcal{T}_t = \{\tau \in \mathcal{T}; t \leq \tau \leq T\}$. Then,

$$Y_t = \text{essup}_{\tau \in \mathcal{T}_t} \mathbb{E}\left[\int_t^\tau f(s, Y_s, Z_s) \, ds + \xi_\tau |\mathcal{F}_t\right] \tag{5.6}$$

Furthermore, the stopping time $D_t = \inf\{u; t \le u \le T; Y_u = \xi_u\}$ is optimal, that is

$$Y_t = \mathbb{E}\left[\int_t^{D_t} f(s, Y_s, Z_s)\, ds + \xi_{D_t} | \mathcal{F}_t\right].$$

Moreover, if the obstacle is an Itô's semimartingale of the form,

$$\xi_t = \xi_0 + \int_0^t U_s ds + \int_0^t V_s^* dW_s,$$

then the increasing process K is absolutely continuous with respect to the increasing process $\int_0^t 1_{\{Y_s = \xi_s\}} [f(s, \xi_s, V_s) + U_s]^- ds$

PROOF: The proof of the first point is very similar to the one for the American options. So we only prove the last point.

It follows from the definition of the RBSDE, that

$$d(Y_t - \xi_t) = -((f(t, Y_t, Z_t) + U_t)dt - dK_t + (Z_t - V_t)^* dW_t.$$

If we denote by $\{L_t; t \in [0, T]\}$ the local time at 0, of the semimartingale $Y - \xi$, the Itô-Tanaka formula gives that

$$d(Y-\xi)_t^+ = -1_{\{Y_t > \xi_t\}}(f(t, Y_t, Z_t) + U_t)dt + 1/2dL_t + 1_{\{Y_t > \xi_t\}}(Z_t - V_t)^* dW_t; Y_T^+ = \xi_T.$$

But $(Y - \xi)^+$ and $(Y - \xi)$ are the same. Hence the two above differentials coincide, and so do the martingale and bounded variation parts. Consequently,

$$1_{\{Y_t = \xi_t\}}(Z_t - V_t) = 0 \qquad dt \times d\mathbb{P}, a.s.$$
$$-1_{\{Y_t = \xi_t\}}[f(t, Y_t, Z_t) + U_t]dt - dK_t = 1/2dL_t$$

Since $K + 1/2L$ is increasing, it follows that $-1_{\{Y_t = \xi_t\}}[f(t, Y_t, Z_t) + U_t]$ is non negative, and dK_t is absolutely continuous with respect to $1_{\{Y_t = \xi_t\}}[f(t, \xi_t, V_t) + U_t]^- dt$.
□

Value function of the optimal stopping problem and RBSDE

In particular, for a given driver, the solution of the RBSDE is the value function of an optimal stopping time problem. In this lecture, we give a direct proof of the existence without referring to this property, by using penalized solutions of the RBSDE. More precisely,

Theorem 5.6 *Let f be a given square integrable driver and ξ a bounded in $\mathbb{L}^{2,1}$ obstacle satisfying the above assumptions.*
For each $n \in \mathbb{N}$, let $\{(Y_t^n, Z_t^n); 0 \le t \le T\}$ denote the unique square integrable solution of

$$Y_t^n = \xi_T + \int_t^T f_s\, ds + n \int_t^T (Y_s^n - \xi_s)^- ds - \int_t^T (Z_s^n)^* dW_s, \qquad (5.7)$$

Then, the increasing sequence Y^n converges a.s and in S^2 to a continuous process Y, such that

$$Y_t = \operatorname{esssup}_{\tau \in T_t} E\left[\int_t^\tau f_s \, ds + \xi_\tau | \mathcal{F}_t\right] \tag{5.8}$$

Moreover, there exist two square integrable processes (Z, K) such that (Y, Z, K) is a solution of the RBSDE,

$$Y_t = \xi_T + \int_t^T f_s \, ds + K_T - K_t - \int_t^T Z_s^* \, dW_s,$$

$$Y_t \geq \xi_t, \qquad \int_0^T (Y_t - \xi_t) \, dK_t = 0.$$

PROOF: Let us denote by $(f^n(t, y, z,) = f_t + n(y - \xi_s)^-)_{n \in \mathbb{N}}$ the sequence of penalized standard drivers. This sequence is increasing, and by the comparison theorem, the sequence of the processes $(Y^n)_{n \in \mathbb{N}}$ is also increasing. Notice that the limit process Y is not necessarily continuous. Moreover, since these drivers are convex and more precisely since,

$$f^n(t, y, z,) = f_t + \sup_{0 \leq u \leq n} u(\xi_t - y),$$

the results of Section 3 give that the solutions Y^n are value functions of control problems, that is

$$Y_t^n = \operatorname{esssup}_{0 \leq u_t \leq n} E[\int_t^T e^{-\int_s^T u_\alpha d\alpha}(f_s + u_s \xi_s) ds + e^{-\int_t^T u_\alpha d\alpha} \xi_T | \mathcal{F}_t]$$

where $\{u_t; t \in [0, T]\}$ is a progressively measurable process.

For the continuation of the proof, let us introduce the continuity modulus of the process ξ, defined as $\omega(h) = \sup_{t \in [0, T-h[}, |\xi_{t+h} - \xi_t|$. From the continuity of ξ on $[0, T[$, $\omega(h)$ converges a.s. to 0. Since ξ is bounded in $\mathbb{L}^{2,1}$, the convergence also holds in $\mathbb{L}^{2,1}$. The notation $\xi_T^* = \sup_{t \in [0, T]} |\xi_t|$ is useful.

For any stopping time $\tau \geq t$, define the process $u^{n,\tau} = n 1_{[\tau, T]}$. The sequence of random variables

$$U_t^n = \int_t^T e^{-\int_t^s u_\alpha^{n,\tau} d\alpha}(f_s + u_s^{n,\tau} \xi_s) ds + e^{-\int_t^T u_\alpha^{n,\tau} d\alpha} \xi_T$$

$$= \int_t^T e^{-n(s-\tau)^+} f_s ds + \int_\tau^T e^{-n(s-\tau)} n \xi_s ds + e^{-n(T-\tau)} \xi_T$$

converges a.s. to $U_t^\tau = \int_t^\tau f_s ds + 1_{\{\tau < T\}} \xi_\tau + 1_{\{\tau = T\}} \xi_T$. Actually, the convergence is uniform. In fact, by using the following representation of U_t^τ,

$$U_t^\tau = \int_t^\tau f_s ds + \int_\tau^T e^{-n(s-\tau)} n \xi_\tau ds + e^{-n(T-\tau)} \xi_\tau,$$

and by using the fact that the jump of ξ at maturity is non-negative, $(\xi_{T-} \leq \xi_T)$, we prove

$$
\begin{aligned}
U_t^\tau - U_t^n &\leq \int_\tau^T e^{-n(s-\tau)}[-f_s + n(\xi_\tau - \xi_s)]ds + e^{-n(T-\tau)}(\xi_\tau - \xi_{T-}) \\
&\leq -\int_\tau^T e^{-n(s-\tau)} f_s ds + \int_\tau^{(\tau+h)\wedge T} e^{-n(s-\tau)} n(\xi_\tau - \xi_s)ds \\
&\quad + 1_{\{T-h<\tau\leq T\}}(\xi_\tau - \xi_{T-})e^{-n(T-\tau)} \\
&\quad + \int_{(\tau+h)\wedge T}^T e^{-n(s-\tau)} n(\xi_\tau - \xi_s)ds + 1_{\{\tau\leq T-h\}}(\xi_\tau - \xi_{T-})e^{-n(T-\tau)}, \\
&\leq n^{-1/2}\left(\int_0^T |f_s|^2 ds\right)^{1/2} + 2\omega(h) + 4\xi_T^* e^{-nh}.
\end{aligned}
$$

Because $\mathbb{E}[U_t^n|\mathcal{F}_t] \leq Y_t^n$, the variable $X_t^* = \text{esssup}_{\tau\in\mathcal{T}_t} \mathbb{E}\left[\int_t^\tau f_s\,ds + \xi_\tau|\mathcal{F}_t\right] = \text{esssup}_{\tau\in\mathcal{T}_t}\mathbb{E}[U_t^\tau|\mathcal{F}_t]$ satisfies

$$
X_t^* \leq Y_t^n + \mathbb{E}[n^{-1/2}\left(\int_0^T |f_s|^2 ds\right)^{1/2} + 2\omega(h) + 4\xi_T^* e^{-nh}|\mathcal{F}_t].
$$

Suppose for the moment that we have proved that $X_t^* \geq Y_t^n$, a.s.. By the martingale inequalities, we have

$$
\mathbb{E}[\sup_{t\in[0,T]} |X_t^* - Y_t^n|^2] \leq \mathbb{E}\left(n^{-1/2}\left(\int_0^T |f_s|^2 ds\right)^{1/2} + 2\omega(h) + 4\xi_T^* e^{-nh}\right)^2.
$$

Taking $h_n = n^{-1/2}$ such that nh_n goes to infinity, the right side goes to 0. By passing to the limit, we have that $X_t^* = Y_t$ but also, the uniform convergence gives that the paths of the process Y are continuous and that this process is bounded in $\mathbb{L}^{2,1}$.

Now, in order to show that $X_t^* \geq Y_t^n$, a.s., let us introduce the stopping time $D_t^n = \inf\{s \in [t,T]; Y_s^n \leq \xi_s\}$, such that $Y_{D_t^n}^n = \xi_{D_t^n}$. So, we have

$$
Y_t^n = \mathbb{E}[\int_t^{D_t^n} f_s ds + Y_{D_t^n}^n|\mathcal{F}_t] = \mathbb{E}[\int_t^{D_t^n} f_s ds + \xi_{D_t^n}|\mathcal{F}_t] \leq X_t^*,
$$

and the inequality is proved. Moreover, the sequence of stopping times D_t^n is increasing with limit a stopping time D_t^*. By passing to the limit in the previous relation, it follows that

$$
Y_t = X_t^* = \mathbb{E}[\int_t^{D_t^*} f_s ds + Y_{D_t^*}|\mathcal{F}_t] = \mathbb{E}[\int_t^{D_t^*} f_s ds + \xi_{D_t^*}|\mathcal{F}_t]
$$

Also, from the continuity of $Y > \xi$, D_t^* is the first time after t where $Y = \xi$, that is $D_t^* = \inf\{s \in [t,T]; Y_s = \xi_s\}$.

Since the processes $\{Y_t^n + \int_0^t f_s ds; t \in [0,T]\}$ are supermartingales, the limit process $Y_t + \int_0^t f_s ds$ is also a continuous supermartingale, bounded in $\mathbb{L}^{2,1}$. From the Doob-Meyer decomposition, and the assumption of a Brownian filtration, there exist a continuous increasing process K and a square integrable martingale $M_t = Y_0 + \int_0^t Z_s^* dW_s$, such that

$$Y_t + \int_0^t f_s ds = M_t - K_t$$

The previous property shows that $K_{D_t^*} - K_t = 0$, or equivalently that $\int_0^T (Y_s - \xi_s) dK_s = 0$.□

Actually, this point of wiew may be used as in El Karoui, Kapoudjian, Pardoux, Peng, and Quenez (1995) to prove the existence of a solution in the general case. However, the uniform convergence of the different approximating processes is more difficult to be obtained.

Proposition 5.7 *For each* $n \in \mathbb{N}$, *let* $\{(Y_t^n, Z_t^n); 0 \leq t \leq T\}$ *denote the unique square integrable solution fo the BSDE,*

$$Y_t^n = \xi + \int_t^T f(s, Y_s^n, Z_s^n) ds + n \int_t^T (Y_s^n - \xi_s)^- ds - \int_t^T Z_s^{n,*} dW_s. \quad (5.9)$$

Then, the increasing sequence Y^n *converges a.s and in* $\mathcal{S}^2$ *to* Y, *the sequence* Z^n *converges to* Z *in* $\mathbb{H}^2$ *and* K^n *converges to* K *in* $\mathcal{S}^2$ *as n tends to infinity, where* K^n *is defined for* $t \in [0, T]$ *by*

$$K_t^n = n \int_0^t (Y_s^n - \xi_s)^- ds, \; 0 \leq t \leq T.$$

Linear RBSDE

The case of RBSDE associated with a linear driver is not very different, and the solution $\{Y_t; 0 \leq t \leq T\}$ of the RBSDE is the value function of an optimal stopping time problem (which does not depend on the solution (Y, Z)). This is also the price process of the American claim ξ in a complete market.

Proposition 5.8 *Let* (β_t, γ_t) *be a bounded* $(\mathbb{R}, \mathbb{R}^n)$-*valued predictable vector-process, and* φ_t *be an element of* $\mathbb{H}_T^{2,1}$. *Let* f *be the standard driver defined by*

$$f(t, y, z) = \varphi_t + \beta_t y + \gamma_t^* z.$$

Let $(\Gamma_{t,s}; t \leq s \leq T)$ *be the adjoint process satisfying the linear SDE :*

$$d\Gamma_{t,s} = \Gamma_{t,s}(\beta_s \, ds + \gamma_s^* \, dW_s); \quad \Gamma_{t,t} = 1.$$

Then the unique solution (Y, Z, K) *of the RBSDE with driver* f *satisfies, for each* $0 \leq t \leq T$,

$$Y_t = \text{ess} \sup_{\tau \in \mathcal{T}_t} \mathbb{E}\left[\Gamma_{t,\tau} \xi_\tau + \int_t^\tau \Gamma_{t,s} \varphi_s \, ds | \mathcal{F}_t\right], a.s..$$

Furthermore, the stopping time $D_t = \inf\{s; s \in [t, T]; Y_s = \xi_s\}$ *is optimal.*

PROOF: Classical calulations using Itô's formula show the supermartingale property of the process $\Gamma_t Y_t + \int_0^t \Gamma_s \phi_s ds$. The continuation of the proof is similar to the one where the driver is known.
□.

5.4 Non-linear Pricing of American contingent claims

Using the solutions of RBSDE's, we develop a theory of non-linear pricing of American square integrable contingent claims satisfying the above assumptions. Let us denote by $\mathcal{C}$ the set of admissible payoffs.

Given a non-linear driver f, we denote by $Y(\xi)$ the non-linear American price process of the claim ξ, and by $X(T, \xi_T)$ the non-linear European price process of the value at time T of the claim. We first prove that, as in complete markets, the American non-linear price is the supremum of the European price processes of all "possible" payoffs.

Theorem 5.9 *The non-linear American price $Y_t(\xi)$ at time t is the maximal price to finance an early exercise that is*

$$Y_t(\xi) = \text{esssup}_{\tau \in T_t} X_t(\tau, \xi_\tau) = X_t(D_t, \xi_{D_t}) \qquad (5.10)$$

where D_t is defined by $D_t = \inf\{s; s \in [t, T], \; Y_s = \xi_s\}$.
The American price system $\xi \in \mathcal{C} \mapsto Y(\xi)$ is increasing, time-consistent and sub-linear if the driver is sub-linear.

PROOF: Let $(Y(\xi), Z, K)$ be the solution of the RBSDE associated with the American price process $Y(\xi)$. Since K is flat on $[t, D_t[$, the pair $(Y_s, Z_s; s \in [t, D_t])$ is solution of the BSDE with terminal condition (D_t, ξ_{D_t}). By uniqueness, the following equality holds,

$$Y_s(\xi) = X_s(D_t, \xi_{D_t}), \qquad s \in [t, D_t].$$

Moreover, for any stopping time τ, $Y_t(\xi)$ may be considered as a generalized solution of non-reflected BSDE's on $[0, \tau]$, with driver $f(t, y, z) + "\frac{dK_t}{dt}" \geq f(t, y, z)$ and terminal value $Y_\tau(\xi) \geq \xi_\tau$. By the comparison theorem for BSDE's, it results that, for $t \in [0; \tau]$, $Y_t(\xi) \geq X_t(\tau, \xi_\tau)$, and the proof is ended.

To show the properties of the American price systems, we extensively use the comparison theorem for RBSDE's. The proofs are the same as for the European price systems except for the no-arbitrage property, because a strict comparison theorem does not hold for RBSDE's. □

Remark : For the American strategies, the no arbitrage property holds only in a weak sense: more precisely, let ξ, ξ' be two payoffs and let Y, Y' be the associated American prices (with the same driver f). Let us denote by D and D' the associated optimal stopping times starting from time 0. Suppose that $\{\xi_t \geq \xi'_t; t \in [0, T]\}$, and that $Y_0 = Y'_0$. Then, $D \leq D'$, the payoffs are the same at the stopping time D' that is, $\xi_{D'} = \xi'_{D'}$, and the prices are equal between

time 0 and time D' that is, a.s. $Y_s = Y'_s$ for $0 \leq s \leq D'$.
Indeed, since the drivers are the same,

$$Y_0 = Y'_0 = X_0(D', \xi'_{D'}) \leq X_0(D', \xi_{D'}) \leq Y_0.$$

Also, by the strict comparison theorem for classical BSDEs, it follows that

$$\xi'_{D'} = \xi_{D'}.$$

Moreover, the equality $Y_0 = X_0(D', \xi_{D'})$ implies that D' is an optimal stopping time for the payoff ξ. It follows that

$$
\begin{aligned}
Y_0 &= \mathbb{E}[\int_0^{D'} f(s, Y_s, Z_s)ds + K_{D'} + Y_{D'}] \\
&= \mathbb{E}[\int_0^{D'} f(s, Y_s, Z_s)ds + \xi_{D'}],
\end{aligned}
$$

Because $Y_{D'} \geq \xi_{D'}$, it follows that a.s. $K_{D'} = 0$ and $Y_{D'} = \xi_{D'}$.
Hence, Y and Y' are both equal to the price of the European claim with exercise time D' and payoff ξ'_D. Hence, by uniqueness, $Y_s = Y'_s$ for $0 \leq s \leq D'$. This property of no arbitrage still holds at any time $t \in [0, T[$. $\square$

American Price system as maximum or maxmin price systems

Let us consider what happens when the standard driver is an infimum of standard drivers. As for the European non-linear price processes, the properties of the American price system are easily deduced from the comparison theorem and uniqueness of RBSDE.

Theorem 5.10 *Let (f, f^α) be a family of standard drivers and (ξ, ξ^α) be a family of contingent claims in C. Let (Y, Z, K) and $(Y^\alpha, Z^\alpha, K^\alpha)$ be the associated (price, portfolio, cost)-strategies. Suppose that there exists a parameter $\bar{\alpha}$ such that*

$$
\begin{aligned}
f(t, Y_t, Z_t) &= \operatorname*{ess\,inf}_\alpha f^\alpha(t, Y_t, Z_t) = f^{\bar{\alpha}}(t, X_t, Z_t) && d\mathbb{P} \otimes dt \, a.s. \\
\xi_t &= \operatorname*{ess\,inf}_\alpha \xi_t^\alpha = \xi_t^{\bar{\alpha}} && \mathbb{P} \, a.s. \qquad (5.11)
\end{aligned}
$$

Then the American price processes Y and Y^α satisfy:

$$Y_t = \operatorname*{ess\,inf}_\alpha Y_t^\alpha = Y_t^{\bar{\alpha}}, \quad \forall t \in [0, T], \, \mathbb{P} \, a.s. \qquad (5.12)$$

The following minimax relation holds at any time t between American price and European price processes,

$$\operatorname*{ess\,sup}_\tau \operatorname*{ess\,inf}_\alpha X_t^\alpha(\tau, \xi_\tau^\alpha) = Y_t(\xi) = X_t^{\bar{\alpha}}(D_t, \xi_{D_t}) = \operatorname*{ess\,inf}_\alpha \operatorname*{ess\,sup}_\tau X_t^\alpha(\tau, \xi_\tau^\alpha) \quad (5.13)$$

The same equalities hold when the driver is the supremum of standard drivers.

PROOF: The first equality is as for the European price systems a simple consequence of the comparison theorem and uniqueness for RBSDE.

The right equality in the minimax relation results from the previous relation between American prices and European prices, since

$$Y_t = \operatorname*{ess\,inf}_{\alpha} Y_t^{\alpha} = \operatorname{essinf}_{\alpha} \operatorname{esssup}_{\tau} X_t^{\alpha}(\tau, \xi_{\tau}^{\alpha}).$$

From the equality $Y_t = X_t^{\overline{\alpha}}(D_t, \xi_{D_t})$, and the similar result for European price processes, we deduce

$$Y_t(\xi) = X_t^{\overline{\alpha}}(D_t, \xi_{D_t}) = \operatorname{essinf}_{\alpha} X_t^{\alpha}(D_t, \xi_{D_t}^{\alpha}) \leq \operatorname{esssup}_{\tau} \operatorname{essinf}_{\alpha} X_t^{\alpha}(\tau, \xi_{\tau}^{\alpha})$$

But we have still the reversed inequality, $\operatorname{essinf}_{\alpha} \operatorname{esssup}_{\tau} X_t^{\alpha}(\tau, \xi_{\tau}) \geq \operatorname{esssup}_{\tau} \operatorname{essinf}_{\alpha} X_t^{\alpha}(\tau, \xi_{\tau}^{\alpha})$ and so the inequalities are in fact equalities □

When the driver is convex or concave, we have the following extension of the European case (Notation is introduced in Section 3).

Corollary 5.11 *Let f be a convex standard generator and $\mathcal{M}^{\beta,\gamma}$ be the dual markets associated with the running cost F. Then the American f-price system is the superprice of the American contingent claim with respect to the $\mathcal{M}^{\beta,\gamma}$-markets and the running cost F. Moreover, the f-American price is the standard American price in an optimal fictitious market associated with $(\overline{\beta}, \overline{\gamma})$ which achieves the supremum in the conjugacy relation,*

$$f(t, Y_t, Z_t) = F(t, \overline{\beta}_t, \overline{\gamma}_t) - \overline{\beta}_t, Y_t - \overline{\gamma}_t^* Z_t.$$

5.5 Relation between a RBSDE and an obstacle problem for a non linear parabolic PDE.

In this subsection, we will show that the reflected BSDE studied in the previous subsections allows us to give a probabilistic representation of solutions of some obstacle problems for PDEs. In terms of pricing of American contingent claims, this property means that the American price of a payoff only depending on the prices of the underlying asset has the same property. In other words, the market is efficient. The second consequence for the pricing problem is to obtain a computational method by solving these PDE's variational inequalities.

For that purpose, we will put the pricing problem in a Markovian framework. The framework is the same as in the case of Markovian BSDE's and the assumptions on the state process and the coefficients b, σ, f and g are the same. Furthermore, the obstacle ξ_t is supposed to satisfy

$$\xi_t = h(t, S_t^{r,y})$$

where $h : [0, T] \times \mathbb{R}^d \to \mathbb{R}$ satisfies

$$h(t, x) \leq K(1 + |x|^p), \ t \in [0, T], \ x \in \mathbb{R}^d. \tag{5.14}$$

We assume moreover that $h(T, x) \leq g(x)$, $x \in \mathbb{R}^d$.

We shall denote by $(Y_s^{r,y}, Z_s^{r,y}, K_s^{r,y} \; r \leq s \leq T)$ the solution of the RBSDE associated with obstacle $h(t, S_t^{r,y})$, driver $f(s, S_s^{r,y}, y, z)$. As in the case of a classical BSDE, we know that $Y_r^{r,y}$ is a deterministic function of (r, y) denoted by $u(r, y)$. We have the following theorem:

Theorem 5.12 *Suppose that the coefficients f, b, σ, h are jointly continuous in t and x. Then, the function $u(r, y)$ is solution of the unique viscosity solution of the obstacle problem :*

$$
\begin{cases}
\min\left(u(t, x) - h(t, x), -\dfrac{\partial u}{\partial t}(t, x) - \mathcal{L}_t u(t, x) - f(t, x, u(t, x), (\nabla u \sigma)(t, x)) \right) = 0, \\
\quad (t, x) \in (0, T) \times \mathbb{R}^d \\
u(T, x) = h(T, x), \; x \in \mathbb{R}^d \; ;
\end{cases}
$$

$$(5.15)$$

where $\mathcal{L}_t$ is the infinitesimal operator associated with the Itô process $S^{t,x}$. Such systems are also called variational inequalities.
The function u gives the non-linear American price for a contingent claim with payoff $h(t, S_t^{r,y})$.

SKETCH OF THE PROOF : We are going to use the approximation of the RBSDE (5.6) by penalization.
For each $(r, y) \in [0, T] \times \mathbb{R}^d$, $n \in \mathbb{N}^*$, let $\{(^nY_s^{r,y}, {}^nZ_s^{r,y}), t \leq s \leq T\}$ denote the solution of the BSDE :

$$
{}^nY_s^{r,y} = g(S_T^{r,y}) + \int_s^T f(t, S_t^{r,y}, {}^nY_t^{r,y}, {}^nZ_t^{r,y}) \, dr
$$

$$
+ n \int_s^T ({}^nY_t^{r,y} - h(t, S_t^{r,y}))^- - \int_s^T {}^nZ_t^{r,y} \, dW_t, \; r \leq s \leq T.
$$

By the previous results on classical BSDE in the markovian case, we know that

$$
u_n(r, y) = {}^nY_t^{r,y}, \; 0 \leq r \leq T, \; x \in \mathbb{R}^d,
$$

is the viscosity solution of the parabolic PDE :

$$
\begin{cases}
\dfrac{\partial u_n}{\partial t}(t, x) + L_t u_n(t, x) + f_n(t, x, u_n(t, x), (\nabla u_n \sigma)(t, x)) = 0, \; 0 \leq t \leq T, \; x \in \mathbb{R}^d ; \\
\qquad\qquad\qquad\qquad\qquad\qquad\qquad\qquad u(T, x) = g(x), \; x \in \mathbb{R}^d ;
\end{cases}
$$

where $f_n(t, x, r, p\sigma(t, x)) = f(t, x, r, p\sigma(t, x)) + n(r - h(t, x))^-$.
But from the results on the approximation of the RBSDE (Theorem 5.6) by penalization, for each $0 \leq t \leq T$, $x \in \mathbb{R}^d$,

$$
u_n(r, y) \uparrow u(r, y) \text{ as } n \to \infty.
$$

Since u_n and u are continuous, it follows from Dini's theorem that the above convergence is uniform on compacts. Using this property and classical technics of the theory of viscosity solutions, it is possible to obtain that $u(r, y)$ is a viscosity solution of the obstacle problem. Concerning the proof of the uniqueness of the viscosity solution of the obstacle problem, one is referred to El Karoui, Kapoudjian, Pardoux, Peng, Quenez (1995). □

References

Antonelli, F. (1992) "Backward-Forward Stochastic Differential Equations", *Annals of Applied Probability* 3. 777-793 (1993).

Barles,G. (1994) "Solutions de viscocité des équations de Hamilton-Jacobi du premier ordre et Applications", *Mathématiques et Applications* 17, Springer 1994.

Barles,G & Kobylansky,M.(1996): "Existence and uniqueness results for backward stochastic differential equationswhen the generator has a quadratic growth." *Université de Tours, preprint.*

Benes, V.E. (1971) "Existence of Optimal Stochastic Control Law'" *SIAM J. of Control,*9, 446-472.

Bergman, Y. (1991) "Option Pricing with Divergent Borrowing and Lending Rates," *Working Paper*, Dpt of Economics Brown University.

Bensoussan,A :"On the Theory of Option Pricing" *Acta Applicandae Mathematicae* 2, 139-158, 1984.

Bensoussan,A, & Lions,J.L.(1978) : "Applications des Inéquations Varitionnelles en Contrôle Stochastique", Dunod, Paris .

Bismut, J.M. (1973) "Conjugate Convex Functions in Optimal Stochastic Control," *J.Math. Anal. Apl.,*44, 384-404.

Bismut, J.M. (1978) "Contrôle des systèmes linéaires quadratiques : applications de l'intégrale stochastique," *Sémin.Proba. XII.*, Lect. Notes in Math., 649, 180-264,Springer.

Buckdahn, R. (1993) "Backward Stochastic Differential Equations driven by a Martingale," *Preprint.*

Cinlar, E., Jacod, J., Protter, P., & Sharpe, M.J. (1980) "Semimartingale and Markov Processes," *Z.f.W.*, 54, 161-219.

Crandall,M., Ishii,H., &Lions P.L.(1992): "User's guide to the viscosity solutions of second order partial differential equations", *Bull. A.M.S.* 27, 1-67,

Cvitanic, J.(1996) "Optimal Trading under constraints" *this volume.*

Cvitanic, J., & Karatzas, I. (1992) "Convex duality in Constrained Portfolio Optimization," *Annals of Applied Probability* 2, *pp.767-818* .

Cvitanic, J., & Karatzas, I. (1993) "Hedging Contingent Claims with Constrained Portfolios," *Annals of Applied Probability* 3, *pp.652-681* .

Cvitanic, J., & Ma, J. (1996) "Hedging Options for a Large Investor and Forward-Backward SDE's," To appear in *Annals of Applied Probability* .

Delbaen, F., & Schachermayer, W. (1994) "A General Version of the Fundamental Theorem of Asset Pricing," *Math.Annal*,**123**.

Dellacherie, C.(1977) "Sur l'existence de certains essinf et essup de familles de processus mesurables," *Sem.Proba.XII.* Lectures. Notes in Math.**649**. Springer Verlag

Dellacherie, C.& Meyer, P.A.(1980) " Probabilités et Potentiel" *Chap V to VIII, Théorie des martingales.* Hermann

Dellacherie, C.& Meyer, P.A.(1980) " Probabilités et Potentiel" *Chap XII to XVI, Théorie du potentiel, Processus de Markov.* Hermann

Duffie, D. (1992) *Dynamic Asset Pricing Theory* Princeton University Press.

Duffie, D., & Epstein, L. (1992) "Stochastic Differential Utility," *Econometrica*, **60**, n.2, 353-394.

Duffie, D., Ma, J., & Yong, J. (1994) "Black's Consol Rate Conjecture", *Working paper*, Graduate School of Business, Stanford University.

Duffie, D., & Skiadias, C. (1991) "Continuous-time Security Pricing : A Utility Gradient Approach", *Working paper*, Graduate School of Business, Stanford University.

Ekeland, I., & Turnbull, T. (1979) *Infinite Dimensional Optimization and Convexity* Chicago Lectures in Math.

El Karoui, N. (1981) "Les aspects probabilistes du contrôle stochastique" *Lectures Notes in Math.* **816**.Springer Verlag.

El Karoui, N. & Huang S.J. (1996) " A General Result of Existence and Uniqueness of Backward Stochastic Differential Equations" to appear *Pitman Series , ed :El Karoui,N and Mazliak,L.*

El Karoui, N., Kapoudjian, C., Pardoux, E., Peng, S. & Quenez, M.C. (1995) "Reflected solutions of backward SDE's, and Related Obstacle Problems for PDE's," *to appear Annals of Probability.*

El Karoui, N., Peng,S., & Quenez, M.C. (1994) "Optimization of Utility Functions" *Working Paper*, Paris VI University.

El Karoui, N., Peng,S., & Quenez, M.C. (1997) "Backward Stochastic Differential Equations in Finance" *Mathematical Finance*, to appear in January 1997.

El Karoui, N., & Quenez, M.C. (1991) "Programmation dynamique et évaluation des actifs contingents en marché incomplet ," *C.R.Acad.Sci.Paris* , **t.313**,pp 851-854.

El Karoui, N., & Quenez, M.C. (1995) "Dynamic Programming and Pricing of Contingent Claims in Incomplete Market," *Siam J.of Control and Opti.*, **33**, n.1.

Epstein, L., & Zin, S. (1989) "Substitution, Risk Aversion and the Temporal Behavior of Consumption and Asset Returns: A Theorical Framework", *Econometrica*, **57**,n.4, 837-969.

Fleming, W.F., & Soner, M. (1993) *Controlled Markov processes and Viscosity Solutions* Springer-Verlag, New-York, Heidelberg, Berlin.

Föllmer, H., & Schweizer,M. (1990) "Hedging of Contingent Claims under Incomplete Information," (1990), *Applied stochastic analysis* eds. M.H.A. Davis and R.J. Elliot, Gordon and Breach, London.

Geoffard, P.Y. (1995) "Discounting and Optimizing, Utility maximization : a Lagrange Variational Formulation as a Minmax Problem," *Journal of Economic Theory*.

Hamadene, S. ,& Lepeltier, J.P. (1995) "Zero-Sum Stochastic Differential Games and Backward Equations", *Systems and Control Letters*, **24**, 259-263.

Harrison, M., & Kreps, D. (1979) "Martingales and Arbitrage in Multiperiod Securities Markets", *Journal of Economic Theory*, **20**, 381-408.

Harisson, M. ,& Pliska, S.P. (1981) "Martingales and Stochastic Integrals in the Theory of Continuous Trading," *Stochastic Processes and their Applications*, **11**, 215-260.

Harisson, M., & Pliska, S.P. (1983) "A Stochastic Calculus Model of Continuous Trading : Complete Markets," *Stochastic Processes and their Applications*, **15**, 313-316.

He,H., & Pearson,N.D. (1991): "Consumption and Portfolio Policies with Incomplete Markets and Short-Shale Constraints ; the infinite dimensional case", *Journal of Economic Theory* **54** 259-304

Hu & Peng,S (1995): "Solution of Forward-Backward Stochastic Differential Equations" *Proba. Theory. Rel. Fields,***103** pp.273-283.

Jacka, S. (1993) : "Local times, optimal stopping and semimartingales," *The Annals of Probability* **21**, 329–339.

Jouini, E., & Kallal, H. (1992) "Arbitrage and Equilibrium in Securities Markets with Shortsale Constraints," *Working Paper*, Univ. Paris I.

Karatzas,I. (1988) "On the pricing of American Options" *Appl. Math. Optimization,***17**, pp. *37-60,*

Karatzas,I. (1989) ", Optimization problems in the theory of continuous trading", *SIAM. J.Control. Optimization,***27**, pp. *1221-1259*

Karatzas, I., & Shreve, S.E. (1996) "Methods on Mathematical Finance ", Book to appear

Karatzas,I., & Shreve, S.E. (1987) "Brownian Motion and Stochastic Calculus", Springer Verlag. New York.

Korn, R. (1992) "Option Pricing in a Model with a Higher Interest Rate for Borrowing than for Lending," Working paper.

Kramkov, D.O. (1996) "Optional decomposition of supermartingales and hedging contingent claims in incomplete security markets," *Probability Theory and Related Fields* **105**, *459-479 (1996)*.

Krylov, N. (1980) *Controlled Diffusion Processes*, New York : Springer Verlag.

Lepeltier J-P. & San Martin J.(1996) " Backward stochastic differential equations with continuous generator " to appear in *Statistics and Probability Letters*

Ma, J., Protter, P., & Yong, J. (1994) "Solving forward-backward stochastic differential equations explicitly - a four step scheme" *Proba. Theory Relat. Fields* **98**, *339-359 (1994)*.

Merton, R. (1971) "Optimum Consumption and Portfolio Rules in a Continuous Time Model", *Journal of Economic Theory.* **3**, *pp. 373-413*.

Müller,S.(1987) "Arbitrage pricing of contingent claims" *Lecture Notes in Economics and Mathematical Systems,*254 Springer Verlag.

Musiela,M & Rutkowoski (1997) "Arbitrage Pricing of Derivative Securities. Theory and Aplplications" *Book to appear* Springer-verlag.

Pardoux, E.(1996) "BSDE's and semilinear PDE's ", *Working paper, Geilo Lectures 1996.*

Pardoux, E., & Peng, S.(1990) "Adapted Solution of a Backward Stochastic Differential Equation", *Systems and Control Letters, bf 14, 55-61.*

Pardoux, E., & Peng, S. (1992) " Backward Stochastic Differential equations and Quasilinear Parabolic Partial Differential Equations" *Lecture Notes in CIS* **176**, *200-217, Springer.*

Pardoux, E., & Peng, S. (1995) "Some backward SDEs with non Lipschitz coefficients", *Proc. Conf. Metz, to appear.*

Peng, S. (1991) "Probabilistic Interpretation for Systems of Quasilinear Parabolic Partial Differential Equations", *Stochastics,* **37**; 61–74.

Peng, S. (1992a) " A Generalized Dynamic Programming Principle and Hamilton-Jacobi-Bellman equation," *Stochastics, Vol. 38, 119-134.*

Peng, S. (1992b) " A Nonlinear Feynman–Kac Formula and Applications," *Proceedings of Symposium of System Sciences and Control theory, Chen & Yong ed. 173-184, World Scientific, Singapore.*

Peng, S. (1993) "Backward Stochastic Differential Equation and It's Application in Optimal Control," *Appl. Math. & Optim.* 27:125–144.

Quenez, M.C. (1993) " Méthodes de contrôle stochastique en Finance," *Thèse de doctorat de l'Université Pierre et Marie Curie.*

Revuz,D. & Yor, M. (1994) " Continuous Martingales and Brownian Motion", Springer Verlag 1994.

Schweizer, M. (1992) "Mean-Variance hedging for general claims, " *The Annals of Applied Probability,* **2**, *pp.171-179.*

Svensson, L.E.O., & Werner, J. (1990) "Portfolio Choice and Asset Pricing with non-traded Assets, " *Research Paper,* **2005**, *pp.*.Graduate School of Business, Stanford University.

Uzawa, H. (1968) "Time preference, the consumption-function, and optimal asset holdings ", in *Value, capital, and Growth : Papers in Honor of Sir John Hicks.* J.N.Wolfe ed., Edinburgh University, Edinburgh.

MARKET IMPERFECTIONS,
EQUILIBRIUM AND ARBITRAGE

Elyès Jouini

CREST-ENSAE, CERMSEM-Université de Paris I

and Ecole Polytechnique (Paris and Tunis)

Contents

Introduction

The theory of asset pricing, which takes its roots in the Arrow-Debreu model (Theory of value [1959, chap. 7]), the Black and Scholes formula (1973) and Cox and Ross (1976 a and b), has been formalized in a general framework by Harrison and Kreps (1979), Harrison and Pliska (1979) and Kreps (1981). In these models, securities markets are assumed to be frictionless. The main result is that a price process is arbitrage free (or, equivalently, compatible with some equilibrium) if and only if it is, when appropriately renormalized, a martingale for some equivalent probability measure. The theory of pricing by arbitrage follows from there. Contingent claims can be priced by taking their expected value with respect to an equivalent martingale measure. If this value is unique, the claim is said to be priced by arbitrage. The new probabilities can be interpreted as state prices (the prices of 1 dollar tomorrow in each state of the world) or as the intertemporal marginal rates of substitution of an agent maximizing his expected utility.

In this work, we will propose a general model that takes frictions into account. There is an important body of literature on this subject concerned with optimal portfolio selection problems. Among others we can cite: Magill and

Constantinides (1976), Constantinides (1986), Taksar and al. (1988), Duffie and Sun (1990), Grossman and Laroque (1990), Fleming and al. (1990), Davis and Norman (1990), and Dumas and Luciano (1991). In these studies the bid and ask price processes of a risky asset are exogenously given: they are usually diffusions of constant ratio (i.e. transaction costs are proportional). Typically, it is then found that portfolios are not rebalanced by maximizing agents as long as their share of wealth invested in the risky asset remains in a certain interval.

Some authors have also studied hedging strategies in the presence of transaction costs. Early investigations are Gilster and Lee (1984) and Leland (1985). In a continuous time model they find hedging strategies, for call options, that are revised at a finite number of times only and that are asymptotically exact (when the number of revisions goes to infinity). The strategies consist in following Black and Scholes ratios with an adjusted (upward) volatility. Dybvig and Ross (1986), Prisman (1986) and Ross (1987) have studied the case of a two-period economy with taxes. Figlewski (1989) has performed numerical simulations to evaluate the importance of transaction fees in hedging strategies. More recently, Bensaid and al. (1992) have developped an algorithm to hedge any contingent claim in the presence of proportional transaction costs in a binomial model. Their method elaborates on the idea that perfect duplication may be a suboptimal way of hedging if transaction costs are large enough.

In the spirit of Harrison and Kreps (1979) we start by characterizing bid and ask securities price processes that are arbitrage free (or, equivalently, viable, i.e. compatible with some equilibrium for a certain class of maximizing agents) in a fairly general finite horizon model of securities market. It turns out that all such processes can be obtained as a perturbation of a price process that is arbitrage free in a frictionless economy (the bid price lying above and the ask price lying below this process). Indeed, we find that a bid-ask price process is arbitrage free if and only if there exists an equivalent probability measure that transforms some process between the bid and the ask price processes into a martingale or a super/sub-martingale(after a normalization). Such a probability measure will be called a martingale measure by analogy with the frictionless case. In particular, we find that the bid and the ask price processes do not have to be arbitrage free in an economy without frictions in order to be viable. As in the perfect markets case, the martingale probabilities can be interpreted as possible intertemporal marginal rates of substitution of an agent maximizing his expected utility. They can also be interpreted as state prices and we show how to compute arbitrage bounds on the bid-ask prices of any contingent claim using the set of martingale measures. These bounds are respectively the minimum cost necessary to hedge the contingent claim, and the maximum amount one could borrow against it, using traded securities. They define a possible range for the bid and the ask price at which a new security could be traded if it were to be introduced in the market. Indeed, nobody would buy the security for more than the upper bound since there would be a way of getting at least the same payoff for cheaper, and nobody would sell it for less than the lower bound since there would be a way of borrowing a larger amount against it, using traded securities (see Bensaid and

al. [1992] for an extensive discussion of this point). We find that the interval
defined by these bounds is equal (modulo the boundary) to the set of expectation
of the claim with respect to all the martingale measures. There is also a sense,
made explicit in the work, in which these are the tightest bounds one can find
without knowledge about agents' preferences. Other potential applications of
our analysis are those of the martingale approach in the frictionless case.

Two simple examples can illustrate intuitively our results. Consider, to start
with, a deterministic world where agents consume at date 0 and T and can trade
two securities (1 and 2). Security 1 (the numeraire) is assumed to be always
worth 1. At any date t, security 2 can be bought at its ask price $Z(t)$, and can
be sold at its bid price $Z'(t)$. Note that in this deterministic world a martingale
is merely a constant process. Now suppose that there is no constant process
lying between the bid price Z' and the ask price Z of security 2. It is easy to see
that this means that there exist two dates t and t' such that $Z'(t') > Z(t)$. In
this case, buying an arbitrarily large amount A of security 2 at date t and selling
[1] it at date t' one ends up with an arbitrarily large net profit $A(Z'(t') - Z(t))$ at
the final date, without spending anything at date 0. Therefore, the bid-ask price
processes Z and Z' are not arbitrage free in this case. Conversely, if there is a
constant process between Z and Z', security 2 can never be bought at a lower
price than it can ever be sold at, and there are no opportunities of arbitrage
regardless of the behavior of the bid-ask price processes. In particular, the bid
and the ask prices do not have to be constant, as in a frictionless economy. In
order to make sense of the more general results recall that a martingale is the
stochastic analog of a constant.

Another simple example with uncertainty can also illustrate the link between
the martingale measures and the arbitrage bounds on the price of a contingent
claim: the minimum cost to hedge it and the maximum amount that can be
borrowed against it through securities trading. Consider an economy where
there are two dates, 0 and 1, and two possible states of the world at date 1 :
"up" and "down". Two securities can be traded: a bond that is always worth
1 (i.e. the riskless rate is equal to zero), and a stock that is worth $S_u = 110$
in state "up" and $S_d = 90$ in state "down". We assume that there is a bid-ask
spread in trading the stock at date 0 : it can be bought for S_0 and it can be sold
for S_0'. It is easy to see that there are no arbitrage opportunities in this economy
as long as $S_0 \geq S_0'$, $S_0 > S_d$, and $S_u > S_0'$. It is also easy to see that these three
inequalities are satisfied if and only if there exists a probability measure that
puts a strictly positive weight p^* on state "up" and a strictly positive weight
$1 - p^*$ on state "down", and for which we have $S_0 \geq p^* S_u + (1 - p^*)S_d \geq S_0'$.
Consider now a call option on the stock with exercise price equal to 100, that
pays 10 in state "up" and 0 in state "down". If there is no bid-ask spread on the
stock and its price is $S_0 = 100$, it is easy to see that there is only one risk-neutral
(martingale) probability, 0.5 for the "up" and the "down" state. The call is then
worth its expected payoff with respect to this probability: 5. The same result

---

[1]If $t < t'$ the purchase is financed by going short in security 1 and if $t > t'$ the proceeds
from the sale are invested in security 1.

could be obtained by duplicating the call, i.e. buying 0.5 shares of stock and selling 45 in bonds, since this portfolio generates the same payoff as the call and its value is precisely 5. Now suppose that there is a bid-ask spread on the stock: the stock can still be sold for $S_0' = 100$ but can be bought for $S_0 > 100$. It is then easy to see that the minimum cost to hedge the call is $0.5 S_0 - 45$ if $S_0 \leq 110$ and 10 otherwise. Indeed, in the first case the optimal strategy is to buy 0.5 shares of stock and to sell 45 in bonds. In the second case however the transaction costs on the stock are too large and it is optimal to buy 10 in bonds, although this strategy does better than duplicate the call since it yields a payoff of 10 in every state of the world. Moreover, the maximum amount that can be borrowed against the call is 5, since the optimal strategy consists in selling 0.5 shares of stock and buying 45 in bonds. On the other hand, the processes that lie between the bid and the ask price take the values $S_u = 110$ in state "up", $S_d = 90$ in state "down" and any value between $S_0' = 100$ and S_0 at date 0. The set of positive probabilities that transform some of these processes into a martingale are of the form $1 > p^* > 0$ for state "up", with $\frac{S_0 - 90}{20} \geq p^* \geq 0.5$. It is then easy to check that the interval defined by the expected values of the payoff of the call with respect to these probabilities is equal to the interval defined by the bounds computed using optimal hedging strategies.[2]

There is no doubt that shortselling and borrowing costs are another salient feature of financial markets. For instance, investors usually do not have full use of the proceeds from short sales of stocks (see Cox and Rubinstein [1985] p. 98-103), and this effectively represents a shortselling cost that is proportional to the holding period. In the Treasury bond market, short sales are performed through reverse repurchase agreements (reverse repos) in which the shortseller lends money at the reverse repo rate and takes the bond as a collateral. The shortselling cost is then the spread between the repo rate (at which an owner of the bond can borrow money collateralized by the bond through a repurchase agreement) and the reverse repo rate. If the bond is "on special", i.e. if it is particularly difficult to borrow, its repo rate is lower than the repo rate on general collateral. In this case, the shortselling cost is the sum of the spread between the repo rate and the reverse repo rate and of the spread between the repo rate and the repo rate on general collateral. Stigum (1983) reports typical costs between 0.25% and 0.65%, but they can be much larger for specific bonds. Amihud and Mendelson (1991) show that costs of this magnitude are substantial enough to wipe out the profits from the arbitrage between Treasury bills and Treasury notes with less than six months to maturity (with identical cashflows) that are relatively cheap, even taking trading costs into account. On the other hand, borrowing costs vary substantially with the size and the credit rating of the market participant, and since they can be quite large, their significance is even less questionable.

Our framework permits to study a market with short sales constraints and different borrowing and lending rates. More precisely we can consider two sorts of securities. Shortselling the first type of securities is not allowed, i.e. they can

---

[2] Note that when $S_0 > 110$ they coincide up to the boundary only.

only be held in nonnegative amounts, whereas the second type of securities can only be held in nonpositive amounts. In particular, we do not assume that the borrowing rate is equal to the lending rate. However this model includes the case where some (or all) securities are not subject to any constraints; we include these securities twice: in the first type and in the second type so that they can be held in nonnegative and nonpositive amounts. We show that this type of economy is arbitrage-free if and only if there exists a numeraire and an equivalent probability measure that transforms the normalized (by the numeraire) price processes of traded securities that cannot be sold short into a supermartingale, and the normalized price processes of the securities that can only be held short into a submartingale.[3] In such an economy, even if a contingent claim can be duplicated by dynamic trading, it is not necessarily possible to price such a contingent claim by arbitrage. This comes from the fact that the underlying securities cannot be sold short. However, arbitrage bounds can be computed for arbitrary contingent claims: they are the minimum amount it costs to hedge the claim and the maximum amount that can be borrowed against it using dynamic securities trading (see Bensaid et al. [1992] for an extensive discussion of this point). These are the tightest bounds that can be inferred on the price of a claim without knowing preferences. We find that these arbitrage bounds on the bid-ask prices of a contingent claim are respectively equal to the smallest and the largest expectations of its future normalized cashflows - with respect to all the numeraire processes and supermartingale probability measures. This model includes situations where holding negative amounts of a security is possible but costly (in the form of a lower expected return per unit of time), and where the riskless borrowing and lending rates differ. In this case, and when the underlying security price follows a diffusion process, we use the previous results to characterize arbitrage-free economies and we determine the set of super/submartingale probability measures.

To interpret these results, note that a martingale is a process that is constant on average: the expectation of its future value at any time is equal to its current value. Therefore, if securities prices are martingales investors cannot enjoy the possibility of a gain without the risk of a loss, and cannot suffer the risk of a loss without enjoying the possibility of a gain. This prevents arbitrage opportunities, i.e. sure gains, that could be generated by buying securities or by shortselling them. On the other hand, if short sales are prohibited a security (or a portfolio of securities) may provide the risk of a loss without providing the possibility of a gain; this does not constitute an arbitrage opportunity since the arbitrage would consist in shortselling the security (or the portfolio of securities) and this is not permitted. As a result, price processes only need to be nonincreasing on average (i.e. supermartingales) to prevent arbitrage opportunities.

Another possible interpretation is that a probability measure that transforms price processes into martingales corrects for the risk aversion of the agents involved in trading: it puts more weight on "bad" (low consumption) states than

---

[3] A martingale is a process that is constant on average, a supermartingale is a process that is nonincreasing on average, and a submartingale is a process that is nondecreasing on average.

the subjective probability measure and as a result the price of an asset can be computed by taking the expectation, weighted by the new probabilities, of its future cashflows. In other words, these new probabilities make the securities price processes compatible with expected utility maximization by risk-neutral agents (hence the often used name of "risk-neutral probabilities"). Indeed, with these probabilities, securities appear to be fair bets to risk-neutral agents. On the other hand, a supermartingale is a process that is nonincreasing on average: the expectation of its future value at any time is less than or equal to its current value. Hence a security that has a supermartingale as a price process does no longer appear as a fair bet to a risk-neutral agent. He would like to sell the security short in unlimited amounts. The short sales constraint, however, prevents him from doing so.

In Section 1 we analyze a two period economy where agents trade in the first period contingent claims to consumption in the second period, belonging to a convex cone of a larger space of payoffs. The prices of these claims are assumed to be given by a sublinear functional (positively homogeneous and subadditive, i.e. such that the price of the sum of two claims is at most equal to the sum of their prices). Enlarging the set of possible price functionals from the linear to the sublinear functionals allows us to take frictions into account: in particular, a long position in a claim costs more than one gets by going short in the same claim. We find that a model is arbitrage free (or equivalently, viable, i.e. compatible with some equilibrium) if and only if there exists a strictly positive linear functional that lies below the sublinear price functional on the set of marketed claims. We show how to compute bounds on the bid and the ask prices of contingent claims: the minimum amount it costs to hedge the claim and the maximum that can be borrowed against it, using securities trading. These bounds can be related to equilibrium analysis and there is a sense in which they are the tightest bounds that can be derived without knowledge about preferences. We show that for any contingent claim the interval defined by these "arbitrage" bounds is equal (modulo its boundary) to the set of expectations of the payoff of the claim with respect to all the martingale measures.

Our choice of a sublinear price functional is justified by the multiperiod model introduced in Section 2, where consumption takes place at dates 0 and T, and where consumers can buy a finite number of securities at their ask price and sell them at their bid price at any time under some constraints. In this section, we apply the results of Section 1 to characterize arbitrage free (or equivalently, viable, i.e. compatible with some equilibrium) securities bid-ask price processes. We find that a securities bid-ask price process is arbitrage free if and only if, after a normalization, there exists an equivalent probability measure and a martingale or a super/sub-martingale, with respect to this probability, that lies between the bid and the ask price processes. This result allows us to associate to each economy with frictions a family of possible underlying frictionless economies, and every economy with frictions can be seen as the perturbation of a frictionless one, and conversely. Note that we do not impose any particular form on the price processes (diffusion or others) and then the martingale property is not contained in our assumptions; Furthermore, we do not impose proportionality between the bid and the ask price and the spread can evolve arbitrarily.

In Section 3, we apply the general framework introduced in Section 2 to transaction costs, incomplete markets and shortselling costs including different borrowing and lending rate. We establish in these contexts valuation formulas for derivative assets. In the shortselling costs case, this allows us to derive a (nonlinear) partial differential equation that must be satisfied by the arbitrage bounds on the prices of derivative securities and to determine the optimal hedging strategies. This partial differential equation is similar to that obtained by Black and Scholes (1973), with two additional nonlinear terms proportional to the spread between the borrowing and lending rates, and to the shortselling cost. Numerical results suggest that the arbitrage bounds can be quite sharp; and substantially sharper than those obtained by using the Black and Scholes hedge ratios.

In Section 4, we characterize efficient consumption bundles in dynamic economies with uncertainty, taking market frictions into account. We define an efficient consumption bundle as one that is an optimal choice of at least a consumer with increasing, state-independent and risk-averse Von Neumann-Morgenstern preferences. In an economy with frictions, a consumption bundle (i.e. a contingent claim to consumption) is available, through securities trading, for a minimum cost equal to its largest price in the underlying linear economies defined by the underlying pricing rules. The linear pricing rule for which this value is attained is said to "price" the contingent claim. We show that a consumption bundle is efficient if and only if it does not lead to lower consumption in cheaper - for the linear pricing rule that "prices" it - states of the world. We also characterize the size of the inefficiency of a consumption bundle, i.e. the difference between the investment it requires and the smallest investment needed to make every maximizing agent at least as well off.

These results allow us to define a measure of portfolio performance that does not rely on mean-variance analysis (and avoids the problems associated with it : see Dybvig [1988 a] and Dybvig and Ross [1985 a and b]), taking market frictions into account. These results can also be used to evaluate the efficiency of a hedging strategy. In the perfect market case hedging and investment decisions can be separated into two distinct stages : duplicate the contingent claim to be hedged and invest optimally the remaining funds. Hence the efficiency of a hedging strategy is not really an issue since hedging amounts to duplication (or perfect hedging). In the presence of market frictions, however, hedging and investment decisions are intimately related and cannot be separated. Hence the efficiency of hedging strategies becomes an issue that can be handled with the previous results. We also evaluate the inefficiency of investment strategies followed by practitioners (such as stop-loss strategies) as in Dybvig (1988 b), and we find that the presence of market frictions may rationalize trading strategies otherwise inefficient, or at least reduce substantially their inefficiency.

In Section 5, we consider a model in which agents face investments opportunities (or investments) described by their cash flows as in Gale (1965), Cantor and Lippman (1983,1995), Adler and Gale (1993) and Dermody and Rockafellar (1991,1995). These cash flows can be at each time positive as well as negative. It is easy to show that such a model is a generalization of the classical one with financial assets. As in Cantor and Lippman (1983,1995) and Adler and Gale (1993), we will show that the absence of arbitrage opportunities is equivalent to the existence of a discount rate such that the net present value of all projects is nonpositive. We will extend this result in three directions : allowing our model to contain an infinite number of investments, allowing the cash flows to be continuous as well as discrete and finally, considering risky cash flows which is never the case for all the mentioned references. We impose that the opportunities can not be sold short and that the opportunities available today are those that are available tomorrow and the days after...(stationarity). In fact our framework permits to encompasses short sales constraints and transaction costs as well. In such a model we prove that the set of arbitrage prices is smaller than the set obtained without stationarity. Under some assumptions, we prove that there is a unique price for an option compatible with the no arbitrage condition : the Black and Scholes price even if there is transaction costs.

1 Arbitrage and equilibrium in a two period model

Let $(\Omega, \mathcal{F}, P)$ be a probability space, $X = L^2(\Omega, \mathcal{F}, P)$ the space of square integrable random variables[4] on $(\Omega, \mathcal{F}, P)$, that we assume to be separable, and X_+ the set of random variables $x \in X$ such that $P(x \geq 0) = 1$ and $P(x > 0) > 0$. If B is an element of $\mathcal{F}$, we denote by 1_B the element of X which is equal to 1 on B and to 0 elsewhere. We also denote by R the real line and by $\bar{R}$ the extended real line $R \cup \{-\infty, +\infty\}$, and if $a = (a_i)$ and $b = (b_i)$ are vectors of R^N we denote by $a \cdot b$ the dot product in R^N of a and b, by a^+ the vector $(max(0, a_i))$ and by a^- the vector $(-min(0, a_i))$. If M is a subset of X, we denote by $cl(M)$ the closure of M, and we say that M is a convex cone if for all x, y in M and all λ in R_+ we have $x + y \in M$ and $\lambda x \in M$. For instance, a linear subspace of X is a convex cone. If $\pi : M \to R$ is a functional defined on M, π is sublinear if for all x, y in M and all λ in R_+ we have $\pi(x+y) \leq \pi(x) + \pi(y)$ and $\pi(\lambda x) = \lambda \pi(x)$. Note that a sublinear functional is convex. A functional $f : X \to R$ is positive if for all $x \in X_+$ we have $f(x) > 0$. We denote the set of positive linear functionals on X by Ψ.

Consider a two period economy where consumers can purchase, at date zero, claims to consumption at date one denoted $m \in M$, where M is a convex cone of $X = L^2(\Omega, \mathcal{F}, P)$. The claim m is available to consumers at a price $\pi(m)$, in terms of today's consumption, where π is a sublinear functional defined on M. We consider this class of functionals, that includes the linear functionals, in order to take frictions and more precisely bid-ask spreads into account. For instance, consumers pay, at date 0, $\pi(m)$ to take a long position in m and receive $-\pi(-m)$ when taking a short position in m (if it is available). Since π is sublinear, we

---

[4] We shall identify random variables that are equal everywhere except on a set of probability zero, and consider X as a space of classes of random variables.

have $\pi(m) + \pi(-m) \geq 0$ and consumers pay more to buy the claim m than they receive when selling it. A more detailed justification of this choice will be given in the context of the multiperiod model of Section 2, where π is the result of a dynamic hedging process and appears naturally to be sublinear. In the next we shall characterize the price systems (M, π) that are compatible with equilibrium for the class of consumers that have continuous, convex and strictly increasing preferences, and show that they coincide with the price systems that are arbitrage free.

More precisely, the consumption space of our consumers is supposed to be $R \times X$ and we assume that every consumer is defined by a preorder of preferences $\preceq$ satisfying the following assumption [5]:

Assumption (C) :

(i) for all $(r^*, x^*) \in R \times X$, $\{(r, x) \in R \times X : (r^*, x^*) \preceq (r, x)\}$ is convex,
(ii) for all $(r^*, x^*) \in R \times X$, $\{(r, x) \in R \times X : (r^*, x^*) \preceq (r, x)\}$ and $\{(r, x) \in R \times X : (r, x) \preceq (r^*, x^*)\}$ are closed.
(iii) for all $(r^*, x^*) \in R \times X$, for all $r > 0$ and all $x \in X_+$ we have $(r^*, x^*) \prec (r^* + r, x^*)$ and $(r^*, x^*) \prec (r^*, x^* + x)$.

The class of such preferences is denoted by $\mathcal{C}$. We can now define the following notion of viability for a price system (M, π) which is identical to the definition in Harrison and Kreps (1979), except that because of the frictions induced by the sublinearity of the price functional, our consumers have budget sets that are convex cones instead of being half spaces.

Definition 1.1 *A price system (M, π), where M is a convex cone of X and π is a sublinear functional on M, is said to be viable if there exist a preorder $\preceq$ in the class $\mathcal{C}$ and a pair $(r^*, m^*) \in R \times M$ such that:*

$$r^* + \pi(m^*) \leq 0$$

and $(r, m) \preceq (r^*, m^*)$ for all $(r, m) \in R \times M$ satisfying $r + \pi(m) \leq 0$.

This says that a price system (M, π) is viable if some consumer in the class $\mathcal{C}$ can find an optimal net trade (r^*, m^*) in $R \times M$, subject to his budget constraint $r^* + \pi(m^*) \leq 0$. In fact, a viable pair (M, π) is one for which we can find equilibrium plans for some consumers belonging to the class $\mathcal{C}$ who can trade claims in $R \times M$, and this is the interpretation that we shall keep in mind.

Indeed, it is obvious that if some consumers are in equilibrium, they all have found an optimal net trade. Conversely, assume that a pair (M, π) is viable in the sense of Definition 2.1 and note $\preceq \in \mathcal{C}$ a preorder of preferences for which an optimal net trade (r^*, m^*) can be found. Define a new preorder $\preceq^*$ on $R \times X$ by: $(r, x) \preceq^* (r', x')$ if $(r + r^*, x + m^*) \preceq (r' + r^*, x' + m^*)$. It is then easy to show that $\preceq^*$ belongs to $\mathcal{C}$. Moreover $(0, 0)$ is an optimal net trade for a consumer

[5] Although this assumption looks quite general, it actually excludes typical examples. It might be possible to extend the results below to a more general class of preferences, but we have not investigated this issue yet, as we shall emphasize the no arbitrage condition in this work.

with preorder of preferences $\preceq^*$. If otherwise let $(r', m') \in R \times M$ such that $(0,0) \prec^* (r', m')$, i.e. $(r^*, m^*) \prec (r' + r^*, m' + m^*)$, and $r' + \pi(m') \leq 0$. We then have $r' + r^* + \pi(m' + m^*) \leq r' + r^* + \pi(m') + \pi(m^*)$ by sublinearity of π and hence we must have $r' + r^* + \pi(m' + m^*) \leq 0$, in addition to $(r^*, m^*) \prec (r' + r^*, m' + m^*)$, which contradicts the optimality of (r^*, m^*) in the budget set of our consumer with preorder of preferences $\preceq$. Therefore, in an economy populated by agents with preferences given by the preorder $\preceq^* \in \mathcal{C}$ and who can trade claims in $R \times M$, the prices given by the functional π are equilibrium prices since all agents weakly prefer the net trade $(0,0)$ to any other trade available to them in their budget set.

The following theorem characterizes the viability of a price system (M, π) in terms of comparisons of the price functional π with positive and continuous[6] linear functionals. We find that (M, π) is viable if and only if there exists a (frictionless) linear price functional that is viable and lies below π on M.

Theorem 1.1 *A price system (M, π), where M is a convex cone of X and π is a sublinear functional on M, is viable if and only if there exists a positive and continuous linear functional ψ defined on X such that $\psi\,|_M \leq \pi$. In particular, a necessary condition of viability is that π is positive.*

The proof of this Theorem is adapted from the proof of the similar one in Harrison and Kreps (1979) and is given in Jouini and Kallal (1996b).

The price system (M, ψ) can be interpreted as a price system of a frictionless economy where the price of every claim lies between its bid and ask price in the economy with frictions. Such a price system will be called an underlying frictionless price system of (M, π). This terminology can be justified if we think of our transaction costs economy as a perturbation of a frictionless one. The underlying frictionless price systems are then candidates for such initial frictionless economies.

Also, if M is a linear subspace of X and π is a linear functional as in the frictionless case, we find the classical result of Harrison and Kreps (1979): (M, π) is viable if and only if there exists a strictly positive linear extension of π to the whole space X.

We are now going to study the link between viability and the absence of opportunities of arbitrage in our model. Strictly speaking, an opportunity of arbitrage is a positive claim to consumption tomorrow available for nothing (or less) today. Although a viable system obviously cannot admit such opportunities, the converse does not hold in general: we also need to eliminate the possibility of getting arbitrarily close to an arbitrage opportunity in order to obtain the viability of the price system (see Kreps [1981] for an example). Hence, the following

Definition 1.2 *A free lunch is a sequence of real numbers r_n that converges to some $r^* \geq 0$, a sequence of contingent claims x_n in X that converges to some*

---

[6] Note that continuity is redundant since positive linear functionals on a Banach lattice (and hence on X) are continuous (see Jameson [1974, prop. 33.14]).

$x^* \geq 0$ such that $r^* + x^* \in X_+$, and a sequence of claims m_n in M such that $m_n \geq x_n$ and $r_n + \pi(m_n) \leq 0$ for all n[7].

Hence a free lunch is a way to get a payoff arbitrarily close to a given positive claim at no cost, or to get a payoff arbitrary close to a nonnegative claim for a negative cost. [8] This definition is fairly natural, given that we have assumed that agents have continuous preferences.

Theorem 1.2 *A price system* (M, π), *where* M *is a convex cone of* X *and* π *is a sublinear functional on* M, *admits no free lunch if and only if it is viable.*

In Jouini and Kallal (1995a) we prove that the no free-lunch assumption is equivalent to the existence of an underlying frictionless economy. Theorem 1.2 is then a direct consequence of this last result with Theorem 1.1. A direct proof of Theorem 1.2 is given in Jouini and Kallal (1996b). In the following, the terms viable and arbitrage free will be synonymous.

If we now consider a claim x that does not necessarily belong to M we can, as in Harrison and Kreps (1979), ask what prices would be reasonable (in the sense of viability) for x. We introduce the notion of bid-ask price consistent with the economy (M, π). Possible interpretations are discussed later on. We shall assume in the whole section that the price functional π is l.s.c..

Definition 1.3 *Let* (M, π) *be a viable price system. Let* $x \in X$ *and* $(q, p) \in \bar{R}^2$, *we say that* (q, p) *is a bid-ask price system of* x *consistent with* (M, π) *if there exists a l.s.c. sublinear functional* π' *defined on the convex cone* $M' = \{m + \lambda x : (m, \lambda) \in \dot{M} \times R\}$ *such that* $(-\pi'(-x), \pi'(x)) = (q, p)$, $\pi'\,|_M = \pi$ *and* (M', π') *is viable.*[9]

Note that the sublinearity of π implies that if (q, p) is a bid-ask price system of x consistent with (M, π) then $p \geq q$. In Jouini and Kallal (1996b) we prove that in a viable economies, any claim admits a consistent bid-ask price system.

Let us also define for all $x \in X$ the set

$$\Pi(x) = \{\psi(x) : \psi \in \Psi \text{ and } \psi\,|_M \leq \pi\}.$$

This set is precisely the set of prices of x in all underlying frictionless economies and it is easy to see that $\Pi(x) = -\Pi(-x)$. Note that if (M, π) is viable then

---

[7] I.e. (r_n, m_n) is in the budget set for all n.

[8] Note that if $M \cap X_+ \neq \emptyset$ one implies the other. Hence in this case we could define a free lunch as a sequence of claims $x_n \in X$ that converges to some claim $x^* \in X_+$, and a sequence of claims $m_n \in M$ such that $m_n \geq x_n$ and $\pi(m_n) \leq 0$, and get the same results in what follows. If M is closed and is such that $M = M - X_+$, and if π is nondecreasing lower semicontinuous the condition: $\pi(x) > 0$ for all $x \in M \cap X_+$, is equivalent to the absence of our free lunches.

[9] In fact, to be consistent with our framework we should not allow for infinite prices. The extension to this case, however, is straightforwood. In particular, the viability of (M', π') implies that $p > -\infty$ and $q < +\infty$. If $p = +\infty$ this means that the claim x is not marketed and if $q = -\infty$ it means that a short position in the claim x is not marketed.

$\Pi(x)$ is a nonempty interval of R, since the set $\{\psi \in \Psi : \psi \mid_M \leq \pi\}$ is nonempty and convex.

As it is shown in Jouini and Kallal (1996b), in the multiperiod securities price model a pleasant fact is that the set of consistent prices $C(x)$ is equal, modulo its boundary, to $\Pi(x)$, the set of prices of x in all the underlying frictionless economies.

As pointed out in Kreps (1981), one can take the view that $x \notin M$ is marketed somewhere and that π prices claims in M that are also marketed at the same time. A consistent price is one for which this situation is viable. Another view, perhaps more useful in practice, is to consider a market where claims in M are traded, before a market for a claim $x \notin M$ opens. A price of x consistent with (M, π) is then a price for which x might sell in this new market. However, as the new claim x is introduced, new opportunities are available to the agents and the old equilibrium might collapse. In a frictionless economy, this does not happen if x is priced by arbitrage (see Kreps [1981, Theorem 5]): as you introduce a new claim x that is priced by arbitrage, an equilibrium remains so.

In our case, as one introduces a new claim $x \notin M$, an extended price system (M', π') does not necessarily admit $(0,0)$ as an equilibrium for the preorder $\preceq$ if (M, π) does. This is nonetheless true for some extended price system, namely the supremum of all the underlying frictionless price functionals.

Proposition 1.3 *Let (M, π) be a viable price system such that $-X_+ \subset M$ and $\pi(-y) \leq 0$ for all $y \in X_+$, and let $\preceq$ be a preoder of preferences in the class $\mathcal{C}$ such that: $(r, m) \preceq (0, 0)$ for all $(r, m) \in R \times M$ satisfying $r + \pi(m) \leq 0$. Let $x \in X$, and let the price system (M', π'), defined by $M' = \{m + \lambda x : (m, \lambda) \in M \times R\}$, and $\pi'(m') = \sup_{\psi' \in \Psi : \psi' \mid_M \leq \pi} \{\psi'(m')\}$, for all $m' \in M'$. Then (M', π') is viable and we have $(r', m') \preceq (0, 0)$ for all $(r', m') \in R \times M'$ satisfying $r' + \pi'(m') \leq 0$.*

Proof : Separate strictly the convex sets J and K as in the proof of Theorem 1.1. ∎

This supremum gives in fact, as it is shown in Jouini and Kallal (1996b) the only prices that could lead to the same equilibrium if the claim were to be introduced in the market, for every possible preferences in the general class $\mathcal{C}$.

Therefore, without knowledge about preferences, it is not possible to infer tightest bounds than those given by $C(x) = cl(\Pi(x))$ for the newly introduced claim x. In our multiperiod securities price model these as tight as possible bounds are in fact the minimum amount it costs to hedge the claim x and the maximum that can be borrowed against it using securities trading.

2 The multiperiod model and the martingale approach

We consider a multiperiod economy where consumers can trade a finite number of securities at all dates $t \in \mathcal{T}$, with $\mathcal{T} \subset [0, T]$. Although we impose a finite horizon there is no other restriction on market timing: our framework includes discrete as well as continuous time models. Without loss of generality we shall assume that $\{0, T\} \subset \mathcal{T}$. These securities have a bid and an ask price at any time $t < T$, and deliver a contingent amount of consumption units at date T. Moreover, we assume that there are two sorts of securities: those which cannot be held short (i.e. in negative amounts) and those that can only be held short. The first set of securities might include a riskless asset in which consumers can invest, and the second set of securities might include a riskless asset that they can use to borrow funds. Of course, this model encompasses the case where a security can be held both in positive and negative amounts : when this is the case we include its price process twice in the model, once in each set of securities.

More precisely, let a filtration $\{\mathcal{F}_t\}_{t \in \mathcal{T}}$ model the information structure of our economy. Let $Z'(t) = (Z'_1(t), \ldots, Z'_K(t))$ and $Z(t) = (Z_1(t), \ldots, Z_K(t))$ be two K dimensional positive processes that denote the bid and the ask prices of the K securities traded in the market. Consumers can buy security k at a price $Z_k(t)$ and sell it at a price $Z'_k(t)$ at any date $t \in \mathcal{T}$. Moreover, we assume that Z and Z' are adapted to $\{\mathcal{F}_t\}_{t \in \mathcal{T}}$, i.e. that bid-ask prices at any date depend on past and current information only, and that $E((Z_k)^2(t)) < \infty$ and $E((Z'_k)^2(t)) < \infty$ for all $t \in \mathcal{T}$ and $k = 1, \ldots, K$, which means that bid and ask prices have finite second moments. We also assume that the bid-ask price processes Z and Z' and the filtration $\{\mathcal{F}_t\}_{t \in \mathcal{T}}$ are right-continuous.[10]

For the same reasons as in Harrison and Kreps (1979), consumers are only allowed to use simple strategies, i.e. they are only allowed to trade at a finite (but arbitrarily large) number of (arbitrary) dates, that do not depend on the path followed by prices. This restriction eliminates in particular doubling strategies.[11] Furthermore since the bid and the ask prices are possibly different, we separate trading strategies into a cumulative long position and a cumulative short position (which are consequently nondecreasing). The difference between these two gives the net position of our consumer in the K securities at any date. Formally, a simple trading strategy is a pair (θ, θ') of nonnegative and nondecreasing K dimensional processes such that there exists a set of trading dates $0 = t_0 \leq \ldots \leq t_N = T$ for which $(\theta(t, \omega), \theta'(t, \omega))$ is constant over the intervals of time $[t_{n-1}, t_n)$, for $n = 1, \ldots, N$. We also assume that (θ, θ') is adapted to $\{\mathcal{F}_t\}_{t \in \mathcal{T}}$, which means that investors can only use current and past information to trade, and we impose the integrability conditions $E((\theta_k Z_k)^2(t)) < \infty$, $E((\theta'_k Z'_k)^2(t)) < \infty$, $E((\theta_k Z'_k)^2(t)) < \infty$, and $E((\theta'_k Z_k)^2(t)) < \infty$ for all $t \in \mathcal{T}$ and $k = 1, \ldots, K$, which guarantee that the payoffs of the strategies have finite second moments.

We incorporate short sales constraints into the model by restricting investors to hold nonnegative net positions in securities $k = 1, \ldots, S$, and nonpositive net positions in securities $k = S + 1, \ldots, K$. We say that a trading strategy is admissible if it satisfies these constraints.

Definition 2.1 *A simple strategy is said to be admissible if it does not require*

---

[10] A filtration $\{\mathcal{F}_t\}_{t \in \mathcal{T}}$ is right-continuous if for all $t \in [0, T]$, $\mathcal{F}_t$ is the intersection of the σ-algebras $\mathcal{F}_s$, where $s > t$.

[11] Harrison and Pliska (1979) and Dybvig and Huang (1988) suggest that imposing a lower bound on wealth instead would lead to similar results even allowing continuous trading. However, these papers assume the absence of free lunch, and one of the aims of this section is to characterize the absence of free lunch.

selling short securities $k = 1, \ldots, S$, i.e. if $\theta_k(t, \omega) \geq \theta'_k(t, \omega)$ for all $k = 1, \ldots, S$, and if it does not require holding positive quantities of securities $k = S+1, \ldots, K$, i.e. if $\theta_k(t, \omega) \leq \theta'_k(t, \omega)$ for all $k = S + 1, \ldots, K$.

Also, we normalize the price of one of the securities that cannot be sold short to 1, i.e. we assume that $Z_1(t) = Z'_1(t) = 1$, for all $t \in \mathcal{T}$. This essentially amounts to express securities prices in terms of a numeraire (or unit of account) that can be stored. On the other hand, we assume that it is possible to have negative balances in the unit of account (i.e. to borrow) but that it can be costly to do so. In order to model this point, we assume that one of the securities that can be held in nonpositive quantities only has a price process that is nondecreasing and bounded, i.e. we assume that the price process $Z_{S+1}(t) = Z'_{S+1}(t)$ is positive, nondecreasing[12] and bounded. These assumptions amount to set a zero interest rate and to allow for a spread between the borrowing and lending rate. If an interest rate is actually earned[13] on positive balances of the numeraire, we can normalize all securities prices by the value of the accumulation process, i.e. the value of a portfolio where one unit of numeraire is invested at the initial date, and we are back to our model. Therefore, this assumption is made without much loss of generality.

Consumers are assumed not to have external sources of financing, they consume only at dates 0 and T, and hence sell (or short) some securities in order to purchase others. This is formalized by the notion of self-financing strategy which characterizes the feasible (in the sense of the budget constraints) trading strategies.

Definition 2.2 *A simple strategy (θ, θ') is said to be self-financing if for $n = 1, \ldots, N$ we have*

$$(\theta(t_n) - \theta(t_{n-1})) \cdot Z(t_n) \leq (\theta'(t_n) - \theta'(t_{n-1})) \cdot Z'(t_n).$$

The set of simple self-financing admissible strategies is denoted by Θ. It is easy to see that Θ is a convex cone of the set of trading strategies. In fact, a self-financing trading strategy (θ, θ') can be seen as a way of transferring wealth from the initial date 0 to the final date T : it guarantees, at date T,

$$(\theta - \theta')^+(T) \cdot Z'(T) - (\theta - \theta')^-(T) \cdot Z(T)$$

units of consumption contingent on the state of the world, and costs $\theta(0) \cdot Z(0) - \theta'(0) \cdot Z'(0)$ units of date 0 consumption.

In the subsequent analysis we shall characterize economies where there are no opportunities of arbitrage profits, i.e. where it is not possible to obtain a payoff that is nonnegative and strictly positive in some states of the world, at no cost. To be more precise, we shall examine economies in which there are no free

---

[12] We could actually only assume that it has bounded variation and derive the fact that it has to be nondecreasing from the absence of arbitrage.

[13] The interest rate can be negative in real terms.

lunches, which include arbitrage opportunities : roughly speaking, a free lunch is the possibility of getting payoffs arbitrarily close to a given positive contingent claim at an arbitrarily small cost. Back and Pliska (1990) provide an example of an economy where there are no arbitrage opportunities and yet the "fundamental theorem of asset pricing" fails : prices do not have the "martingale property" and there is no linear pricing rule. Since our aim is to find the counterpart of the "fundamental theorem of asset pricing" in an economy with market frictions, we shall study the implications of the absence of free lunches in our model.

Definition 2.3 *A free lunch is a sequence of contingent claims x_n in X converging to some x^* in X_+ and such that there exists a sequence of strategies (θ^n, θ'^n) in Θ satisfying $(\theta^n - \theta'^n)^+(T) \cdot Z'(T) - (\theta^n - \theta'^n)^-(T) \cdot Z(T) \geq x_n$ for all n, and*

$$\lim_n \{\theta'^n(0) \cdot Z(0) - \theta'^n(0) \cdot Z'(0)\} \leq 0.$$

An immediate consequence of the absence of free lunch is that the ask price of any security must lie above its bid price: $P(\{\omega : Z'(t, \omega) \leq Z(t, \omega)\}) = 1$, for all $t \in \mathcal{T}$ (see Jouini and Kallal [1995a]).

In order to describe the investment opportunities in our economy, we consider the set M of marketed claims, i.e. of payoffs that can be hedged by an admissible self-financing simple trading strategy. Formally, a claim $x \in X$ belongs to M if there exists an admissible self-financing simple trading strategy (θ, θ') in Θ satisfying $(\theta - \theta')^+(T) \cdot Z'(T) - (\theta - \theta')^-(T) \cdot Z(T) \geq x$. It is easy to see that M is a convex cone.

When there are shortsale constraints and/or transaction costs it is not true that the cheapest way to achieve at least a given contingent payoff at date T is to duplicate it.[14] However, the set of available contingent rates of return in our economy can be represented by the price functional π defined, for every $x \in X$, by

$$\pi(x) = \inf\{ \liminf_n \{\theta^n(0) \cdot Z(0) - \theta'^n(0) \cdot Z'(0)\} : (\theta^n, \theta'^n) \in \Theta,$$
$$(\theta^n - \theta'^n)^+(T) \cdot Z'(T) - (\theta^n - \theta'^n)^-(T) \cdot Z(T) \geq x_n$$
$$\text{and } (x_n) \subset X \text{ converges to } x\}.$$

In words, $\pi(x)$ represents the infimum cost necessary to get at least a payoff arbitrarily close to x at date T. This means that π indeed summarizes the investment opportunities in our dynamic economy. It turns out that this price functional is sublinear, which means that it is less expensive to hedge the sum $x + y$ of two claims than to hedge x and y separately. It is easy to see why: the sum of the trading strategies that hedge x and y hedges $x + y$ but some orders to buy and sell the same security at the same date might cancel out, generating some savings on transaction costs and/or making the short sales constraints nonbinding. We are now able to appreciate the main differences between our model and the frictionless case: in a frictionless economy the set of marketed claims M is a *linear subspace* and the price functional π that characterizes the

---

[14] This fact has been underlined by Bensaid et al. (1992) in a discrete time and states framework.

opportunity set of returns is *linear*. Instead we find that in securities market models with frictions, M is a convex cone and π is a sublinear functional.

At this point, we are in a position to prove our main result that characterizes arbitrage free[15] bid and ask price processes. We find that the absence of free lunch [16] is equivalent to the existence of a numeraire process Z_0 and an equivalent probability measure that transforms some processes lying between the normalized bid-ask price processes of the securities that cannot be sold short into *supermartingales,* and some processes lying between the normalized bid-ask price processes of the securities that can only be held short into *submartingales.* [17]

Let us first define the set $\Theta'(t,k) = \{(\theta,\theta') \in \Theta : (\theta,\theta')(s) = 0$ for all $s < t, \theta'_k(T) = 1, \theta_j(T) = 0$ for $j \neq 0, \theta'_j(T) = 0$ for $j \notin \{0,k\}$ and $\theta(t) \cdot Z(t) = \theta'(t) \cdot Z'(t)\}$, and the set $\Theta(t,k) = \{(\theta,\theta') \in \Theta : (\theta,\theta')(s) = 0$ for all $s < t, \theta_k(T) = 1, \theta'_j(T) = 0$ for $j \neq 0, \theta_j(T) = 0$ for $j \notin \{0,k\}$, and $\theta(t) \cdot Z(t) = \theta'(t) \cdot Z'(t)\}$.

In words, $\Theta'(t,k)$ is the set of the strategies that consist in going short in one security k (and investing the proceeds in security 0) between time t and the final date T (but not necessarily at the same time in different events). The set $\Theta(t,k)$ is defined symmetrically.

The following theorem characterizes arbitrage free bid-ask price processes with shortsales constraints and generalizes the main result of Jouini and Kallal (1995a). This last one will be derived as a consequence of the following theorem in Section 3.

Theorem 2.1 *Assume that the borrowing and the lending rates are equal (i.e. $Z_1 = Z_{K+1} = 1$) then :*
(i) The securities price model admits no multiperiod free lunch if and only if there exist at least a probability measure P^* equivalent[18] to P with $E((\frac{dP^*}{dP})^2) < \infty$ and a process Z^* satisfying[19] $Z' \leq Z^* \leq Z$ such that Z^*_k is a supermartingale for $k = 1, \ldots, K$ and a submartingale for $k = K+1, \ldots, S$ with respect to the filtration $\{\mathcal{F}_t\}$ and the probability measure P^*.
(ii) Moreover, if we denote by E^* the expectation operator associated to the "(super/sub)-martingale measure" P^*, there is a one-to-one correspondence between the set of such expectation operators and the set of linear functionals $\psi \in \Psi$ such that $\psi \mid_M \leq \pi$. This correspondence is given by the following formulas:

$$P^*(B) = \psi(1_B), \text{ for all } B \in \mathcal{F} \text{ and } \psi(x) = E^*(x), \text{ for all } x \in X.$$

---

[15] We shall use the terms "arbitrage free" and "absent of free lunches" interchangeably.

[16] We could also relate the absence of free lunch to the viability as in Section 1.

[17] A stochastic process Y is a martingale with respect to the probability measure P^* and the filtration $\{\mathcal{F}_t\}$ if $E^*(Y(s) \mid \mathcal{F}_t) = Y(t)$ for all $s \geq t$, where E^* is the expectation operator associated to P^*. A process Y is a supermartingale with respect to P^* and $\{\mathcal{F}_t\}$ if $E^*(Y(s) \mid \mathcal{F}_t) \leq Y(t)$ for all $s \geq t$. A process Y is a submartingale with respect to P^* and $\{\mathcal{F}_t\}$ if $E^*(Y(s) \mid \mathcal{F}_t) \geq Y(t)$ for all $s \geq t$.

[18] I.e. P and P^* have exactly the same zero measure sets.

[19] This means that for all t, $P(\{\omega : Z'(t,\omega) \leq Z^*(t,\omega) \leq Z(t,\omega)\}) = 1$.

Furthermore for such an expectation operator we can take

$$Z_k^*(t,\omega) = \sup_{(\theta,\theta')\in\Theta'(t,k)} \{E^*((\theta_0 - \theta_0')(T) \mid \mathcal{F}_t)(\omega)\},\ for\ k = 1,\ldots,K$$

$$and\ Z_k^*(t,\omega) = \inf_{(\theta,\theta')\in\Theta(t,k)} \{E^*(-(\theta_0 - \theta_0')(T) \mid \mathcal{F}_t)(\omega)\},\ for\ k = K+1,\ldots,S.$$

(iii) Furthermore, for all $m \in M$ *we have*

$$[-\pi(-m),\pi(m)] = cl\{E^*(m) : P^*\ is\ a\ (super/sub)\text{-}martingale\ measure\ \}.$$

The proof of this Theorem is at the end of this section. One consequence of this result is that *any* arbitrage-free (or equivalently, viable) bid ask price process (Z',Z) can be seen as a perturbation of an arbitrage-free frictionless economy price process Z^*, with $Z' \leq Z^* \leq Z$. Conversely, it is obvious that if Z^* is an arbitrage-free frictionless price process, then any perturbed price system process (Z',Z), with $Z' \leq Z^* \leq Z$, is arbitrage-free and Z^* defines one of the underlying frictionless economies.

In particular, neither Z' nor Z needs to be a martingale (i.e. arbitrage-free by themselves in a frictionless model) for the securities price model to be arbitrage-free. This has been illustrated in the introduction by the example of a deterministic economy. Let us now briefly examine two stochastic examples. Consider, to start with, an economy where a stock bid and ask price processes follow diffusions with a singularity (zero volatility) at a point where their drift is not equal to the riskless rate. In a frictionless model this would lead to arbitrage opportunities. However, if the bid price is strictly lower than the ask price, according to our result the model with transaction costs is arbitrage-free. Now let us turn to a two-period economy with two possible states of the world ("up" and "down") at date 1. Suppose that the interest rate is equal to zero, and that a stock is available for trading at date 0 and pays off (at date 1) $S_u = \$110$ in state "up" and $S_d = \$90$ in state "down". We denote by S_0 its ask price, and by S_0' its bid price at date 0. According to Theorem 2.1 (part (i)), the model is arbitrage-free if and only if $S_0 \geq S_0'$, $S_0 > 90$, and $S_0' < 110$ (which do not imply that S or S' is a martingale). Also note that if the model is arbitrage-free, there are infinitely many equivalent martingale measures: they are of the form $p^* = \frac{S_0^*-90}{110-90}$ where S^* is any process satisfying $S_u^* = S_u$, $S_d^* = S_d$, and $S_0^* \in [S_0',S_0] \cap [90,110]$. Nonetheless, it is easy to see that in this model any contingent claim can be obtained by trading in the bond and in the stock, i.e. that markets are *complete*.[20]

Another consequence of this Theorem (part (iii)) is that it allows a representation, in terms of the set of equivalent martingale measures, of the set of investment opportunities available in this economy. A contingent claim x is available at a cost (in terms of today's consumption) $\pi(x)$ which is equal to

[20] It would be possible to have infinitely many processes S^* between the bid and the ask price processes, with each of the processes S^* having infinitely many martingale measures. To see this, add a third state of the world (e.g. "middle") to this example, with the stock paying $S_m = 100$ in that state. In this case, markets would be incomplete.

the *largest* expected value with respect to the martingale measures. Similarly, the maximum amount $-\pi(-x)$ that can be borrowed (for today's consumption) against the claim x is equal to its *smallest* expected value with respect to the martingale measures. This means that the martingale measures can be interpreted as possible stochastic discount factors, i.e. today's prices of \$1 to be paid in a given state of the world. This also means that the interval $[-\pi(-x), \pi(x)]$ defines arbitrage bounds on the price of the contingent claim x. Indeed, x would not be traded for a price outside this interval since this would either lead to an arbitrage opportunity or be dominated by trading in the underlying securities. On the other hand, it is possible to show that any bid-ask price pair in this interval can be supported by some strictly increasing continuous linear (and hence convex) preferences. Therefore, without any further knowledge about preferences, $-\pi(-x)$ and $\pi(x)$ are the tightest bounds that can be inferred on the price of the claim x.

Let us illustrate this point by going back to our two-periods-and-states-of-the-world example. Suppose that $S_0' = 100$ and $S_0 \geq 100$. It is easy to see that the minimum cost to hedge the call is equal to $0.5S_0 - 45$ (i.e. the cost of duplication) if $S_0 \leq 110$, and is equal to 10 otherwise. Indeed, if $S_0 > 110$ the transaction costs on the stock are too large and duplication is no longer optimal. Moreover, the maximum amount that can be borrowed against the call is \$5. On the other hand, the martingale measures are of the form $(p^*, 1 - p^*)$, with $1 > p^* > 0$ and $\frac{S_0 - 90}{20} \geq p^* \geq 0.5$. It is easy to check that the interval defined by the expected values of the payoff of the call with respect to these probabilities is equal to the interval defined by the bounds computed using the optimal hedging strategies (note that when $S_0 > 110$ they coincide up to the boundary only).

Also, in our securities price model a maximizing agent will equate his marginal utility to one of the martingale measures (one that "prices" his optimal consumption bundle). Indeed, it is easy to see that if (r^*, m^*) is an optimal consumption bundle satisfying the budget constraint $r^* + \pi(m^*) \leq 0$, then some hyperplane separates feasible bundles from bundles that are strictly preferred to (r^*, m^*). This hyperplane is given by one of the positive linear functionals ψ that lie below π on M and is tangent to the indifference surface that goes through (r^*, m^*). Moreover, it is easy to see that we must have $\psi(m^*) = \pi(m^*)$, which means that ψ "prices" m^*. And by Theorem 2.1 (part (ii)) such a linear pricing rule ψ corresponds to a martingale measure. Conversely, any equivalent martingale measure E^* could be the vector of intertemporal marginal rates of substitution of a maximizing agent in our economy. Indeed, by Theorem 2.2 (part (ii)) the linear pricing rule ψ defined by $\psi(x) = E^*(x)$ for all $x \in X$ lies below π on M. Moreover it is easy to see that an agent with a utility function defined by $u(r, x) = r + \psi(x)$ for all $(r, x) \in R \times X$ and with an endowment $(0, 0)$ will be happy not to trade.

In the following Theorem we will focus our attention to the shortselling constraints and for this purpose we will assume that there is no bid-ask spreads.

Theorem 2.2 *Assume that there is no bid-ask spreads on the marketed securities (i.e. $Z = Z'$) and that the lending rate can be written on the form*

$Z_{K+1}(t,\omega) = exp(\int_0^t r(t,\omega)dt)$ where r is some nonnegative bounded (by some $\bar{r}$) and right continuous adapted instantaneous rate process then :

(i) The securities price model admits no free lunches if and only if there exist a probability measure P^* equivalent to P with $E((\frac{dP^*}{dP})^2) < \infty$, and a positive numeraire process $Z_0^* = exp(\int_0^t r_0(s,\omega)ds)$ with $0 \leq r_0(t,\omega) \leq r(t,\omega)$, such that ($\frac{Z_k}{Z_0^*}$ is a supermartingale for $k = 1,\ldots,K$ and a submartingale for $k = K+1,\ldots,S$ with respect to the filtration $\{\mathcal{F}_t\}$ and the probability measure P^*.

(ii) If we denote by E^* the expectation operator associated to P^*, there is a one-to-one correspondence between the set of such expectation operators and the set of positive linear functionals $\psi \in \Psi$ such that $\psi \mid_M \leq \pi$. This correspondence is given by :

$$P^*(B) = \psi(1_B), \text{ for all } B \in \mathcal{F} \text{ and } \psi(x) = Z_0^*(0)E^*(\frac{x}{Z_0^*(T)}), \text{ for all } x \in X.$$

(iii) For all $x \in M$ we have,
$$[-\pi(-x), \pi(x)] = cl\{\psi(x) : \psi \mid_M \leq \pi \text{ and } \psi \in \Psi\}$$
$$= cl\{Z_0^*(0)E^*(\frac{x}{Z_0^*(T)}) : Z_0^* \text{ and } P^* \text{ are as in } (i)\}.$$

The proof is at the end of this section. In words, we find that the absence of free lunch is equivalent to the existence of (1) a numeraire process Z_0^* that increases at a rate smaller than or equal to the borrowing rate and larger than or equal to the lending rate, and (2) an equivalent probability measure that transforms the (normalized by the numeraire) price processes of the securities that cannot be sold short into *supermartingales,* and the (normalized by the numeraire) price processes of the securities that can only be held in nonpositive quantities into *submartingales.* This super/submartingale measure is in fact what is sometimes called a "risk-neutral" measure. In particular, if a security is not subject to any constraint, it follows that its (normalized) price process $\frac{Z_i^*}{Z_0^*}$ must be a supermartingale and a submartingale, i.e. a martingale, under the "risk-neutral" measure. Hence, if there are no constraints at all in the economy, and if there is no spread between the borrowing and the lending rates, the numeraire must be equal to 1 and there must exist an equivalent probability measure for which securities prices are martingales (as shown by Harrison and Kreps [1979]).

While part (i) of the Theorem characterizes the absence of arbitrage in terms of a numeraire process and of supermartingale/submartingale measures, part (ii) states that these probability measures and numeraire define all the positive linear functionals that lie below the sublinear functional π (that describes the random returns available in our dynamic economy). These positive linear functionals can be interpreted as representing the price functionals of underlying frictionless economies. This turns out to be useful to compute arbitrage bounds on the price of any contingent claim. Indeed, part (iii) states that for every contingent claim x, the interval $[-\pi(-x), \pi(x)]$ is equal to the closure of the set of prices of the claim x in all the underlying frictionless economies. Recall that $\pi(x)$ is the smallest amount necessary to reach (or dominate) the payoff x by trading in the underlying securities. Hence, if the claim x were to be introduced in the market at date 0 it would not be bought for more than $\pi(x)$ or sold for a price under $-\pi(-x)$, since a better deal could be achieved through securities trading in both cases. Therefore the bid and the ask prices of x must fall in the interval $C(x) = [-\pi(-x), \pi(x)]$. Note that this does not mean that if the bid price of x is below $-\pi(-x)$ or if the ask price of x is above $\pi(x)$ then there are opportunities of arbitrage; it only means that there are no transactions at these prices[21]. However, if the bid price of x were to be above $\pi(x)$ or its ask price were to be below $-\pi(-x)$ then selling the claim x at the bid price or buying it at the ask price and hedging the position through securities trading would constitute a free lunch. Also, Theorem 2.2 characterizes the set $\Pi(x)$ using the set of martingale measures. Hence, the interval $C(x) = [-\pi(-x), \pi(x)]$ defined by the arbitrage bounds on the bid-ask prices of the claim x is equal to the closure of the set of expectations of x with respect to all the equivalent probability measures that transform some process between the bid and the ask price processes of each traded security into a martingale.

We conclude this section by the proofs of Theorem 2.1 and 2.2

Proof of Theorem 2.1 : First, let P^* be a probability measure equivalent to P and let Z^*, with $Z' \leq Z^* \leq Z$, be a super/submartingale with respect to P^* and $\{\mathcal{F}_t\}$ (as in the theorem). Define the linear functional ψ by $\psi(x) = E^*(x)$ for all $x \in X$. Since $\rho = \frac{dP^*}{dP} \in X$ we have by the Riesz representation Theorem that $\psi(x) = E^*(x) = E(\rho x)$ is continuous. Since P and P^* are equivalent, we have $\psi \in \Psi$.

Let $m \in M$ and let $(\theta, \theta') \in \Theta$ with trading dates $0 = t_0 \leq t_1 \leq \ldots \leq t_N = T$, such that $(\theta - \theta')^+(T) \cdot Z'(T) - (\theta - \theta')^-(T) \cdot Z(T) \geq m$. Since $Z' \leq Z^* \leq Z$ and (θ, θ') is nondecreasing and self-financing, we have, for $n = 1, \ldots, N$,

$$E^*((\theta(t_n) - \theta(t_{n-1})) \cdot Z^*(t_n) - (\theta'(t_n) - \theta'(t_{n-1})) \cdot Z^*(t_n) \mid \mathcal{F}_{t_{n-1}})$$
$$\leq E^*((\theta(t_n) - \theta(t_{n-1})) \cdot Z(t_n) - (\theta'(t_n) - \theta'(t_{n-1})) \cdot Z'(t_n) \mid \mathcal{F}_{t_{n-1}}) \leq 0.$$

and then

$$E^*((\theta - \theta')(t_n) \cdot Z^*(t_n) \mid \mathcal{F}_{t_{n-1}}) \leq E^*((\theta - \theta')(t_{n-1}) \cdot Z^*(t_n) \mid \mathcal{F}_{t_{n-1}}).$$

Using the fact that Z_k^* is a supermartingale when the security k can not be sold short (i.e. $\theta - \theta' \geq 0$) and a submartingale when security k can only be held in nonpositive amount (i.e. $\theta - \theta' \leq 0$) we obtain

$$E^*((\theta - \theta')(t_n) \cdot Z^*(t_n) \mid \mathcal{F}_{t_{n-1}}) \leq (\theta - \theta')(t_{n-1} \cdot Z^*(t_{n-1}.$$

By iteration, $E^*((\theta - \theta')(T) \cdot Z^*(T)) \leq (\theta - \theta')(0) \cdot Z^*(0) \leq \theta(0) \cdot Z(0) - \theta'(0) \cdot Z'(0)$. Since $(\theta - \theta')^+(T) \cdot Z'(T) - (\theta - \theta')^-(T) \cdot Z(T) \leq \{(\theta - \theta')^+(T) - (\theta - \theta')^-(T)\} \cdot Z^*(T)$ we have $E^*((\theta - \theta')^+(T) \cdot Z'(T) - (\theta - \theta')^-(T) \cdot Z(T)) \leq \theta(0) \cdot Z(0) - \theta'(0) \cdot Z'(0)$. It is easy to see that this implies that there cannot be any multiperiod free lunch.

---

[21] Note that we are assuming the opening of a spot market for the contingent claim x at date 0 only.

Moreover, taking the infimum over the strategies $\theta \in \Theta$ such that $(\theta - \theta')^+(T) \cdot$ $Z'(T) - (\theta - \theta')^-(T) \cdot Z(T) \geq m$, we obtain that $\psi(m) = E^*(m) \leq \tilde{\pi}(m)$, where $\tilde{\pi}(m) = \inf\{\theta(0) \cdot Z(0) - \theta'(0) \cdot Z'(0) : (\theta, \theta') \in \Theta$ and $(\theta - \theta')^+(T) \cdot Z'(T) - (\theta - \theta')^-(T) \cdot Z(T) \geq m\}$, for all $m \in M$. Using the fact that ψ is continuous, we have by Lemma 1 below that $\psi \mid_M \leq \pi$.

Lemma 1 (Jouini and Kallal (1995a)) π *is the largest l.s.c. functional that lies below $\tilde{\pi}$.*

Conversely, assume that the securities price model, or equivalently (M, π), admits no free lunch and let $\psi \in \Psi$ such that $\psi \mid_M \leq \pi$, as guaranteed by Theorem 1.1. Define P^* from ψ by $P^*(B) = \psi(1_B)$ for all $B \in \mathcal{F}$. By linearity and strict positivity of ψ it is clear that P^* is a measure equivalent to P. Using the fact that $Z_0 = Z'_0 = 1$ it is easy to show that $P^*(1_\Omega) = 1$.

Since ψ is continuous, by the Riesz representation Theorem there exists a random variable $\rho \in X$ such that $\psi(x) = E(\rho x)$, for all $x \in X$. Thus, $P^*(B) = E(\rho 1_B)$ for all $B \in \mathcal{F}$ and $\frac{dP^*}{dP} = \rho$ is square integrable. It remains to show that there exists a process Z^*, with $Z'_k \leq Z^*_k \leq Z_k$, and such that Z^*_k is a supermartingale for $k = 1, \ldots, K$ and a submartingale for $k = K+1, \ldots, S$ with respect to P^* and $\{\mathcal{F}_t\}$.

Consider the processes Z^* defined in (ii). In words, when the security k can be sold short, Z^*_k is the supremum of the conditional expected value of the proceeds from the strategies that consist in going short in one security k (and investing the proceeds in security 0) between time t and the final date T (but not necessarily at the same time in different events). When the security k can only be held in nonpositive quantities, the process Z^* is defined symmetrically. We obviously must have $Z^*_k(t) \leq Z_k(t)$ for $k = 1, \ldots, K$ and for all t. Indeed, one of the strategies in $\Theta(t, k)$ consists in going long at once at date t in one unit of security k. Moreover we also must have $Z'(t) \leq Z^*(t)$ for $k = 1, \ldots, K$ and all t. Indeed, assume instead that there exists $t \in [0, T]$ and $B \in \mathcal{F}_t$ such that $P(B) > 0$ and $Z'(t, \omega) > Z^*(t, \omega)$ for all $\omega \in B$. Then there would exists a strategy $(\alpha, \alpha') \in \Theta(t, k)$ satisfying $E^*(Z'_k(t)1_B) > E^*(-(\alpha_0 - \alpha'_0)(T)1_B)$. At the end of this strategy we have one unit of security k and we can sale it at a price $Z'_k(T)$. This operation defines a new trategy (θ, θ') with $\theta_0(T) = \alpha_0(T) + Z'_k(T)$, $\theta_k(T) = 0$ and $\theta = \alpha$ in all other cases. We then have $E^*((\theta_0 - \theta'_0)(T)1_B) > 0$, and hence $E^*([(\theta - \theta')^+(T) \cdot Z'(T) - (\theta - \theta')^-(T) \cdot Z(T)]1_B) > 0$ which contradicts the fact that $E^*([(\theta - \theta')^+(T) \cdot Z'(T) - (\theta - \theta')^-(T) \cdot Z(T)]1_B) \leq \pi([(\theta - \theta')^+(T) \cdot Z'(T) - (\theta - \theta')^-(T) \cdot Z(T)]1_B) \leq 0$. Hence $Z'_k(t) \leq Z^*_k(t)$ for all t and Z^*_k lies between Z_k and Z'_k for $k = 1, \ldots, K$. A symetric argument gives the result for $k = K+1, \ldots, S$.

Moreover we must have $E^*(Z^*_k(s) \mid \mathcal{F}_t) \leq Z^*_k(t)$ for $k = K+1, \ldots, S$ and $E^*(Z^*_k(s) \mid \mathcal{F}_t) \geq Z^*_k(t)$ for $k = 1, \ldots, K$ for all $s \geq t$, since $\Theta'(s, k) \subset \Theta'(t, k)$ and $\Theta(s, k) \subset \Theta(t, k)$ for all $s \geq t$ and all $\omega \in \Omega$, and by the law of iterated expectations.

For part (iii) see Jouini and Kallal (1996b)   ∎

Proof of Theorem 2.2:   First note that if there is no multiperiod free lunch, then the set M is a convex cone and the price functional π is a lower semicontinuous[22] sublinear functional on M which takes value in R. It is then easy to see that our multiperiod model admits no free lunches if and only if the induced two-period model (M, π) has the same property.

Then let P^* be a probability measure equivalent to P such that $\frac{dP^*}{dP} \in X$ and let Z_0^* be a bounded (and bounded away from zero) absolutely continuous positive numeraire process such that $\frac{Z_k}{Z_0^*}$ is a supermartingale for $k = 1, \ldots, K$ and a submartingale for $k = K + 1, \ldots, S$. Without loss of generality, we shall assume that $Z_0^*(0) = 1$. Define the linear functional ψ by $\psi(x) = E^*(\frac{x}{Z_0^*(T)})$ for all $x \in X$. Since $\rho = \frac{dP^*}{dP} \in X$ we have that (by the Riesz representation Theorem) ψ is continuous. Since P and P^* are equivalent, ρ is strictly positive. Thus, ψ is also strictly positive and we have $\psi \in \Psi$.

Since there is no bid-ask spreads on the traded securities, the cost and the result of a strategy (θ, θ') only depend from the net position $\theta - \theta'$. In the next, we will denote by θ this net position and we will consider that the strategy is entirely described by θ. Let $m \in M$ and let θ be a simple self-financing strategy that is admissible (i.e. that satisfies the short sales constraints), with trading dates $0 = t_0 \leq t_1 \leq \ldots \leq t_N = T$, such that $\theta(T) \cdot Z(T) \geq m$. Since the strategy θ is self-financing, we must have

$$E^*\left((\theta(t_n) - \theta(t_{n-1})) \cdot \frac{Z(t_n)}{Z_0^*(t_n)} \mid \mathcal{F}_{t_{n-1}}\right) \leq 0,$$

which can also be written

$$E^*\left(\theta(t_n) \cdot \frac{Z(t_n)}{Z_0^*(t_n)} \mid \mathcal{F}_{t_{n-1}}\right) \leq E^*\left(\theta(t_{n-1}) \cdot \frac{Z(t_n)}{Z_0^*(t_n)} \mid \mathcal{F}_{t_{n-1}}\right).$$

Since θ is adapted to the filtration $\{\mathcal{F}_t\}$ we must have

$$E^*\left(\theta(t_{n-1}) \cdot \frac{Z(t_n)}{Z_0^*(t_n)} \mid \mathcal{F}_{t_{n-1}}\right) = \theta(t_{n-1}) \cdot E^*\left(\frac{Z(t_n)}{Z_0^*(t_n)} \mid \mathcal{F}_{t_{n-1}}\right)$$

Moreover using the supermartingale and the submartingale properties with respect to P^*, together with the fact that $\theta_i \geq 0$ for all $i = 1, \ldots, K$, $\theta_j \leq 0$ for all $j = K + 1, \ldots, S$, we obtain

$$\theta(t_{n-1}) \cdot E^*\left(\frac{Z(t_n)}{Z_0^*(t_n)} \mid \mathcal{F}_{t_{n-1}}\right) \theta(t_{n-1}) \cdot \frac{Z(t_{n-1})}{Z_0^*(t_{n-1})},$$

which can be rewritten

$$E^*\left(\theta(t_n) \cdot \frac{Z(t_n)}{Z_0^*(t_n)} \mid \mathcal{F}_{t_{n-1}}\right) \leq \theta(t_{n-1}) \cdot \frac{Z(t_{n-1})}{Z_0^*(t_{n-1})}.$$

---

[22] I.e. Such that $\{(m, \lambda) \in M \times R : \lambda \geq \pi(m)\}$ is closed in $M \times R$, or equivalently such that $\{m \in M : \lambda \geq \pi(m)\}$ is closed in M for all $\lambda \in R$, or equivalently such that $\liminf_{n}\{\pi(m_n)\} \geq \pi(m)$ whenever the sequence $(m_n) \subset M$ converges to $m \in M$.

Iterating this reasoning, and using the law of iterated expectations, we obtain that

$$E^*(\theta(T) \cdot \frac{Z(T)}{Z_0^*(T)} \leq \theta(0) \cdot Z(0).$$

This inequality precludes the existence of a free lunch. Taking the infimum over the strategies $\theta \in \Theta$ it is then easy to show that $\psi \mid_M \leq \pi$.

Conversely, assume that the multiperiod model (or equivalently the induced two-period model (M, π)) admits no free lunch, and let $\psi \in \Psi$ such that $\psi \mid_M \leq \pi$, as guaranteed by Theorems 1.1 and 1.2. Since ψ is continuous, by the Riesz representation Theorem there exists some $\rho \in X$ such that $\psi(x) = E(\rho x)$, for all $x \in X$.

Suppose, to start with, that $Z_{K+1} = Z_1 = 1$. Define P^* from ψ by $P^*(B) = \psi(1_B)$ for all $B \in \mathcal{F}$. By linearity and strict positivity of ψ it is clear that P^* is a measure equivalent to P. As in Jouini and Kallal (1995a) it is easy to show that $\pi(1_\Omega) = 1$, $\pi(-1_\Omega) = -1$ and hence that $\psi(1_\Omega) = P^*(1_\Omega) = 1$. It follows that P^* is a probability measure. Moreover, $P^*(B) = E(\rho 1_B)$ for all $B \in \mathcal{F}$ and $\frac{dP^*}{dP} = \rho$ is square integrable. It remains to show that Z_k is a supermartingale for $k = 1, \ldots, K$ and a submartingale for $k = K + 1, \ldots, S$ with respect to P^*. In order to do this, consider the strategy $\theta \in \Theta$ that consists in buying security Z_i, for some $i = 1, \ldots, K$ at time t in event $B_t \in \mathcal{F}_t$, and borrowing the funds by selling short security Z_{K+1}. This strategy does not cost anything and its payoff at the final date T is $1_{B_t}(Z_i(T) - Z_i(t))$, therefore we must have $\pi(1_{B_t}(Z_i(T) - Z_i(t)) \leq 0$, and hence $\psi(1_{B_t}(Z_i(T) - Z_i(t)) \leq 0$ which means $E^*(1_{B_t}(Z_i(T) - Z_i(t)) \leq 0$. This implies that $E^*(Z_i(T) \mid \mathcal{F}_t) \leq Z_i(t)$, which means that Z_i is a supermartingale with respect to P^*. Similarly, in order to show that Z_j is a submartingale with respect to P^* consider the startegy that consists in selling short security Z_j, for some $j = K+1, \ldots, S$ at time t in event $B_t \in \mathcal{F}_t$, and in investing the proceeds in security Z_1.

Let us now turn to the general case where $Z_{K+1}(t) = exp(\int_0^t \tilde{r}(s)ds)$ for some nonnegative, bounded (by some constant $\bar{r}$), and right-continuous adapted process $\tilde{r}$. Consider the set of dates $0 = t_0 < t_1 < \ldots < t_N = T$ and the sequence of real numbers $\bar{R}_n = exp(\int_{t_n}^{t_{n+1}} \bar{r})$ for $n = 0, \ldots, N - 1$. Let K be the set of sequences $R = (R_0, R_1, \ldots, R_{N-1})$ of essentially bounded random variables such that R_n is $\mathcal{F}_{t_n}$ measurable and takes a.e. values in $\prod_0^{N-1}[1, \bar{R}_n]$ for all $n = 0, \ldots, N-1$. For any sequence of random variables $R = (R_0, R_1, \ldots, R_{N-1}) \in K$, and with the convention $R_N = 1$, let us define

$$a_n(R) = \sup_{i=1,\ldots,K} E[\frac{Z_i(t_{n+1})}{Z_i(t_n)} \frac{R_{n+1}E(\rho R_{n+2} \ldots R_N | \mathcal{F}_{t_{n+1}})}{E(\rho R_{n+1} \ldots R_N | \mathcal{F}_{t_n})} \mid \mathcal{F}_{t_n}],$$

$$b_n(R) = \inf_{j=K+1,\ldots,S} E[\frac{Z_j(t_{n+1})}{Z_j(t_n)} \frac{R_{n+1}E(\rho R_{n+2} \ldots R_N | \mathcal{F}_{t_{n+1}})}{E(\rho R_{n+1} \ldots R_N | \mathcal{F}_{t_n})} \mid \mathcal{F}_{t_n}].$$

Note that by choosing $i = 1$ and $j = K + 1$, it is easy to see that $a_n(R) \geq 1$ and $b_n(R) \leq \bar{R}_n$. This implies that $(a_n(R), b_n(R)) \cap [1, \bar{R}_n] \neq \emptyset$, where for any real numbers a and b the set (a, b) denotes the closed interval $[a, b]$ if $a \leq b$, and the closed interval $[b, a]$ otherwise.

Let us denote $[\alpha_n(R), \beta_n(R)] = (a_n(R), b_n(R)) \cap [1, \bar{R}_n]$ and let us define the set-valued map ϕ for all $R \in K$ by

$$\phi(R) = (\phi_0(R), \ldots, \phi_{N-1}(R)) = \prod_0^{N-1} [\alpha_n(R), \beta_n(R)].$$

It is easy to see that K is convex and compact for the weak-star topology $\sigma(L^\infty, L^1)$ of L^∞, that $\phi(R) \subset K$ for every $R \in K$, that ϕ takes convex, nonempty and compact values for this topology, and that its graph is closed (which implies that it is upper-semicontinuous).

Let us now modify ϕ in the following way:
(1) If there exists $n \geq 0$ such that $b_n(R)$ is not larger than or equal to $a_n(R)$, let us denote by n_0 the smallest index for which this happens. This means that there exist a nonnull event $B_{n_0} \in \mathcal{F}_{t_{n_0}}$, and i_0 and j_0 such that $b_{n_0}(R) < a_{n_0}(R)$ on B_{n_0}. For every $n > n_0$, let

$\phi_n^*(R) = \{\inf(a_n(R), \beta_n(R))\}$ if $\frac{Z_{i_0}(t_{n_0+1})}{Z_{i_0}(t_{n_0})} - \frac{Z_{j_0}(t_{n_0+1})}{Z_{j_0}(t_{n_0})} \geq 0$, and

$\phi_n^*(R) = \{\sup(b_n(R), \alpha_n(R))\}$ otherwise.
(2) Let $\gamma_0(R) = \frac{1}{E(\rho R_1 \ldots R_N)}$. Then let

$\phi_0^*(R) = \{\gamma_0(R)\}$ if $\gamma_0(R) \in [\alpha_0(R), \beta_0(R)]$,

$\phi_0^*(R) = \{\beta_0(R)\}$ if $\gamma_0(R) > \beta_0(R)$, and

$\phi_0^*(R) = \{\alpha_0(R)\}$ if $\gamma_0(R) < \alpha_0(R)$.
(3) Everywhere else we let

$\phi_n^*(R) = \phi_n(R)$.

It is easy to show that the graph of ϕ^* is included in the graph of ϕ, and that ϕ^* is also convex, nonempty and compact valued and that it is upper-semicontinuous. By the Kakutani-Fan Theorem (see Fan [1952]), this implies that ϕ^* admits a fixed point, that we shall denote by R^*. We shall now proceed in three steps.

Step 1: For all $n = 0, \ldots, N$ we have $a_n(R^*) \leq b_n(R^*)$.

Indeed, suppose that this property is not satisfied, and consider the smallest index n_0 for which it fails to be true. This means that there exist a nonnull event $B_{n_0} \in \mathcal{F}_{t_{n_0}}$, and i_0 and j_0 such that, with the convention $R_N^* = 1$, $E((\frac{Z_{i_0}(t_{n_0+1})}{Z_{i_0}(t_{n_0})} - \frac{Z_{j_0}(t_{n_0+1})}{Z_{j_0}(t_{n_0})}) \frac{\rho R_{n_0+1}^* \ldots R_N^*}{E(\rho R_{n_0+1}^* \ldots R_N^* | \mathcal{F}_{t_{n_0}})} \mid \mathcal{F}_{t_{n_0}}) > 0$ on B_{n_0}. For every $n > n_0$, let $i(n)$ be the ($\mathcal{F}_{t_n}$ measurable) index that realizes the supremum in $a_n(R^*)$ and $j(n)$ be the ($\mathcal{F}_{t_n}$ measurable) index that realizes the supremum in $b_n(R^*)$. Consider the self-financing trading strategy that is triggered at time t_{n_0} in the event B_{n_0}, and consists in buying one dollar of security i_0 and selling short one dollar of security j_0, liquidating the position at time t_{n_0+1} and investing the proceeds in security $i(n_0 + 1)$ or financing the losses by selling security $j(n_0 + 1)$ depending on the state of the world, and rolling over the position at each date n by buying security $i(n)$ or selling short security $j(n)$. The payoff of such a strategy is equal to

$$1_{B_{n_0}}\left(\frac{Z_{i_0}(t_{n_0+1})}{Z_{i_0}(t_{n_0})} - \frac{Z_{j_0}(t_{n_0+1})}{Z_{j_0}(t_{n_0})}\right)\frac{Z_{i(n_0+1)}(t_{n_0+2})}{Z_{i(n_0+1)}(t_{n_0+1})}\cdots\frac{Z_{i(N-1)}(t_N)}{Z_{i(N-1)}(t_{N-1})}$$

if $\frac{Z_{i_0}(t_{n_0+1})}{Z_{i_0}(t_{n_0})} - \frac{Z_{j_0}(t_{n_0+1})}{Z_{j_0}(t_{n_0})} \geq 0$, and to

$$1_{B_{n_0}}\left(\frac{Z_{i_0}(t_{n_0+1})}{Z_{i_0}(t_{n_0})} - \frac{Z_{j_0}(t_{n_0+1})}{Z_{j_0}(t_{n_0})}\right)\frac{Z_{j(n_0+1)}(t_{n_0+2})}{Z_{j(n_0+1)}(t_{n_0+1})}\cdots\frac{Z_{j(N-1)}(t_N)}{Z_{j(N-1)}(t_{N-1})}$$

otherwise.

Denote by Δ this payoff. Since this strategy is self-financing, satisfies the short sales constraints, and does not require any investment, we must have $E(\rho\Delta) \leq 0$. Note that

$$E(\rho\Delta) = E(1_{B_{n_0}}E((\frac{Z_{i_0}(t_{n_0+1})}{Z_{i_0}(t_{n_0})} - \frac{Z_{j_0}(t_{n_0+1})}{\tilde{Z}_{j_0}(t_{n_0})})A_{n_0+1} \mid \mathcal{F}_{t_{n_0}})),$$

where k is equal to i or j depending on the state of the world and

$$A_{n_0+1} = E(\rho\frac{Z_{k(n_0+1)}(t_{n_0+2})}{Z_{k(n_0+1)}(t_{n_0+1})}\cdots\frac{Z_{k(N-1)}(t_N)}{Z_{k(N-1)}(t_{N-1})} \mid \mathcal{F}_{t_{n_0+1}}).$$

Let us focus on A_{n_0+1}. At date t_{n_0+1} we know what values are taken by k. Let us suppose at first that we have $k = i$ with $i \leq K$. We then have

$$E(\rho\frac{Z_{k(n_0+1)}(t_{n_0+2})}{Z_{k(n_0+1)}(t_{n_0+1})}\cdots\frac{Z_{k(N-1)}(t_N)}{Z_{k(N-1)}(t_{N-1})} \mid \mathcal{F}_{t_{n_0+1}})$$

$$= E(\frac{Z_{i(n_0+1)}(t_{n_0+2})}{Z_{i(n_0+1)}(t_{n_0+1})}\cdots\frac{Z_{i(N-2)}(t_{N-1})}{Z_{i(N-2)}(t_{N-2})}E(\rho\frac{Z_{i(N-1)}(t_N)}{Z_{i(N-1)}(t_{N-1})} \mid \mathcal{F}_{t_{N-1}}) \mid \mathcal{F}_{t_{n_0+1}}).$$

Note that by construction we have $E(\rho\frac{Z_{i(N-1)}(t_N)}{Z_{i(N-1)}(t_{N-1})} \mid \mathcal{F}_{t_{N-1}}) = a_{N-1}(R^*)E(\rho \mid \mathcal{F}_{t_{N-1}})$. Moreover, since $i \leq K$ we must have $R^*_{N-1} = \inf(a_{N-1}(R^*), \beta_{N-1}(R^*))$ which implies $R^*_{N-1} \leq a_{N-1}(R^*)$ and hence we must have $A_{n_0+1} \geq$
$E(\frac{Z_{i(n_0+1)}(t_{n_0+2})}{Z_{i(n_0+1)}(t_{n_0+1})}\cdots\frac{Z_{i(N-3)}(t_{N-2})}{Z_{i(N-3)}(t_{N-3})}E(\frac{Z_{i(N-2)}(t_{N-1})}{Z_{i(N-2)}(t_{N-2})}R^*_{N-1}E(\rho \mid \mathcal{F}_{t_{N-1}}) \mid \mathcal{F}_{t_{N-2}}) \mid \mathcal{F}_{t_{n_0+1}})$.

By construction, we have

$$E(\frac{Z_{i(N-2)}(t_{N-1})}{Z_{i(N-2)}(t_{N-2})}R^*_{N-1}E(\rho \mid \mathcal{F}_{t_{N-1}}) \mid \mathcal{F}_{t_{N-2}}) = a_{N-2}(R^*)E(\rho R^*_{N-1} \mid \mathcal{F}_{t_{N-2}}),$$

and we have $R^*_{N-2} \leq a_{N-2}(R^*)$, which after a sufficiently large number of iterations leads to

$$A_{n_0+1} \geq R^*_{n_0+1}E(\rho R^*_{n_0+2}\cdots R^*_{N-1} \mid \mathcal{F}_{t_{n_0+1}}).$$

Similarly, in the other case where $k = j$ with $j \geq K + 1$ we find that

$$A_{n_0+1} \leq R^*_{n_0+1} E(\rho R^*_{n_0+2} \ldots R^*_{N-1} \mid \mathcal{F}_{t_{n_0+1}}).$$

Altogether, this implies that

$$E(\rho\Delta) \geq E(1_{B_{n_0}} E((\frac{Z_{i_0}(t_{n_0+1})}{Z_{i_0}(t_{n_0})} - \frac{Z_{j_0}(t_{n_0+1})}{Z_{j_0}(t_{n_0})}) E(\rho R^*_{n_0+1} \ldots R^*_{N-1} \mid \mathcal{F}_{t_{n_0+1}}) \mid \mathcal{F}_{t_{n_0}})).$$

Given the hypothesis made this implies that $E(\rho\Delta) > 0$, which is a contradiction.

Step 2: $\gamma_0(R^*) \in [a_0(R^*), b_0(R^*)]$.

Consider the payoff (which clearly defines the strategy that generates it)

$$\Delta' = \frac{Z_{i(0)}(t_1)}{Z_{i(0)}(t_0)} \cdots \frac{Z_{i(N-1)}(t_N)}{Z_{i(N-1)}(t_{N-1})}.$$

Since this strategy consists in rolling over an initial investment of one dollar at date 0, we must have $E(\rho\Delta') \leq 1$. As in step 1 we can show that $E(\rho\Delta') \geq E(\frac{Z_{i(0)}(t_1)}{Z_{i(0)}(t_0)} E(\rho R^*_1 \ldots R^*_{N-1} \mid \mathcal{F}_{t_1}))$. Multiplying both sides by $\gamma_0(R^*)$ we obtain $E(\frac{Z_{i(0)}(t_1)}{Z_{i(0)}(t_0)} \frac{E(\rho R^*_1 \ldots R^*_{N-1} \mid \mathcal{F}_{t_1})}{E(\rho R^*_1 \ldots R^*_{N-1})}) \leq \gamma_0(R^*)$, which means that $a_0(R^*) \leq \gamma_0(R^*)$. In a similar way, we can show that $b_0(R^*) \geq \gamma_0(R^*)$.

Step 3:

By the fixed-point property, we must have
(i) $R^*_0 = \gamma_0(R^*)$
Indeed, since $\gamma_0(R^*) \in [a_0(R^*), b_0(R^*)]$ by Step 2, and since $a_0(R^*) \geq \alpha_0(R^*)$ and $b_0(R^*) \leq \beta_0(R^*)$ we must have $\gamma_0(R^*) \in [\alpha_0(R^*), \beta_0(R^*)]$.
(ii) $a_n(R^*) \leq R^*_n \leq b_n(R^*)$ for all $n = 0, \ldots, N - 1$.
Indeed, we have shown that $a_n(R^*) \leq b_n(R^*)$ for all $n = 0, \ldots, N - 1$ in Step 1.

Let $Z_0(t_0) = 1$ and $Z_0(t_n) = R^*_0 \ldots R^*_{n-1}$, for $n = 1, \ldots, N$. Since by the fixed-point property we must have $R^*_0 = \gamma_0(R^*)$, we have $E(\rho R^*_0 \ldots R^*_{N-1}) = 1$ and hence $E(\rho Z_0(t_N)) = 1$. Moreover since $R^*_n \geq a_n(R^*)$ we must have

$$E(\frac{Z_i(t_{n+1})/Z_0(t_{n+1})}{Z_i(t_n)/Z_0(t_n)} \frac{\rho Z_0(t_N)}{E(\rho Z_0(t_N) \mid \mathcal{F}_{t_n})} \mid \mathcal{F}_{t_n}) \leq 1$$

for $n = 0, \ldots, N - 1$, and $i = 1, \ldots, K$. And, since $R^*_n \leq b_n(R^*)$, we must have $E(\frac{Z_j(t_{n+1})/Z_0(t_{n+1})}{Z_j(t_n)/Z_0(t_n)} \frac{\rho Z_0(t_N)}{E(\rho Z_0(t_N)\mid\mathcal{F}_{t_n})} \mid \mathcal{F}_{t_n}) \geq 1$, for all $n = 0, \ldots, N - 1$, and all $j = K + 1, \ldots, S$.

By letting $E^*(x) = E(\rho Z_0(T)x)$, E^* defines an equivalent probability measure (with $\frac{dP^*}{dP} = \rho Z_0(T) \in X$) and satisfies $E^*(\frac{Z_i(t_{n+1})}{Z_0(t_{n+1})} \mid \mathcal{F}_{t_n}) \leq \frac{Z_i(t_n)}{Z_0(t_n)}$ and $E^*(\frac{Z_j(t_{n+1})}{Z_0(t_{n+1})} \mid \mathcal{F}_{t_n}) > \frac{Z_j(t_n)}{Z_0(t_n)}$, for all $n = 0, \ldots, N - 1$, all $i = 1, \ldots, L$, and all $j = 1, \ldots, S$, which are the desired super/submartingale properties.

To conclude, we extend the numeraire process Z_0 found this way for each time grid $t_0, \ldots, t_N$ into a continuous time process by defining $Z_0(t) = exp(\int_0^t r_N(s, \omega)ds)$, with $r(s) = y_n(s)$ for every $t_n \leq s < t_{n+1}$, where $R_n^* = exp(\int_{t_n}^{t_{n+1}} y_n(s)ds)$ for all $n = 0, 1, \ldots, N-1$ (which implies $0 \leq r \leq \bar{r}$). Note that Z_0 is absolutely continuous (see Billingsley [1986, chap. 6]) and can be written $Z_0(t) = \int_0^t g_N(t, \omega)$, where $g_N(t, \omega) = exp(\int_0^t r_N(s, \omega)ds)r_N(t, \omega)$ belongs to $L^\infty([0, T] \times \Omega)$, and satisfies $0 \leq g_N \leq exp(\int_0^T \bar{r})\bar{r}$. We then do this for finer and finer grids $t_0 \leq \ldots \leq t_N$ of union equal to some dense subset D of $[0, T]$. Since the sequence of slopes $g_N(t, \omega)$ constructed this way is bounded in $L^\infty([0, T] \times \Omega)$, it admits a convergent (for the weak-star topology $\sigma(L^\infty, L^1)$) subsequence. Let us denote by $g^*(t, \omega)$ its limit, and consider the numeraire process $Z_0^*(t) = \int_0^t g^*(s, \omega)ds$. It is easy to see that $Log(Z_0^*(t))$ is also absolutely continuous, which allows us to write $Z_0^*(t) = exp(\int_0^t r_0(s, \omega)ds)$, for some bounded process r_0. By letting $E^*(x) = E(\rho Z_0^*(T)x)$, it is easy to see that E^* defines an equivalent probability measure (with $\frac{dP^*}{dP} = \rho Z_0^*(T) \in X$) and satisfies (using the right-continuity of the filtration and the price processes) $E^*(\frac{Z_i(t)}{Z_0(t)} \mid \mathcal{F}_s) \leq \frac{Z_i(s)}{Z_0(s)}$ and $E^*(\frac{Z_j(t)}{Z_0(t)} \mid \mathcal{F}_s) \geq \frac{Z_j(s)}{Z_0(s)}$ for all $T \geq t \geq s \geq 0$, all $i = 1, \ldots, K$, and all $j = K+1, \ldots, S$, which are the desired super/submartingale properties. The property $0 \leq r_0 \leq \tilde{r}$ then follows.

Since M is a convex cone and π is a lower semi-continuous sublinear functional, statement (iii) is a consequence of Section 1.   ∎

Note that by assuming in our multiperiod securities price model that contingent claims that can be hedged at a finite cost are marketed, we implicitly treat $-\pi(-x)$ and $\pi(x)$ as the bid and the ask price of x in the economy. Another possible interpretation is to consider claims that can be hedged as nonmarketed derivative securities and ask what could be equilibrium prices for them. Following the same reasoning as in Section 1, it is easy to see that any bid-ask price system (q, p), with $p \geq q$, in the interval $[-\pi(-x), \pi(x)]$ is compatible with the securities price process. However, without knowledge about preferences it not possible to infer tightest bounds than $-\pi(-x)$ and $\pi(x)$ on the price of the claim x.

3   Valuation of derivatives

In this section we apply the theoretical results of sections 1 and 2 to some particular cases of economies with frictions.

3.1   Pricing derivatives with transaction costs

In this section we assume that there is no short sales constaints and that the borrowing and the lending rate are equal (to 1 after normalization). In this case, following Theorem 2.1, there exists a probability measure P^* and two processes between Z and Z' such that the first one is a supermartingale and the second one a submartingale with respect to P^*. In fact we can prove the following stronger result.

Theorem 3.1 (Jouini and Kallal, 1995 a) *(i) The securities price model admits no multiperiod free lunch if and only if there exist at least a probability measure P^* equivalent to P with $E((\frac{dP^*}{dP})^2) < \infty$ and a process Z^* satisfying $Z' \leq Z^* \leq Z$ such that Z^* is a martingale with respect to the filtration $\{\mathcal{F}_t\}$ and the probability measure P^*.*

(ii) Moreover, if we denote by E^* the expectation operator associated to P^*, there is a one-to-one correspondence between the set of such expectation operators and the set of linear functionals $\psi \in \Psi$ such that $\psi \mid_M \leq \pi$. This correspondence is given by the following formulas:

$$P^*(B) = \psi(1_B), \text{ for all } B \in \mathcal{F} \text{ and } \psi(x) = E^*(x), \text{ for all } x \in X.$$

(iii) Furthermore, for all $m \in M$ we have

$$[-\pi(-m), \pi(m)] = cl\{E^*(m) : P^* \text{ is a martingale measure }\}.$$

Proof of Theorem 3.1 :

Using Theorem 2.1, we only have to prove that there exists a martingale relatively to P^* between Z and Z'. Since there is no constraints, Theorem 2.1 (ii) permits us to construct two processes $\overline{Z}^*$ and $\underline{Z}^*$ between Z and Z' such that $\overline{Z}^*$ is a supermartingale relatively to P^* and $\underline{Z}^*$ a submartingale.

Note that we must have $\overline{Z}^*(t) \leq \underline{Z}^*(t)$ for all t. Indeed, assume instead that there exists $t \in [0, T]$ and $B \in \mathcal{F}_t$ such that $P(B) > 0$ and $\overline{Z}^*(t, \omega) > Z^*(t, \omega)$ for all $\omega \in B$. Then there would exist strategies $(\theta, \theta') \in \Theta'(t, k)$ and $(\alpha, \alpha') \in \Theta(t, k)$ satisfying $E^*((\theta_0 - \theta'_0)(T)1_B) > E^*(-(\alpha_0 - \alpha'_0)(T)1_B)$, i.e. $E^*(((\theta_0 + \alpha_0) - (\theta'_0 + \alpha'_0))(T)1_B) > 0$, and hence $E^*([\{(\theta + \alpha) - (\theta' + \alpha')\}^+(T) \cdot Z'(T) - \{(\theta + \alpha) - (\theta' + \alpha')\}^-(T) \cdot Z(T)]1_B) > 0$ which contradicts the fact that $E^*([\{(\theta + \alpha) - (\theta' + \alpha')\}^+(T) \cdot Z'(T) - \{(\theta + \alpha) - (\theta' + \alpha')\}^-(T) \cdot Z(T)]1_B) \leq \pi([\{(\theta + \alpha) - (\theta' + \alpha')\}^+(T) \cdot Z'(T) - \{(\theta + \alpha) - (\theta' + \alpha')\}^-(T) \cdot Z(T)]1_B) \leq 0$. Hence $\overline{Z}^*(t) \leq \underline{Z}^*(t)$ for all t. To conclude, we use the following

Lemma 2 *If $Z'(t) \leq \overline{Z}^*(t) \leq \underline{Z}^*(t) \leq Z(t)$, where $\overline{Z}^*(t)$ is a supermartingale and $\underline{Z}^*(t)$ is a submartingale with respect to $\{\mathcal{F}_t\}$ and a probability $P^* = \rho P$, with $\rho \in X$, then there exists a martingale Z^* with respect to $\{\mathcal{F}_t\}$ and P^* such that $Z' \leq Z^* \leq Z$.*

The proof of this Lemma is given in Jouini and Kallal (1995a).

∎

Note that Theorem 2.1 is not in this last reference and Theorem 3.1 is then proved directly.

In the next, we will show how we can use this approach to price contingent portfolios. The main difference with the previous approach is that, in the next, we will not compare liquidation values for different strategies but the final quantity of each security for each strategy. We replace then the main definitions of previous sections as follows.

Definition 3.1 *A contingent claim C is defined by $(C_0, \ldots, C_K) \in X^{K+1}$ the contingent portfolio guaranteed by C*

Note that a contingent claim C is not necessarily attainable by a strategy belonging to Θ. Thus, we consider the following set M of marketed claims, i.e., claims that can be dominated (or hedged) by the payoff of a simple self-financing trading strategy.

Definition 3.2 *A claim C is said to be marketed if there exists a self-financing simple strategy (θ, θ') in Θ such that $(\theta_k - \theta'_k)(T, \omega) \geq C_k(\omega)$. for $k = 0, \ldots, K$ and for almost every ω.*

As we have already seen, when there are transaction costs it is not true that the cheapest way to obtain a given minimal contingent payoff at date T is to duplicate it by dynamic trading. A simple example can illustrate it. Assume that a call option on a stock is to be hedged using a riskless bond and the underlying stock only. Also suppose that there are transaction costs in trading the stock at intermediate dates (between now and maturity). It is then easy to see that if transaction costs are prohibitively high it is cheaper to buy the stock and hold it until maturity (which leads to a payoff that is strictly larger than the payoff of the call) than to try to duplicate the call. Hence, we consider the price functional π defined for every marketed claim $m \in M$ by

$$\pi(C) = \inf\{ \liminf_n \{\theta^n(0) \cdot Z(0) - \theta'^n(0) \cdot Z'(0)\} : (\theta^n, \theta'^n) \in \Theta,$$
$$(\theta^n - \theta'^n)(T) \geq C^n$$
$$\text{and } (C^n) \subset M \text{ converges to } C\}.$$

In words, $\pi(C)$ represents the infimum cost necessary to get at least a final contingent portfolio arbitrarily close to C at date T. In the subsequent analysis we shall characterize the situations in which there are no possibilities of arbitrage profits. In fact, as in the previous sections we shall rule out a weaker form of arbitrage opportunities: (portfolio) free lunches. In words, a (portfolio) free lunch is the possibility of getting a contingent portfolio (at date T) with a value arbitrarily close to a given element of X_+ using a self-financing simple strategy at a cost (in terms of date 0 consumption) arbitrarily small. More precisely

Definition 3.3 *A (portfolio) free lunch is a sequence of contingent claims C^n in M, converging to some C and a sequence (θ^n, θ'^n) in Θ such that $C^+ Z'(T) - C^- Z(T) \in X_+$, $(\theta^n - \theta'^n)(T) \geq C^n$ for all n and $\lim_n \{\theta^n(0) \cdot Z(0) - \theta'^n(0) \cdot Z'(0)\} \leq 0$.*

An obvious consequence of the absence of (portfolio) free lunches is that the bid price of any security must lie above the ask price (i.e. for all $t \in \mathcal{T}$, $P(\{\omega : Z'(t, \omega) \leq Z(t, \omega)\}) = 1$).

We shall now see that the price functional π is sublinear: it is less expensive to hedge the sum $x + y$ of two contingent claims than to hedge the claims x and y separately and add up the costs. It is easy to see why: the sum of two strategies that hedge the claims x and y hedges the claim $x + y$ but some orders to buy and sell the same security at the same date might cancel out. Some of the transaction costs might be saved this way.

Proposition 3.2 *If there is no (portfolio) free lunch, then the set of marketed claims M is a convex cone and the price functional π is a lower semicontinuous sublinear functional which takes value in R.*

Proof of Proposition 3.2 : Using the fact that Θ is a convex cone and the fact that $Z \geq Z'$, it is relatively easy to show that M is a convex cone. Since the null strategy permits to dominate the null portfolio at a zero cost we have $\pi(0) \leq 0$. If $\pi(0) < 0$, it is easy to construct a (portfolio) free lunch and then $\pi(0) = 0$. Furthermore we can easily see that $\pi(\lambda C) = \lambda\pi(C)$ for all $C \in M$ and $\lambda \in R_+$. Using a limit argument it is also easy to obtain that for all $C, C' \in M$
$$\pi(C + C') \leq \pi(C) + \pi(C').$$
Let $\lambda \in R$ and C^n be a sequence in M converging to $C \in M$ such that $\pi(C^n) \leq \lambda$, for all n. Then, by a diagonal extraction process, there exist a sequence $\tilde{C}^n \in M$ and a sequence $(\theta^n, \theta'^n) \in \Theta$ such that $\|\tilde{C}^n - C^n\| \leq \frac{1}{n}$, $(\theta^n - \theta'^n)(T) \geq \tilde{C}^n$ and $\theta^n(0) \cdot Z(0) - \theta'^n(0) \cdot Z'(0) \leq \lambda + \frac{1}{n}$. Since $\tilde{C}^n$ converges to C we must then have $\pi(C) \leq \lambda$. Hence, the set $\{C \in M : \pi(C) \leq \lambda\}$ is closed and π is l.s.c.

Let now $C \in M$ such that $\pi(C) = -\infty$, for all $\lambda > 0$, we have $\pi(\lambda C) = -\infty$ and if λ converges to 0 we obtain by the lower semi-continuity property that $\pi(0) = -\infty$. This contradiction implies that π takes its values in R. ∎

We are now in a position to prove our main result that characterizes arbitrage free bid and ask price processes. We find that the securities prices model is arbitrage free if and only if there exists a process, lying between the bid and ask price processes, that is a martingale under some equivalent measure. Moreover, we obtain that the arbitrage price of a contingent portfolio is equal to the supremum of the expectations of its terminal value evaluated with some security price process Z^* in the bid-ask interval and relatively to a probability measure P^* such that Z^* is a P^* martingale. Recall that Theorem 2.1 claims that when we only compare terminal liquidation values, the arbitrage price is equal to the supremum of the expectations of the terminal liquidation value relatively to a probability measure P^* such that there exists some martingale Z^* relatively to P^* in the bid-ask interval. The process Z^* does not appear in the valuation formula it only has an effect on the choice of P^*. In the following Theorem Z^* imposes restrictions on P^* and appears directly in the valuation formula.

Theorem 3.3 *(i) The securities price model admits no (portfolio) free lunch if and only if there exist at least a probability measure P^* equivalent to P with $E((\frac{dP^*}{dP})^2) < \infty$ and a process Z^* satisfying[23] $Z' \leq Z^* \leq Z$ such that Z^* is a martingale with respect to the filtration $\{\mathcal{F}_t\}$ and the probability measure P^*.*
(ii) For all contingent claim C in M we have $\pi(C) = \sup E^*(C \cdot Z^*(T))$ where the supremum is taken over all the expectation operators E^* associated to a probability measure P^* and all the processes Z^* such that (P^*, Z^*) satisfy the conditions of (i).

Proof of Theorem 3.3 : First, let P^* be a probability measure equivalent to P and let Z^*, with $Z' \leq Z^* \leq Z$, be a martingale with respect to P^*

---

[23] This means that for all t, $P(\{\omega : Z'(t, \omega) \leq Z^*(t, \omega) \leq Z(t, \omega)\}) = 1$.

and $\{\mathcal{F}_t\}$. Define the linear functional ψ by $\psi(C) = E^*(Z^*(T) \cdot C)$ for all $C \in X^{K+1}$. Since $\rho = \frac{dP^*}{dP} \in X$ we have by the Riesz representation Theorem that $\psi(C) = E^*(Z^*(T) \cdot C) = E(\rho Z^*(T) \cdot C)$ is continuous.

Let $\tilde{\Omega}$ be the space equal to $\Omega \times \{0, \ldots, K\}$ endowed with $(\tilde{\mathcal{F}}, \tilde{P})$ the natural probability structure defined by $(\mathcal{F}, P)$. Let $\tilde{X}$ be the set defined by $\tilde{X} = L^2(\tilde{\Omega}, \tilde{\mathcal{F}}, \tilde{P})$. The set $\tilde{X}$ can be identified with X^{K+1} and π can be seen as a sublinear functional on a subset M of $\tilde{X}$. In the next $\tilde{\Psi}$ is defined for $\tilde{X}$ as Ψ for X

Since P and P^* are equivalent, we have $\psi \in \tilde{\Psi}$.

Let $C \in M$ and let $(\theta, \theta') \in \Theta$ with trading dates $0 = t_0 \leq t_1 \leq \ldots \leq t_N = T$, such that $(\theta - \theta')(T) \geq C$. Since $Z' \leq Z^* \leq Z$ and (θ, θ') is nondecreasing and self-financing, we have, for $n = 1, \ldots, N$,

$$E^*((\theta(t_n) - \theta(t_{n-1})) \cdot Z^*(t_n) - (\theta'(t_n) - \theta'(t_{n-1})) \cdot Z^*(t_n) \mid \mathcal{F}_{t_{n-1}})$$
$$\leq E^*((\theta(t_n) - \theta(t_{n-1})) \cdot Z(t_n) - (\theta'(t_n) - \theta'(t_{n-1})) \cdot Z'(t_n) \mid \mathcal{F}_{t_{n-1}}) \leq 0.$$

Using the fact that Z^* is a martingale with respect to $\{\mathcal{F}_t\}$ and P^*, we have

$$E^*((\theta - \theta')(t_n) \cdot Z^*(t_n) \mid \mathcal{F}_{t_{n-1}}) \leq E^*((\theta - \theta')(t_{n-1}) \cdot Z^*(t_n) \mid \mathcal{F}_{t_{n-1}})$$
$$\leq (\theta - \theta')(t_{n-1}) \cdot Z^*(t_{n-1}).$$

By iteration, $E^*((\theta - \theta')(T) \cdot Z^*(T)) \leq (\theta - \theta')(0) \cdot Z^*(0) \leq \theta(0) \cdot Z(0) - \theta'(0) \cdot Z'(0)$. Since $(\theta - \theta')^+(T) \cdot Z'(T) - (\theta - \theta')^-(T) \cdot Z(T) \leq \{(\theta - \theta')^+(T) - (\theta - \theta')^-(T)\} \cdot Z^*(T)$ and, by positivity of the price processes, $C^+ Z'(T) - C^- Z(T) \leq (\theta - \theta')^+(T) \cdot Z'(T) - (\theta - \theta')^-(T) \cdot Z(T)$ we have $E^*(C^+ Z'(T) - C^- Z(T)) \leq \theta(0) \cdot Z(0) - \theta'(0) \cdot Z'(0)$. It is easy to see that this implies that there cannot be any (portfolio) free lunch.

Furthermore, $\psi(C) = E^*(Z^*(T) \cdot C) \leq E^*((\theta - \theta')(T) \cdot Z^*(T))\theta(0) \cdot Z(0) - \theta'(0) \cdot Z'(0)$. Taking the infimum over the strategies $(\theta, \theta') \in \Theta$ such that $(\theta - \theta')(T) \geq C$, we obtain that $\psi(C) = E^*(C) \leq \tilde{\pi}(C)$, where $\tilde{\pi}(m) = \inf\{\theta(0) \cdot Z(0) - \theta'(0) \cdot Z'(0) : (\theta, \theta') \in \Theta \text{ and } (\theta - \theta')(T) \geq C\}$, for all $C \in M$. We can prove as in Lemma 1 that π is the largest l.s.c. functional that lies below $\tilde{\pi}$. Using the fact that ψ is continuous, we have then $\psi \mid_M \leq \pi$.

Lemma 3 *The absence of (portfolio) free lunch is equivalent to the absence of free lunch for (M, π).*

Proof of the Lemma : Consider a (portfolio) free lunch as in definition 4.1.3. We have $\theta^n(0) \cdot Z(0) - \theta'^n(0) \cdot Z'(0) \geq \pi((\theta^n - \theta'^n)(T)) \geq \pi(C^n)$ and taking the lim inf on both sides we see that for a subsequence of the sequence C^n, we have $\lim_n \{\theta^n(0) \cdot Z(0) - \theta'^n(0) \cdot Z'(0)\} \leq 0$. If (M, π) admits no free lunch, by Theorems 1.1 and 1.2 there exists $\psi \in \tilde{\Psi}$ such that $\psi \mid_M \leq \pi$ and we have then $\psi(C) \leq 0$ which is impossible.

Conversely, consider a free lunch for (M, π). We have $x_n \in M$ and $r_n + \pi(x_n) \leq 0$ for all n. Using the definition of π we can find, by a diagonal extraction process, sequences of strategies $(\theta^n, \theta'^n) \in \Theta$ and of claims $y_n \in M$ such that

$\|x_n - y_n\| \leq 1/n$, $(\theta^n - \theta'^n)(T) \geq y_n$ and $r^* + \lim_n \{\theta^n(0) \cdot Z(0) - \theta'^n(0) \cdot Z'(0)\} \leq 0$.

If $r^* = 0$, we then have a multiperiod free lunch since y_n converges to $x^* \in \tilde{X}_+$ and then $C^+ Z'(T) - C^- Z(T) \in X_+$. If $r^* > 0$, it suffices to modify the strategies such as the amount r^* is invested in some security to obtain a (portfolio) free lunch. This ends the proof of the Lemma.

Assume now that there is no (portfolio) free lunch or equivalently that there is no free lunch for (M, π) and let $\psi \in \tilde{\Psi}$ such that $\psi \mid_M \leq \pi$, as guaranteed by Theorems 1.1 and 1.2. Since ψ is continuous, by the Riesz representation Theorem there exists a random variable $\rho \in \tilde{X}$ such that $\psi(x) = E(\rho x)$, for all $x \in \tilde{X}$ or equivalently there exists $(\rho_0, \ldots, \rho_K)$ in X^{K+1} such that $\psi(C) = E(\rho \cdot C)$, for all $C \in X^{K+1}$. Define P^* from ψ by $P^*(B) = E(\rho_0 1_B)$ for all $B \in \mathcal{F}$. By linearity and strict positivity of ψ it is clear that P^* is a measure equivalent to P. Using the fact that $Z_0 = Z_0' = 1$ it is easy to show that $P^*(1_\Omega) = 1$ and $\frac{dP^*}{dP} = \rho_0$ is square integrable.

It remains to show that there exists a process Z^*, with $Z_k' \leq Z_k^* \leq Z_k$, and such that Z_k^* is a martingale with respect to P^* and $\{\mathcal{F}_t\}$, for $k = 1, \ldots, K$. In fact, we will prove that the martingale relatively to P^* and $\{\mathcal{F}_t\}$ defined by $Z_k^*(t) = E^*(\rho_k/\rho_0 \mid \mathcal{F}_t)$ lies between Z_k' and Z_k.

Let $k \in \{1, \ldots, K\}$, $t \in \mathcal{T}$ and $B \in \mathcal{F}_t$. Let C the contingent claim defined by $C_k = 1_B$, $C_0 = -Z_k(t)1_B$ and $C_h = 0$ for $h \neq 0, k$. The contingent claim C is duplicable. It suffices to buy at t, if $\omega \in B$, one unit of the security k and to pay with security 0 units. This strategy costs nothing and we have then $E^*((-C_0 Z_k(t) + \rho_k/\rho_0)1_B) = E((-\rho_0 Z_k(t) + \rho_k)1_B) = \psi(C) \leq \pi(C) \leq 0$. Then, we have $E^*(\rho_k/\rho_0 1_B) \leq E^*(Z_k(t)1_B)$, for all t and all $B \in \mathcal{F}_t$. This implies that $Z_k^* \leq Z_k$. By a symetric argument we obtain $Z_k^* \geq Z_k'$ which achieves to prove the point (i) of the theorem.

In fact, we have also proved that every $\psi \in \tilde{\Psi}$ such that $\psi \mid_M \leq \pi$ is equal to $E^*(C \cdot Z^*)$ for some process Z^* between Z' and Z and some probability measure P^* such that Z^* is a martingale relatively to P^* and conversely.

Following Section 1, $\pi(C) = \sup \psi(C)$ where the supremum is taken over all the functionals $\psi \in \tilde{\Psi}$ such that $\psi \mid_M \leq \pi$. This result permits to achieve the proof of (ii). ∎

3.2 Shortselling costs in the diffusion case

We now turn to an illustration of our results, in an economy where traded securities prices follow a continuous time diffusion process.

In this subsection, we consider the continuous time case where the set of trading dates $\mathcal{T}$ is equal to $[0, T]$ and securities prices follow diffusion processes. To keep things simple we shall treat the case where there are only two securities, a bond and a stock that follows a diffusion process. We model the spread between the borrowing and lending rate by having a (locally) riskless security in which investors can only go short and a riskless security in which they can only go long. Similarly, we model the cost of shortselling the underlying stock by having a risky security in which investors can only go short, and a risky security in which they can only go long, with possibly different returns. We shall not consider bid-ask spreads in the traded securities here. Again to simplify, we shall assume that there is a single source of uncertainty in the economy, modeled by a Brownian Motion.[24]   Finally, we shall focus our attention on the pricing of derivative securities that pay a given function of the underlying stock price at maturity.

Formally, let $W = \{W(t) : 0 \leq t \leq T\}$ be a one-dimensional standard Brownian Motion defined on the probability space $(\Omega, \mathcal{F}, P)$, and let $\{\mathcal{F}_t\}$ be the augmented[25] filtration generated by W, with $\mathcal{F}_T = \mathcal{F}$. Consider the risky securities price processes Z_1 and Z_2 satisfying the stochastic integral equations:

$$Z_1(t) = Z_1(0) + \int_0^t \mu_1(Z_1(s), s) Z_1(s) ds + \int_0^t \sigma_1(Z_1(s), s) Z_1(s) dW(s)$$

$$Z_2(t) = Z_2(0) + \int_0^t \mu_2(Z_1(s), s) Z_2(s) dt + \int_0^t \sigma_2(Z_1(s), s) Z_2(s) dW(s)$$

and the locally riskless securities price processes

$$B_1(t) = exp(\int_0^t r_1(Z_1(s), s) ds),$$

$$B_2(t) = exp(\int_0^t r_2(Z_1(s), s) ds),$$

for all $t \in [0, T]$, where the drifts $\mu_i(x, t)$, the volatilities $\sigma_i(x, t) : R \times [0, T] \to R$ and the riskless rates $r_i(x, t)$ are given continuous functions of the state variable x (the price of the stock) and time t, such that $\sigma_i(x, t) > 0$, $r_i(x, t) \geq 0$ and $r_i(x, t)$ is bounded for $i = 1, 2$. Initial values $Z_1(0)$ and $Z_2(0)$ are assumed to be positive so that the stock price processes remain positive.

Investors are assumed to be able to hold only long positions in security Z_1 and only short positions in security Z_2. In fact, the processes Z_1 and Z_2 can model the returns on the same stock, respectively for a long and a short position. In this case, the difference in return between Z_2 and Z_1 models the cost of shortselling the stock[26]. Also, r_1 models the riskless lending rate and r_2 models the riskless borrowing rate, as we assume that investors can only hold nonnegative quantities of B_1 and nonpositive quantities of B_2.

We shall assume that the drift and volatility are sufficiently regular to guarantee the existence and uniqueness of continuous solutions Z_1 and Z_2 with bounded moments to the stochastic integral equations above. We refer to Gihman and Skorohod (1972, Theorem 1, p. 40) for sufficient Lipschitz and growth conditions on $\mu_i(x, t)$ and $\sigma_i(x, t)$.

---

[24] There is no real difficulty in introducing more sources of risk and more traded securities.

[25] This implies in particular that it is right-continuous.

[26] Although we let the volatilities of the long and short positions in the stock be possibly different, the cost of shortselling a security is more likely to be paid in terms of expected return.

Absence of arbitrage and (super/sub)martingale measures

Following Theorem 2.1 consider P^*, a candidate probability measure consistent with the absence of arbitrage, i.e. an equivalent probability measure with $dP^* = \rho dP$ where ρ is strictly positive and square integrable. Using Kunita and Watanabe (1967) Representation Theorem it is easy to show that $E(\rho \mid \mathcal{F}_t)$ can be written as $1 + \int_0^T \xi(s)dW(s)$, where ξ is a measurable adapted process satisfying $\int_0^T E(\xi^2(s))ds < \infty$, and it follows from Ito's Lemma that $E(\rho \mid \mathcal{F}_t) = exp\{\int_0^t \gamma(s)dW(s) - \frac{1}{2}\int_0^t \gamma^2(s)ds\}$ and in particular

$$\rho = exp\{\int_0^T \gamma(s)dW(s) - \frac{1}{2}\int_0^T \gamma^2(s)ds\}$$

where $\gamma(t) = \frac{\xi(t)}{E(\rho|\mathcal{F}_t)}$. Let the process $W^*(t)$ be defined by

$$W^*(t) = W(t) - \int_0^t \gamma(s)ds, 0 \le t \le T.$$

Then according to Girsanov's (1960) Fundamental Theorem, W^* is a Brownian Motion on $(\Omega, \mathcal{F}, P^*)$, and the processes Z_i satisfy, for $i = 1, 2$, the stochastic integral equation

$$Z_i(t) = \int_0^t \{\mu_i(Z_1, s) + \sigma_i(Z_1, s)\gamma(s)\}Z_i(s)ds + \int_0^t \sigma_i(Z_1, s)Z_i(s)dW^*(s).$$

We shall further assume that for each price process Z_i, and every numeraire process $B_0(t) = exp(\int_0^t r_0(s)ds)$ with $r_1(s) \le r_0(s) \le r_2(s)$, there exists an equivalent probability measure P_i, with $E((\frac{dP_i}{dP})^2) < \infty$, that transforms $\frac{Z_i}{B_0}$ into a martingale. A sufficient condition is $E[exp(8 \int_0^T \gamma_i^2(s)ds)] < \infty$, for $i = 1, 2$, where $\gamma_i(t) = -\frac{\mu_i(t) - r_0(t)}{\sigma_i(t)}$ is the instantaneous price of risk associated to the normalized price processes (see the Appendix). Note that if $\gamma_i(t)$ is bounded this condition is satisfied.

We are now in a position to characterize the absence of arbitrage in our model. According to Theorem 2.2, this model is arbitrage free if and only if we can find a numeraire $B_0(t) = exp(\int_0^t r_0(s)ds)$ with a rate of accumulation r_0 that is between the lending rate r_1 and the borrowing rate r_2, and an equivalent probability measure for which $\frac{Z_1}{B_0}$ is a supermartingale and $\frac{Z_2}{B_0}$ is a submartingale. This leads to the following result established in Jouini and Kallal (1995b).

Theorem 3.4 *(i) The model is arbitrage free if and only if there exists an accumulation rate process $r_0(t)$ such that $r_1(t) \le r_0(t) \le r_2(t)$ and $\frac{\mu_1(t) - r_0(t)}{\sigma_1(t)} \le \frac{\mu_2(t) - r_0(t)}{\sigma_2(t)}$ for all $t \in [0, T]$.*
(ii) In this case, for each such accumulation rate process r_0, the set of the (square integrable) Radon-Nikodym derivatives of the equivalent probability measures for

which $\dfrac{Z_1}{exp(\int_0^t r_0(s)ds)}$ *is a supermartingale and* $\dfrac{Z_2}{exp(\int_0^t r_0(s)ds)}$ *is a submartingale is*

$$\Pi = \{\rho : \rho = exp[\int_0^T \gamma(s)dW(s) - \tfrac{1}{2}\int_0^T \gamma^2(s)ds]$$
$$\quad for\ some\ adapted\ process\ \gamma(t)\ such\ that\ -\tfrac{\mu_2(t)-r_0(t)}{\sigma_2(t)} \leq \gamma(t) \leq$$
$$-\tfrac{\mu_1(t)-r_0(t)}{\sigma_1(t)}\}.$$

If the (borrowing and lending) riskless rate is equal to r_0, then $-\gamma_i(t) = \frac{\mu_i(t)-r_0(t)}{\sigma_i(t)}$ represents the instantaneous risk-premium per unit of risk, i.e. the expected return of the security in excess of the riskless rate per unit of volatility. The theorem then says that the price processes are arbitrage free if and only if the price of risk is higher for a short position than for a long position in the stock, i.e. if investors, when long in the stock, get a smaller expected return per unit of risk, than they have to pay per unit of risk when short in the stock. Moreover, the instantaneous price of risk is known to be useful to compute the price of derivative securities in frictionless securities market models. Adding this return to the Brownian Motion we obtain a security that has a unit constant volatility, and an expected return equal to the price of risk. In a world where this basic security is a martingale, i.e. where the adjusted price of risk is equal to zero, all derivative securities should be martingales as well, and their prices should be the expected value of their payoffs (with respect to the new probability). In our case there are two instantaneous prices of risk: the price for a short position and the price for a long position in the stock. As we shall see, bounds can be found on any derivative security price by computing all the expected values of its normalized payoff with respect to the probabilities that make the adjusted price of risk equal to zero, starting from any price of risk between the prices of risk for a long and a short position in the stock. As we would expect, the relevant price of risk is equal to the price of risk for a long position in the stock if the optimal hedging strategy consists in being long in the stock, and it is equal to the price of risk for a short position in the stock if the optimal hedging strategy consists in being short in the stock. If the borrowing and the lending rates differ, the previous steps need to be followed using all the riskless rates that lie between the borrowing and the lending rates, and for which the risk-premium on a long position in the stock is lower than the risk-premium on a short position.

Arbitrage bounds on contingent claims

Let us consider a contingent claim to consumption at the final date T that is a given function of the stock price at that date, i.e. that is of the form $h(Z_1(T))$. Denote by $\pi(h)$ and $-\pi(-h)$ the arbitrage bounds on the bid and the ask price of the claim $h(Z_1(T))$, where $\pi(h)$ is the minimum amount it costs to hedge it and $-\pi(-h)$ the maximum that can be borrowed against it through dynamic securities trading. We have already seen that no investor would agree to pay more than $\pi(h)$ for the claim $h(Z_1(T))$, since there is a way of obtaining (at least) this payoff by trading in the underlying securities. Also, no investor would sell the claim $h(Z_1(T))$ for less than $-\pi(-h)$ since there is a way of obtaining $-\pi(-h)$ against this payoff by trading in the underlying securities.

According to Theorem 2.1, the interval $[-\pi(-h), \pi(h)]$ is equal (modulo its boundary) to the set of expectations $\{E(\rho \frac{h(Z_1(T))}{exp(\int_0^T r_0(t)dt)}) \, : \, r_1(t) \leq r_0(t) \leq r_2(t)$ and $\rho \in \Pi(r_0)\}$, where $\Pi(r_0)$ is the set of probability measures for which $\frac{Z_1(t)}{exp(\int_0^t r_0(s)ds)}$ is a supermartingale and $\frac{Z_2(t)}{exp(\int_0^t r_0(s)ds)}$ is a submartingale. This allows us to express these bounds as the solutions of optimal control problems and to derive a partial differential equation with a boundary condition that they satisfy. These partial differential equations are in fact similar to the partial differential equation derived by Black and Scholes in the frictionless case, with two additional nonlinear terms. One term is proportional to the spread between the borrowing and the lending rates, whereas the other term is proportional to the spread between the risk-premia on long and short positions in the stock.

To derive these results, we consider payoffs that are sufficiently regular and do not grow too fast. More precisely, we assume that the function $h : R \rightarrow R$ is continuous and satisfies the polynomial growth condition $|h(x)| \leq C(1+|x|)^n$ for some constant C and some nonnegative integer n. We denote by $C_p^{2,1}(E)$ the class of functions from a subset E of $R \times [0, T]$ into R that are twice continuously differentiable with respect to the first variable and continuously differentiable with respect to the second variable. We then have

Theorem 3.5 (Jouini and Kallal (1995b)) *(i) If the securities price model is arbitrage free, and if there exists a function $V(x,t) \in C_p^{2,1}(R \times (0,T))$, continuous on $R \times [0, T]$, and satisfying the partial differential equation*

$$r_1 V = \frac{\partial V}{\partial t} + \frac{1}{2}\sigma_1^2 x^2 \frac{\partial^2 V}{\partial x^2} + r_1 x \frac{\partial V}{\partial x}$$

$$+\sigma_1 x (\frac{\mu_2 - r_1}{\sigma_2} - \frac{\mu_1 - r_1}{\sigma_1})[\frac{\partial V}{\partial x}]^- + (r_2 - r_1)[V - x(\frac{\partial V}{\partial x})^+ + x\frac{\sigma_1}{\sigma_2}(\frac{\partial V}{\partial x})^-]^-$$

with boundary condition

$$V(x,T) = h(x), \text{ for all } x \in R$$

then $\pi(h) = V(Z_1(0), 0)$.
(ii) In this case, we have

$$h(Z_1(T)) \;=\; V(Z_1(0), 0) + \int_0^T (\frac{\partial V}{\partial x})^+ dZ_1$$

$$- \int_0^T \frac{\sigma_1 Z_1}{\sigma_2 Z_2}(\frac{\partial V}{\partial x})^- dZ_2 + \int_0^T \frac{[\Delta_0]^+}{B_1} dB_1 + \int_0^T \frac{[\Delta_0]^-}{B_2} dB_2$$

where $\Delta_0 = V - Z_1[(\frac{\partial V}{\partial x})^+ - \frac{\sigma_1}{\sigma_2}(\frac{\partial V}{\partial x})^-]$.

This means that the solution to this partial differential equation V, that gives the arbitrage bound as a function of the stock price and calendar time, can be used to find the optimal hedging strategy in terms of the hedge ratio $\frac{\partial V}{\partial x}$, as in the frictionless case. The optimal hedging strategy, that duplicates the payoff $h(Z_1(T))$ for the lowest possible cost, consists in holding, at any time, $\Delta_1 = (\frac{\partial V}{\partial x})^+$ units of security Z_1, i.e. of stock, and $\Delta_2 = -\frac{\sigma_1 Z_1}{\sigma_2 Z_2}(\frac{\partial V}{\partial x})^-$ units of security Z_2, i.e. going short in $(\frac{\sigma_1}{\sigma_2})(\frac{\partial V}{\partial x})^-$ units of stock, and investing the surplus $(\Delta_0)^+ = [V - \Delta_1 Z_1 - \Delta_2 Z_2]^+ = [V - Z_1(\frac{\partial V}{\partial x})^+ + Z_1\frac{\sigma_1}{\sigma_2}(\frac{\partial V}{\partial x})^-]^+$ in the riskless bond B_1 or financing the deficit $(\Delta_0)^- = [V - \Delta_1 Z_1 - \Delta_2 Z_2]^- = [V - Z_1(\frac{\partial V}{\partial x})^+ + Z_1\frac{\sigma_1}{\sigma_2}(\frac{\partial V}{\partial x})^-]^-$ by selling short the riskless bond B_2. This strategy only requires an initial investment $\pi(h) = V(Z_1(0), 0)$.

It is also possible to determine the equivalent probability measure P^* and the riskless rate that "price" the contingent claim $h(Z_1(T))$, i.e. that are such that $\pi(h) = E^*[\frac{h(Z_1(T))}{exp(\int_0^T r_0(t)dt)}]$. It follows from Theorem 2.2 that the riskless rate is given by

$$r_0 = r_1 1_{\{V - Z_1(\frac{\partial V}{\partial x})^+ + Z_1\frac{\sigma_1}{\sigma_2}(\frac{\partial V}{\partial x})^- \geq 0\}} + r_2 1_{\{V - Z_1(\frac{\partial V}{\partial x})^+ + Z_1\frac{\sigma_1}{\sigma_2}(\frac{\partial V}{\partial x})^- < 0\}}.$$

The equivalent probability measure P^* is given by $\frac{dP^*}{dP} = exp\{\int_0^T \gamma(s)dW(s) - \frac{1}{2}\int_0^T \gamma^2(s)ds\}$, where the risk-premium γ is given by

$$\gamma(t) = -\frac{\mu_1 - r_0}{\sigma_1}1_{\{\frac{\partial V}{\partial x} \geq 0\}} - \frac{\mu_2 - r_0}{\sigma_2}1_{\{\frac{\partial V}{\partial x} < 0\}}.$$

In other words, r_0 and γ are respectively the riskless rate and the risk-premium faced by an investor hedging the contingent claim $h(Z_1(T))$.

It is easy to see that these results apply to the lower bound $-\pi(-h)$ on the price of the claim $h(Z_1(T))$. Indeed, we have $-\pi(-h) = -V(Z_1(0), 0)$ where V satisfies the same partial differential equation as in Theorem 3.5 but with the opposite boundary condition $V(x, T) = -h(x)$. This partial differential equation is nonlinear whenever $r_2 > r_1$ and/or $\frac{\mu_2 - r_1}{\sigma_2} > \frac{\mu_1 - r_1}{\sigma_1}$, i.e. whenever there actually is a cost of borrowing and/or a cost of shortselling the stock. Because of the nonlinearity, the upper and lower bounds $\pi(h)$ and $-\pi(-h)$ do not necessarily collapse to a single arbitrage price as they would in the absence of frictions.

Note that a heuristic reasoning could have led us to these partial differential equations. Indeed, from the sign of $\frac{\partial V}{\partial x}(x, t)$ we can infer whether a hedging strategy needs to be long or short in the stock: if $\frac{\partial V}{\partial x}(x, t) > 0$ it needs to be long since it must rise in value as the stock price moves up and if $\frac{\partial V}{\partial x}(x, t) < 0$ it needs to be short since it must fall in value as the stock price moves up. In the first case, assuming that the riskless rate is r_0, the price of risk for a long position in the stock is $-\frac{\mu_1 - r_0}{\sigma_1}$ and the risk-neutral adjusted expected return on the stock is equal to the riskless rate. In the second case, the price of risk for a short position in the stock is $-\frac{\mu_2 - r_0}{\sigma_2}$ and the risk-neutral adjusted expected return on the stock is then equal to the riskless rate plus the cost (in terms of excess expected return) $\sigma_1(\frac{\mu_2}{\sigma_2} - \frac{\mu_1}{\sigma_1})$ of holding a short position in the stock with the same volatility. The riskless rate r_0 is then set equal to the lending rate r_1 if the hedging portfolio involves lending, and to the borrowing rate r_2 otherwise. Writing Black and Scholes (1973) equation with these risk-neutral adjusted expected returns and riskless rates, we then obtain our partial differential equation.

So far we only have a Verification Theorem, that is to say that if we can find a sufficiently regular solution to the partial differential equations above we can compute the arbitrage bounds on the prices of derivative securities, as well as determine the optimal hedging strategies. Under some additional restrictions on the model, it is possible to show that these partial differential equations indeed have a solution. An existence Theorem is included in the Jouini and Kallal (1995 b).

An example of a model that would satisfy these assumptions is an economy where the stock price Z_1 follows a geometric Brownian motion as in the Black-Scholes economy, with a constant cost (in terms of expected return) of shortselling the stock, and where the riskless borrowing and lending rates are constant. More precisely, let

$$\frac{dZ_1(t)}{Z_1(t)} = \mu dt + \sigma dW(t) \text{ and } \frac{dZ_2(t)}{Z_2(t)} = (\mu + c)dt + \sigma dW(t)$$

$$B_1(t) = exp(rt) \text{ and } B_2(t) = exp((r + s)t)$$

where the drift μ, the volatility σ, the cost of shortselling c, the riskless rate r and the spread s between the borrowing and the lending rates are constant. In this case, the stock price processes can also be written $Z_1(t) = Z_1(0)exp\{(\mu - \frac{1}{2}\sigma^2)t + \sigma W(t)\}$ and $Z_2(t) = exp\{-c(T - t)\}Z_1(t)$. The portfolio of derivative securities to be hedged can be chosen with a payoff of the form

$$h(x) = Max(0, x - K_1) + Max(0, K_2 - x)$$

which corresponds to a call of strike price K_1 plus a put of strike price K_2. It follows from Theorem 3.5 that there are no opportunities of arbitrage if and only if $s \geq 0$ and $c \geq 0$, i.e. if the spread in the riskless rate and the shortselling cost are nonnegative. Moreover, the arbitrage bounds on the claim $h(Z_1(T))$ can then be found by solving the partial differential equation

$$\frac{\partial V}{\partial t} + \frac{1}{2}\sigma^2 x^2 \frac{\partial^2 V}{\partial x^2} + rx\frac{\partial V}{\partial x} + cx[\frac{\partial V}{\partial x}]^- + s[V - x\frac{\partial V}{\partial x}]^- = rV$$

with boundary condition

$$V(x, T) = h(x).$$

Although a closed form solution is not available at this point, this equation can be solved by classical numerical methods, such as the finite difference method. Another possibility is to approximate the model by a binomial model. We have computed some numerical values of the arbitrage bounds of a portfolio formed of a put and a call (both at-the-money, i.e. of strike price equal to the stock price), for reasonable parameter values, and a cost c of the order 1%. We find that the arbitrage bounds are quite sharp, and substantially sharper than the arbitrage bounds obtained by implementing the hedging strategy that consists in adding up the hedge ratios given by the Black and Scholes formula. For instance, for $\sigma = 30\%$, $r = 5\%$, $s = 0$, $T = 6$ months, $Z_1(0) = K_1 = K_2 = 100$, and $c = 1\%$ we find that the arbitrage bounds are approximatively 16.90 and 17.17. Hence, in this case, the spread represents roughly 1.6% of the value of the portfolio. The Black and Scholes hedge ratios would give instead arbitrage bounds equal to 16.78 and 17.29, i.e. an interval twice as wide. These qualitative results are robust to changes in parameter values. More numerical results are given in the next subsection.

3.3   Shortselling costs in the finite periods and states case

In this subsection we consider a model where the set of states of the world Ω and the set of possible trading dates $\mathcal{T}$ are finite. More specifically, we shall assume that the stock price follows a multiplicative binomial process and that the riskless rate is constant, although it will be clear that this is by no means essential to our analysis.

Suppose that there are two securities in the economy: a stock and a riskless bond with a constant return $R = exp(r)$ over each period. The return on a long position in the stock is assumed to take two possible values $u_1 > d_1$ in states "up" and "down" at each date. In order to model the cost in going short in the stock, we assume that a short position in the stock yields different returns $u_2 > d_2$ in states "up" and "down" at each date. [27] As in the continuous time diffusions model, we shall consider two securities price processes Z_1 and Z_2 with these returns, and assume that security 1 cannot be sold short whereas security 2 cannot be held long. Also, we shall restrict our attention to derivative securities with payoffs that are functions of the stock price Z_1, so that the actual values taken by the process Z_2 are irrelevant, and its return matters only. Therefore, we can fix $Z_2(0)$ at an arbitrary (positive) level.

In this economy, a state of the world is the realization of a sequence of T "ups" and "downs" where T is the (finite) horizon of the economy. At any date t we can observe the sequence of "ups" and "downs" that have occured up to that date and there is a certain probability that the next element in the sequence is an "up" or a "down." Therefore, a probability measure on the set of states of the world Ω can be described by an array of conditional probabilities of going "up," at any date t, given each possible sequence of "ups" and "downs" up to time t. The set of probabilities for which the normalized process $\frac{Z_1(t)}{R^t}$ is a supermartingale (if there are any) are of the form α for state "up" tomorrow (and $1 - \alpha$ for state "down" tomorrow), where $\alpha \leq \frac{R-d_1}{u_1-d_1}$, at each date in every event. The probabilities for which the normalized process $\frac{Z_2(t)}{R^t}$ is a submartingale are of the form α' for state "up" tomorrow (and $1 - \alpha'$ for state "down" tomorrow), where $\alpha' \geq \frac{R-d_2}{u_2-d_2}$, at each date in every state. Therefore, according to Theorem

---

[27] I.e. We shall assume that the return on a long position in the stock is negatively correlated with the return on a short position in the stock. This is not required by the mathematics and the other (awkward) case could be treated equally well.

2.1, the model is arbitrage free if and only if $\alpha_2 \leq \alpha_1$, where $\alpha_i = \frac{R-d_i}{u_i-d_i}$ is the probability (for state "up" tomorrow) at each date in every state that transforms $\frac{Z_i(t)}{R^t}$ into a martingale.

The arbitrage bounds on the price of a derivative security that pays a certain function $h(Z_1(T))$ of the stock price at maturity, [28] i.e the minimum cost of hedging it and the maximum amount that can be borrowed against it, can be computed using the set Π of probabilities that transform $\frac{Z_1(t)}{R^t}$ into a super-martingale and $\frac{Z_2(t)}{R^t}$ into a submartingale. Indeed, we have already recalled that the interval defined by these bounds on the price of a derivative security is equal to the set of expectations of the normalized payoff of the security with respect to all these probabilities. These probabilities are of the form α for state "up" tomorrow (and $1 - \alpha$ for state "down" tomorrow), where $\alpha_2 \leq \alpha \leq \alpha_1$, at each date in every state. This suggests the following algorithm to compute the upper bound on the derivative security price, i.e. the minimum cost of hedging it: start from the vector of payoffs of the derivative security at the final date T divided by R^T and for each event at date $T - 1$ compute the maximum conditional expected value of these normalized payoffs, where probabilities are taken in Π. For instance, if in a given event at time $T - 1$ the normalized payoffs at time T are x_u if "up" occurs and $x_d > x_u$ if "down" occurs, the maximum conditional expected value is $\alpha_2 x_u + (1 - \alpha_2)x_d$ since $\alpha_2 \leq \alpha_1$. Once this vector of expected "payoffs" at time $T - 1$ in every event is computed, do the same thing at time $T - 2$ and so on down to date 0. The number obtained at time 0 is hence the upper bound on the ask price of the derivative security that pays $h(Z_1(T))$ at date T. Note that this algorithm also gives the risk-neutral probabilities for which this upper bound is the expectation of the normalized payoff of the derivative security. To compute the lower bound on its bid price, perform the same algorithm but consider minimum conditional expected values at each stage instead.

We shall now turn to the approximation of the Black and Scholes model with short sales costs by a sequence of binomial models of this sort, in the spirit of Cox et al. (1979). If T is the horizon of the economy (in some unit of time: days, weeks, years...) we shall denote by $\frac{T}{n}$ the amount of time between stock price movements, and hence n represents the number of periods. As $n \to +\infty$ we have that $\frac{T}{n} \to 0$ and we must adjust R, u_i and d_i in a way that approximates the Black and Scholes model with short sales costs in the limit. Since the riskless return is R^T over the horizon of the economy, the riskless return $\tilde{R}$ over one period of length $\frac{T}{n}$ must satisfy $\tilde{R}^n = R^T$ and hence $\tilde{R} = R^{\frac{T}{n}} = exp(r\frac{T}{n})$. We shall also choose $u_1 = exp(\sigma\sqrt{\frac{T}{n}})$, $d_1 = exp(\sigma\sqrt{\frac{T}{n}})$, $u_2 = exp(c\frac{T}{n} + \sigma\sqrt{\frac{T}{n}})$, $d_2 = exp(c\frac{T}{n} - \sigma\sqrt{\frac{T}{n}})$, and $q = \frac{1}{2} + \frac{1}{2}\frac{\mu}{\sigma}\sqrt{\frac{T}{n}}$, where q is the probability of the stock price moving "up", $\sigma > 0$ is the volatility of the stock, μ its drift, and $c \geq 0$ the cost (in terms of expected return) of going short in the stock.

---

[28] In fact, it will be clear that any distribution of payoffs in the "tree" defined by the movements of the stock price can be priced using our algorithm.

This leads to the following expectations and variances: $E(ln(\frac{Z_1(T)}{Z_1(0)})) = \mu T$, $E(ln(\frac{Z_2(T)}{Z_2(0)})) = (\mu + c)T$, $Var(ln(\frac{Z_1(T)}{Z_1(0)})) = Var(ln(\frac{Z_2(T)}{Z_2(0)})) = \sigma^2 T - \frac{\mu^2 T^2}{n}$. Therefore, the limits of the means and the variances of the (compounded) returns in a long and a short position in the stock coincide with the means and variances of the returns in the Black Scholes model. It can also be shown using a Central Limit Theorem as in Cox et al. (1979) that the whole distributions coincide as well in the limit.

We shall compute the arbitrage bounds on the prices of two types of financial instruments: straddles and butterfly spreads (see Cox and Rubinstein [1985] for the use of these instruments). A straddle is a combination of a put option and a call option written on the same stock, with the same expiration date T and the same strike price K. Therefore its payoff at date T is of the form:

$$h_s(x) = Max(0, K - x) + Max(0, x - K).$$

On the other hand, a butterfy spread is a combination of a call option of strike price K_1, a call option of strike price K_3 larger than K_1, and a short position in two call options of strike price K_2 with $K_1 < K_2 < K_3$, all written on the same stock and of same expiration date T. Therefore its payoff at date T is of the form:

$$h_b(x) = Max(0, x - K_1) - 2Max(0, x - K_2) + Max(0, x - K_3).$$

In our numerical computations we have adopted the parameters $\sigma = 30\%$ for the volatility of the stock, $r = 5\%$ for the riskless lending rate, $Z(0) = 100$ for the initial value of the stock, and $T = 6$ months for the time to expiration of the puts and calls.

We computed the numerical values of the arbitrage bounds on an at-the-money straddle (i.e. with strike price $K = 100$) when $c = 0\%$ and s ranges between 0% and 5%, when $s = 0\%$ and c ranges between 0% and 5%, and when $s = c$ range between 0% and 5%. We find that the spread between the arbitrage bounds ranges between 0% and 3.4% in the first case, between 0% and 3.5% in the second case, and between 0% and 6.6% in the third case.

We also computed the numerical values of the arbitrage bounds on a butterfly with strike prices $K_1 = 90$, $K_2 = 100$, and $K_3 = 110$ when $c = 0\%$ and s ranges between 0% and 5%, when $s = 0\%$ and c ranges between 0% and 5%, and when $s = c$ range between 0% and 5%. We find that the spread between the arbitrage bounds ranges between 0% and 11.2% in the first case, between 0% and 10.7% in the second case, and between 0% and 20.6% in the third case.

In each case we also computed the arbitrage bounds that would be obtained by adding up the arbitrage bounds given by using the Black and Scholes formula for the puts and calls separately. It appears that the spread between the actual arbitrage bounds is substantially smaller than the spread between these subop-timal bounds. For the straddle, the spread between our bounds is roughly equal to half the spread between the suboptimal bounds. For the butterfly, the effect is even more dramatic and the spread between our bounds is roughly equal to one twentieth of the spread between the suboptimal bounds. The magnitudes of these results are quite robust to changes in parameter values although (as one would expect) they are sensitive to the strike prices of the puts and calls in the portfolios. We may conclude that the arbitrage bounds in the presence of (reasonable) shortselling and borrowing costs are quite sharp, and substantially sharper than the bounds derived from the Black and Scholes formula.

3.4 Incomplete markets

In both Theorems 2.1 and 2.2 we have not assumed that the markets are complete and all our results encompasses this particular case. Nevertheless it seems to be interesting to derive a result concerning incomplete markets without other imperfections. We then have,

Theorem 3.6 (El Karoui-Quenez) *In an incomplete market without other imperfections the arbitrage interval for some contingent claim x is given considering all the normalized expectation values of x relatively to probability measures P^* for which all the price processes of the traded securities are martingales.*

This result is a direct consequence of Theorem 2.1 or Theorem 2.2. Note that, in the general case, we have two sources of indeterminacy for P^* : we have to choose Z^* between the bid and the ask processes and for each Z^* we have many probability measures for which Z^* is a martingale. In the case considered by Theorem 3.6, there is only one source of indeterminacy : Z^* is known (equal to Z and to Z') and we only have to find P^* for which this Z^* is a martingale.

4   Trading strategies with market frictions

We saw in the previous sections that in the presence of market frictions duplication can be suboptimal and conversely some nonoptimal strategies in the frictionless framework become optimal if we introduce frictions in the model. In this section we characterize efficient consumption bundles in dynamic economies with uncertainty, taking market frictions into account. We define an efficient consumption bundle as one that is an optimal choice of at least a consumer with increasing, state-independent and risk-averse Von Neumann-Morgenstern preferences. We incorporate market frictions into the analysis, including dynamic market incompleteness, bid-ask spreads, short sales constraints, different borrowing and lending rates and taxes.

For the perfect market case Dybvig (1988 a) develops a new model, the payoff distribution pricing model (PDPM), and shows that the size of the inefficiency of a contingent claim can be measured by the difference between the investment it requires and the minimum investment necessary to obtain at least the same utility level for all possible agents (utility price). This utility price is equal to the minimum investment necessary to obtain the same distribution of payoffs (its "distributional price"). In the presence of market frictions, however, simple examples show that some distributions of payoffs are inefficient as a whole : there might not exist any efficient consumption bundle with a given distribution of payoffs. Hence, in general the PDPM would ignore a piece of the potential inefficiency of a consumption bundle. On the positive side, we show that the inefficiency of a consumption bundle can be measured by the difference between the investment it requires and the minimum investment necessary to obtain a claim with the same distribution of payoffs *or* a convex combination of such claims (the "utility price"). We also show that the utility price of a consumption bundle is in fact the largest of its distributional prices in the underlying frictionless economies defined by the underlying linear pricing rules.

We consider a multiperiod economy with uncertainty, where consumers can trade at each intermediate date a finite number of securities that give the right to a contingent claim to consumption at the final date. We shall assume that consumption (of a single good, the numeraire) takes place at the initial and the final date only. For expositional purposes we shall also assume that there are a finite number of trading dates and of possible states of the world [29]. The states of the world are numbered by $i = 1$ to n and a contingent claim that gives the right to c_i units of consumption in state i (for $i = 1, \ldots, n$) at the final date is represented by the $n-$ dimensional vector $c = (c_1, \ldots, c_n)$. We do *not* assume that markets are dynamically complete (i.e. that any contingent claim to consumption can be achieved through dynamic securities trading), neither do we assume that short sales are unrestricted, and we allow for bid-ask spreads, different borrowing and lending rates, and possibly other types of market frictions. [30]

As we have already seen, in the presence of market frictions (including market incompleteness, bid-ask spreads, different borrowing and lending rates, short sale constraints and taxes), there exists a convex set K of state price vectors[31] such that the minimum cost today of achieving any contingent claim to consumption $c = (c_1, \ldots, c_n)$ tomorrow is equal to $\pi(c) = \max\{p \cdot c = \sum_{i=1}^{n} p_i c_i : p \in K\}$. Once we normalize prices and consumption bundles by the price of the security that serves as a numeraire[32] the state prices add up to one across states of the world (i.e. $\sum_{i=1}^{n} p_i = 1$) and K is the set of so called "risk-neutral probabilities". The normalized price of any contingent claim is then the largest expected value of its payoff with respect to the risk-neutral probabilities. Also, as we shall see, in such an economy the vectors of intertemporal marginal rates of substitution of maximizing agents can be identified with some element of the set of candidate state price vectors K.

In the perfect market case, Dybvig (1988 a and b) develops the payoff distribution pricing model (PDPM). Assuming a finite number of equiprobable states

---

[29] We shall use tools that make the extension of our results to continuous time economies possible (although technical).

[30] We shall rely on theoretical results of previous sections.

[31] One of which is strictly positive for every state of the world.

[32] This could be a riskless bond.

of the world, he finds that a consumption bundle c is efficient (i.e., choosen by some maximizing agent) if and only if it gives the right to at least as much consumption in states of the world where consumption is strictly cheaper to obtain, i.e if $p_i^* > p_j^*$ implies $c_i \leq c_j$, where p^* is the vector of state prices that represents the linear pricing rule. This is equivalent to the fact that the consumption bundle c minimizes the cost of achieving the distribution of its payoffs, i.e.

$$p^* \cdot c = \min\{p^* \cdot c' : c' \text{ is distributed as } c\}.$$

Dybvig (1988a) then defines the distributional price of an arbitrary contingent claim c (efficient or not) as the minimum cost of achieving the distribution of its payoffs. A claim is then efficient if and only if its market price is equal to its distributional price and the inefficiency of a claim c can be measured by the difference $p^* \cdot c - P(c, p^*)$. This is the difference between the cost of achieving a consumption bundle c and the minimum cost of achieving the same distribution of payoffs (that gives the same utility as c to every agent).

In the presence of market frictions, however, simple examples show that the "distributional approach" does not apply in a straightforward manner. As opposed to the frictionless case, in the presence of market frictions some distributions are in fact inefficient and are never chosen by any maximizing agent. Therefore, the distributional price of a claim does not reveal all its potential inefficiency. In other words, there might be contingent claims that give at least as much utility as c to every agent, are not distributed as c, and are strictly cheaper than the cheapest claim distributed as c.

We shall illustrate this fact by analyzing a two-period economy with two equiprobable states of the world (1 and 2), and where the opportunity set is represented by the set of state price vectors $K_{a,b} = \{(p, 1 - p) : p \in [a, b]\}$ with $0 < a < b < 1$. This is the case, for instance, of an economy with a zero riskless rate and where agents can buy and sell a stock with payoffs $(S_1, S_2), S_1 > S_2$ at an ask price $S^a = aS_1 + (1 - a)S_2$, and a bid price $S^b = bS_1 + (1 - b)S_2$. In this case, the minimum cost to obtain a consumption bundle (c_1, c_2), is $ac_1 + (1-a)c_2$ if $c_1 \leq c_2$ and it is $bc_1 + (1 - b)c_2$ otherwise. Suppose that $c_1 < c_2$, then the distributional price of (c_1, c_2) (i.e. the minimum cost to get a consumption claim distributed as (c_1, c_2)) is $\min\{ac_1 + (1 - a)c_2, bc_2 + (1 - b)c_1\}$. It is then easy to check that if $a < \frac{1}{2}$ then $\min\{ac_1+(1-a)c_2, bc_2+(1-b)c_1\} > \frac{c_1+c_2}{2}$ and since any maximizing agent (with preferences satisfying (i)-(iii) above) weakly prefers the consumption bundle $(\frac{c_1+c_2}{2}, \frac{c_1+c_2}{2})$ to (c_1, c_2) (and to (c_2, c_1)) this shows that the distribution of payoffs of (c_1, c_2) as a whole is inefficient. Moreover, note that this example is not a degenerate one. Both consumption bundles (c_1, c_2) and (c_2, c_1) are in the opportunity set and neither of them is dominated[33] by a consumption bundle that costs the same amount to achieve.[34] Note that, as we shall see, this does not mean that the presence of market frictions makes efficiency a tighter criterion as a general rule.

---

[33] In the sense of a weakly larger payoff in every state of the world and strictly larger in some state.

[34] In the presence of market frictions this does not violate the absence of arbitrage.

In the next subsection we shall characterize efficient portfolios in the presence of market frictions and give a preference-free evaluation of the inefficiency cost. We shall also relate this measure of inefficiency to the measure given by the distributional approach in the underlying frictionless economies. We shall also see how these results apply to the measurement of portfolio performance.

The reader can find all the proofs of this section and more examples in Jouini and Kallal (1996a).

4.1   Efficient trading strategies

Recall that an economy with market frictions in securities trading can be represented by a convex set K of probability measures (state price vectors or linear pricing rules), where at least one element of K assigns a strictly positive weight to every state of the world.[35] In such an economy, achieving at least a consumption bundle c (through securities trading) requires a minimum investment of $\pi(c) = \max\{E(c) : E \in K\}$. We shall say that a linear pricing rule E of K "prices" the consumption bundle c if it satisfies $\pi(c) = E(c)$.

Note that we have implicitly assumed a zero interest rate. This assumption turns out to be innocuous: it only amounts to normalize the state prices by the discount factor (i.e. the sum of the state prices across all the states of the world). In order to analyze an economy with a nonzero interest rate, we only need to multiply all the payoffs by the discount factor in our analysis.

For convenience, we shall assume that there is a finite number n of equiprobable states of the world. The class of weakly concave and strictly increasing Von Neumann-Morgenstern preferences will be denoted by $\mathcal{U}$.

We shall say that a consumption vector is efficient if there exists a risk-averse and strictly increasing Von Neumann - Morgenstern utility function and an initial wealth for which it is an optimal choice. More formally

Definition 4.1 *A contingent claim $c^* \in R^n$ is efficient if there exists $u \in \mathcal{U}$ such that c^* solves* $\max\{u(c) : \pi(c) \leq \pi(c^*)\}$.

This is the same definition as in the frictionless case except that the budget constraint is expressed in terms of a nonlinear pricing operator π, where the nonlinearity comes from the presence of market frictions. However, this pricing operator is of a particular form : it is the supremum of a family of linear positive pricing rules. Hence, the budget constraint can be viewed as a collection of linear budget constraints with linear pricing rules ranging in K. Also, since agents are assumed to have strictly increasing preferences, an efficient claim c^* makes the budget constraint binding and the initial wealth for which it is an optimal choice is necessarily $\pi(c^*)$.

---

[35] It is easy to see that in such an economy, the existence of such a strictly positive vector of state prices is both necessary and sufficient for the absence of arbitrage opportunities. It also means that K, which we assume to be closed, is the closure of the set of its strictly positive elements.

Similarly, we shall say that c^* is *strictly* efficient if it is an optimal choice for an agent with a *strictly* concave and strictly increasing Von Neumann - Morgenstern utility function. We shall denote by $\mathcal{U}_{sc}$ the class of such preferences.

The following theorem characterizes the efficiency of a given contingent claim in terms of a particular state price vector in K : the linear pricing rules that price it.

Theorem 4.1 *A contingent claim $c^* \in R^n$ is (strictly) efficient if and only if there exists a strictly positive probability measure $E^* \in K$ such that*
(i) $E^*(c^*) = \pi(c^*)$,
(ii) c^* is in (strict) reverse order of E^*.

We say that $c^* = (c_1^*, \ldots, c_n^*)$ is in reverse order of $E^* = (e_1^*, \ldots, e_n^*)$ if : $c_i^* > c_j^*$ implies $e_i^* \leq e_j^*$. This means that the payoff is not lower in a "cheaper" state of the world. Similarly, we say that $c^* = (c_1^*, \ldots, c_n^*)$ is in strict reverse order of $E^* = (e_1^*, \ldots, e_n^*)$ if : $c_i^* > c_j^*$ implies $e_i^* < e_j^*$. This means that the payoff is not lower in a "cheaper or as expensive" state of the world. The Theorem then says that a claim is (strictly) efficient if and only if it is in (strict) reverse order of a strictly positive state price vector that "prices" it.

Roughly speaking, this result follows from the first-order conditions: marginal utilities of consumption in each state of the world are proportional to the state price vector representing one of the binding linear budget constraints (that is binding at the cost of the optimal consumption bundle). From the assumption that agents are risk-averse, marginal utilities are decreasing, which implies that payoffs must be higher in cheaper (relative to the binding linear pricing rule) states of the world. The difficulty is that we are dealing with a maximization problem under a continuum of constraints. This Theorem generalizes the price characterizations obtained by Peleg and Yaari (1975) and by Dybvig and Ross (1982) in the incomplete markets case.

The previous result provides us with a diagnostic test: given a contingent claim we are now able to determine whether it is an optimal choice for some maximizing agent. What we need now is an evaluation of the inefficiency cost, i.e. a measure of how far a claim is from being efficient.

A simple measure of the (potential) inefficiency of a consumption bundle c^* is given by $\pi(c^*) - V(c^*)$ where

$$V(c^*) = \sup_{u \in \mathcal{U}} \{\min\{\pi(c) : u(c) \geq u(c^*)\}\}.$$

Indeed, $V(c^*)$ represents the larger amount that is required by rational consumers (with preferences in $\mathcal{U}$) in order to get the same utility level as with c^*. Of course, if c^* is efficient, then $V(c^*) = \pi(c^*)$ and our measure of inefficiency is equal to zero. On the other hand, if c^* is inefficient, the difference $\pi(c^*) - V(c^*)$, which is equal to $\inf_{u \in \mathcal{U}} \{\pi(c^*) - \min\{\pi(c) : u(c) \geq u(c^*)\}\}$, represents the smallest discrepancy, across rational consumers, between the actual cost of c^* and the price at which it would be an optimal choice. Hence our measure of inefficiency $\pi(c^*) - V(c^*)$ does not depend on the choice of a specific utility function.[36]

We shall call $V(c^*) = \sup_{u \in \mathcal{U}}\{\min\{\pi(c) \ : \ u(c) \geq u(c^*)\}\}$ the "utility price of c^*". It turns out that in dynamically complete perfect markets, the utility price of a consumption bundle coincides with its distributional price (see Dybvig [1988a]), i.e. the minimum cost of achieving the same distribution of payoffs. Our example in the introduction clearly shows that this is not the case in the presence of market frictions. In other words, even though an efficient claim obviously minimizes the cost of achieving the distribution of its payoffs (since agents have state-independent preferences), minimizing the cost of a achieving a given distribution of payoffs does not imply efficiency. In order to be efficient a claim c^* also needs to minimize the cost of achieving the distribution of its payoffs in another economy : the frictionless economy represented by a positive linear pricing rule E^* (in K) that prices c^* in the original economy with market frictions.

In the frictionless case, the pricing rule is linear and hence there always exists a minimum cost consumption bundle in the set $\{c : \forall u \in \mathcal{U}, u(c) \geq u(c^*)\}$ that has the same distribution of payoffs as c^* (i.e. that is a permutation of c^*). However, in the presence of market frictions the pricing rule is not linear. Hence, it might be strictly cheaper to obtain a convex combination of consumption bundles that are distributed as c^* than to obtain any claim distributed as c^*. We shall denote by $\Sigma(c^*)$ the set of convex combinations of consumption bundles that are distributed as c^*. We then have,

Theorem 4.2 *for all $c^* \in R^n$, the utility price of c^* is equal to*

$$V(c^*) = \min\{\pi(c) : u(c) \geq u(c^*), \forall u \in \mathcal{U}\} = \min\{\pi(c) : c \in \Sigma(c^*)\}.$$

This says that the utility price of c^* is in fact the cost of the cheapest consumption bundle that is distributed as c^* *or* that is a convex combination of consumption bundles distributed as c^*. Equivalently, according to our Lemma above, the utility price of c^* is then the cost of the cheapest consumption bundle that makes every rational agent (with preferences in $\mathcal{U}$ or in $\mathcal{U}_{sc}$) at least as well off as with c^*. In proving this Theorem we also prove that the utility price $\sup_{u \in \mathcal{U}_{sc}} \min\{\pi(c) \ : \ u(c) \geq u(c^*)\}$ defined relatively to the smaller class of preferences $\mathcal{U}_{sc}$ coincides with the utility price $V(c^*)$.

It also turns out that, even though the utility price does not coincide with the distributional price in the presence of market frictions, there is a link between the utility price and the set of distributional prices in the underlying frictionless economies. Indeed, we find that the utility price of a claim is the largest of its distributional prices in the underlying frictionless economies defined by the underlying linear pricing rules that belong to K. The following Theorem states this result

---

[36] It depends though on the class of preferences that we use. As we shall see, however, it is quite robust to changes in the choice of this class.

Theorem 4.3 *for all $c^* \in R^n$, the utility price of c^* is equal to*

$$V(c^*) = \max\{P(c^*, E) : E \in K\},$$

where $P(c^*, E) = \min\{E(c) : c \text{ is distributed as } c^*\} = \min\{E(c) : c \in \Sigma(c^*)\}$.

Note the analogy with the price at which a consumption bundle is available in this economy (the amount an agent needs to invest to get it) which is the largest of its prices in the underlying frictionless economies with pricing rules belonging to K.

Morever it is shown in Dybvig (1988a) that the distributional price $P(c^*, E)$ can be expressed using the cumulative distribution functions of the payoff c^* (denoted F_{c^*}) and of the state price E (denoted F_E). Recall that $F_{c^*}(x)$ is equal to the probability that the random variable c^* is less than or equal to x (and similarly for F_E). Also, let the inverse of a cumulative distribution function F be defined by $F^{-1}(y) = \min\{x : F(x) \geq y\}$ for all $y \in (0, 1)$ (the values at 0 and 1 will be irrelevant to us). We then have that $P(c^*, E) = \int_0^1 F_{c^*}^{-1}(y) F_E^{-1}(1-y) dy$, which implies that the utility price is equal to

$$V(c^*) = \max\{\int_0^1 F_{c^*}^{-1}(y) F_E^{-1}(1-y) dy : E \in K\}.$$

4.2 Portfolio performance

As in Dybvig (1988a), in measuring performance we follow the tradition of comparing some investment strategy (and the distribution of payoffs it leads to) to the alternative of trading in a market. However, we do not assume that this market is frictionless. This means that we allow it to be (dynamically) incomplete, to have restricted short sales, different borrowing and lending rate and positive bid-ask spreads. Ignoring these frictions would make the benchmark market available to investors more attractive than it actually is, and would lead to an underestimation of the performance of the investment strategy being analyzed. Of course, this effect is mitigated by the fact that the investment strategy itself is subject to transaction costs (and other frictions) and therefore leads to lower payoffs than it would in a perfect market.

An investment strategy is evaluated on the basis of the distribution F_c of its payoff c, where the actual payoff c might depend on information that is not available to the agents (but only to the portfolio manager), allowing for information-trading and private investments outside the benchmark market. The benchmark market itself is described by the set K of linear pricing rules that summarize the investment opportunities available to investors. As far as utility pricing is concerned the relevant characteristic of the benchmark market is the set of cumulative distributon functions of the underlying linear pricing rules $\{F_E : E \in K\}$. The following Corollary is similar to Theorem 4 in Dybvig (1988 a) for the frictionless case, and is a consequence of our Theorem 4.3.

Corollary 4.4 *Suppose that an investment strategy leads from an initial wealth w_0 to a distribution of payoffs F_c. Let $V(c) = \max\{\int_0^1 F_c^{-1}(y)F_E^{-1}(1-y)dy : E \in K\}$. Then,*

(i) If $w_0 < V(c)$, we have superior performance, i.e. there exists a rational agent[37] who prefers receiving the distribution of payoffs F_c to trading in the benchmark market.

(ii) If $w_0 = V(c)$, we have ordinary performance, i.e. every rational agent weakly prefers trading in the benchmark market to receiving the distribution of payoffs F_c.

(iii) If $w_0 > V(c)$, we have inferior performance, i.e. every rational agent strictly prefers trading in the benchmark market to receiving the distribution of payoffs F_c.

Hence, by comparing the initial investment to the utility price of the distribution of payoffs obtained by the investment strategy, one is able to evaluate the performance of the portfolio. If the utility price is lower than the initial investment, then we conclude that the portfolio is not well-diversified and is underperforming. If the utility price is equal to the initial investment, then the portfolio is well-diversified and it is performing as it should. If the utility price is larger than the initial investment, the manager has superior ability and/or information and the portfolio is overperforming.

As argued by Dybvig (1988 a) this provides an alternative to the Security Market Line (SML) in measuring portfolio performance. As opposed to the SML analysis, this alternative gives a correct evaluation even when superior performance is due to private information. Indeed, the SML is based on mean-variance analysis,[38] and even if securities returns are assumed to be jointly normally distributed, they will typically not be normal once conditioned on information (see Dybvig and Ross [1985 a and b]).

4.3   Efficient hedging strategies

We now assume that an agent has some contingent liability x (suppose, for instance, that he has written an option contract) and some initial wealth w_0 to hedge it. A rational agent with a utility function $u \in \mathcal{U}$ will then solve the following maximization problem

$$\max\{u(c - x) : \pi(c) \leq w_0\}.$$

In the case of dynamically complete markets without frictions, this problem can be separated into two different stages: first duplicate the claim x, at a cost $\pi(x)$,

---

[37] I.e. with a utility function in $\mathcal{U}$.

[38] Mean-variance analysis can be justified either by assuming normally distributed returns or by assuming quadratic utility. However, the latter assumption implies undesirable properties such as nonmonotonic preferences and increasing absolute risk aversion.

then solve for the optimal investment with an initial wealth $w_0 - \pi(x)$. Formally, our agent will solve the maximization problem

$$\max\{u(c) : \pi(c) \leq w_0 - \pi(x)\}.$$

Hence, in this case there is no real difference between the efficiency of a hedging strategy and of an investment strategy since optimal hedging consists in duplication of the liability followed by an optimal investment strategy of the remaining funds.

In the presence of market frictions, however, this is no longer true: the hedging problem cannot be separated from the investment problem. This means that the efficiency of hedging strategies is an issue. We propose the following definition of efficient hedging strategies :

Definition 4.2 *A strategy leading to a payoff* $c^* \in R^n$ *is an efficient hedging strategy of a contingent claim* x *if there exists* $u \in \mathcal{U}$ *and* $w_0 \in R$ *such that* c^* *solves* $\max\{u(c - x) : \pi(c) \leq w_0\}$.

This means that we say that a hedging strategy is efficient if it leads to a net payoff that is efficient. As for investment strategies we shall say that a hedging strategy is *strictly* efficient if it is optimal for a rational agent with a strictly concave utility function (i.e. with preferences in $\mathcal{U}_{sc}$). We refer to Jouini and Kallal (1996a) for a characterization of efficient hedging strategies and, in the case of general hedging strategies, for a measurement of the inefficiency.

4.4 Numerical results

Let us consider an economy where there is a riskless bond and a stock that follows a stationary multiplicative binomial model, with an actual probability of 0.5 of going "up" or "down" at each node. This stock can be sold short, paying a constant cost in terms of expected return. In this case we have as it is shown in section 4.3 $K = [\alpha_1, \alpha_2]^{\text{number of nodes}}$ where each component of an element of a measure in K is the conditional probability of going "up" at the corresponding node. Then let us define the scalar $\beta = \max\{([\alpha_1, \alpha_2] \cup [1 - \alpha_2, 1 - \alpha_1]) \cap [0, 0.5]\}$ and the associated probability measure E_β on our tree defined by a constant conditional probability β of going "up" at each node. We then have

Theorem 4.5 *For all* $c^* \in R^n$, *its utility price is equal to* $V(c^*) = E_\beta(\tilde{c})$ *where* $\tilde{c}$ *is distributed as* c^* *and is in reverse order of* E_β.

In Jouini and Kallal (1996a) we present a simple algorithm for computing the utility price of a payoff and evaluating the inefficiency cost of a trading strategy. Of course, if holding the stock is an efficient strategy, going short in the stock is *not* an efficient strategy. Also, shortselling costs would have an impact on the inefficiency of a trading strategy only if this strategy and/or the strategies that dominate it involve some shortselling. We could examine, for instance, a stop-loss strategy in a setup where investors expect a negative return for the stock and go short in it, liquidating their position as soon as the price of the stock reaches a certain level (say 110% of the initial price). This strategy is inefficient (for reasonable parameter values of the stock and bond price processes), but the strategies that dominate it involve more shortselling than the strategy itself, and hence we expect its inefficiency to be smaller than in the absence of shortselling costs. For a high enough cost of shortselling the strategy can even be shown to be efficient.

5   Imperfections and stationarity

In this section, we consider a model in which agents face investments opportunities (or investments) described by their cash flows as in Gale (1965), Cantor and Lippman (1983,1995), Adler and Gale (1993) and Dermody and Rockafellar (1991,1995). These cash flows can be at each time positive as well as negative. It is easy to show that such a model is a generalization of the classical one with financial assets. We impose that the opportunities can not be sold short and that the opportunities available today are those that are available tomorrow and the days after in an infinite horizon model...(stationarity). In such a model we prove in Carassus and Jouini (1996b) that the set of arbitrage prices is smaller than the set obtained without stationarity. This result does not contradicts previous results because we have now an infinite horizon and each security can start at each date (for example, there is new options at each date). Then when we consider some price for a given security, this price can induce arbitrages between this security starting today, tomorrow,... If we impose that there is no transaction costs on the call option, the previous sections imply that the price of the option can be equal to any price in the non arbitrage interval. Dubourg (1994) and Soner and al. (1995) proved that this interval is too large and equal to $[0, S]$. If we impose stationarity in the model there is only one price compatible with the no arbitrage condition : the Black and Scholes one or more generally the frictionless one.

In order to give an idea of the result we will consider in the next the simplest case defined by a deterministic framework.

The model we consider entails the absence of risk, stationarity, and short sales constraints. In the general theory of arbitrage formalized by Harrison and Kreps (1979), Harrison and Pliska (1981), and Kreps (1981), securities markets are assumed to be frictionless, and the main result is that the absence of arbitrage opportunities (or no arbitrage) is equivalent to the existence of an equivalent martingale measure. The existence of state prices follows. In our framework, we will prove that the state prices must have a particular form : e^{-rt}. In fact, our main result is basically that no arbitrage implies the existence of a yield curve, and that the only yield curve process consistent with no arbitrage, in a deterministic and stationary setup, is flat.

In our model we allow short sales constraints, but only in order to give an intuition of our result, let us consider a simple frictionless setup. The absence of arbitrage opportunities implies the existence at any time t of a positive discount function D_t. $D_t(s)$ is the market value at time t of one dollar paid at time $t+s$ (this is in discrete settings just an implication of the separating hyperplane theorem). No arbitrage means no arbitrage even for contracts that may not be present, including forward contracts and zero coupon bonds. Following Cox, Ingersoll, and Ross (1981), the consequence of the no arbitrage condition in a deterministic setting is that the spot bond price is equal to the forward bond price. So the forward price at time t of a bound delivered at time $t+s$ and paying one dollar at time $t+T$, for $s < T$, that is $D_t(T)/D_t(s)$, is equal to the price at time $t+s$ of one dollar paid at time $t+T$, $D_{t+s}(T-s)$. Roughly speaking, stationarity in the model would imply stationarity for D, i.e. $D_t = D_{t+s}$, for all t and s. Hence we get $D(T)/D(s) = D(T-s)$, and the unique solution to this equation is $D(t) = e^{-rt}$, for some constant r. In fact, the stationarity for D is not straightforward and we prove that there exists many discount functions but a unique of this form.

Moreover it is well known that the classical notion of no arbitrage is not always equivalent to the existence of an equivalent martingale measure. Since Kreps (1981), Back and Pliska (1990), and more recently, Delbaen (1992) and Schachermayer (1994), we know that it is necessary to eliminate possibilities of getting arbitrarily close to something positive at an arbitrarily small cost, and therefore we will use this free lunch concept to derive our result.

We assume that every investment is available at each period of the investment horizon, one can subscribe to the investment at each date (stationarity). We will also assume the number invested in each time period to be nonnegative. This requirement here is that no investment can be sold. Notice that this short sale constraint is not a restriction and our model includes the case without constraints. Because the investor wants to become rich in a finite time, we shall also constrain the strategies to end in a finite time.

In the discrete case, an investment project m will be characterized by $(m_0, ..., m_T)$ where the real number m_t represents the cash received from the project in the tth period. If m_t is nonpositive, the investor must pay for the project, and if m_t is nonnegative, the investor is paid by the project. In this formalization, not matter if assets have a price or not. If m_0 is negative it could represent the price to pay in order to assure the cash flow $m_1, ..., m_T$. Here, we choose to include the price in the cash flow sequence, this is to say investments have price zero.

In this section, an investment will be represented as a Radon measure (for example see Bourbaki (1965) or Rudin (1966)). Roughly, for an investment represented by a Radon measure μ, $\int_{t_1}^{t_2} d\mu$ represents the investment payment between times t_1 and t_2. This choice allows us to describe investments with discrete as well as continuous cash flows. The previous discrete payment, $m = (m_0, ..., m_T)$, will be represented by the discrete measure $\mu = \sum_{t=0}^{T} m_t \delta_t$, where δ_t is the Dirac measure in t. But it also allows us to treat investment having continuous payoff, that is investment represented by a function m. In this case, $m(t)dt$ represents the investment payment in the short period dt, and the Radon measure μ associated to this investment will be given by the following measure defined by a density, $d\mu(t) = m(t)dt$.

We allow our model to contain an infinite number of investments. Notice that a continuous rate is modeled by an infinite number of investments, because one should consider all the possible repayment dates. The set of investment income streams is modeled by a family of Radon measure $(\mu_i)_{i \in I}$ with I finite or infinite. We suppose that all investment i have a finite horizon T_i, that is the support of measure μ_i lies in $[0, T_i]$. Otherwise, assuming the existence of an investment with an infinite horizon, it will always be possible to suspend repayment of the debt to infinity. This is not an arbitrage opportunity, because the investor wants to become rich in a finite time, which implies the time horizon to be finite. In this model, the investor is only allowed to choose a finite number of investments. There is an infinite number of possibilities but only a finite number of choices. Let us consider the example of a single investment m. At each time t, we must choose the number of subscriptions to investment m. Let l_t be the chosen number. At time 0, we buy l_0 investments which assures a payoff of $l_0 m_0$. At time 1, the total payoff is $l_0 m_1 + l_1 m_0$, and at time t, it will be $l_0 m_t + l_1 m_{t-1} + ... + l_{t-1} m_1 + l_t m_0$, which can be described by the convolution product $l * m(t)$. In the general case, after selecting a finite subset J of the set I of investments, the investor chooses the number of subscriptions from each element of J. For the same reasons as before, these numbers will be modeled by a family $(l_j)_{j \in J}$ of Radon measures. Roughly, $\int_{t_1}^{t_2} dl_i$ represents the number of investments i bought between times t_1 and t_2. We also require that the support of all measures l_i is in a fixed compact set. Moreover, the no sell assumption requires all the l_i to be positive. The previous payoff calculus is easily generalized and the choice of a finite subset J of I, and a strategy $(l_j)_{j \in J}$ leads to the payoff $\sum_{j \in J} l_j * \mu_j$.

The following example, from Adler and Gale (1993), intended to show whether it is possible to make an arbitrarily large profit in a finite time. Consider an investment which pays \$1 today. The investor must pay \$2 tomorrow and finally receives \$1.01 the day after. We denote this investment by $m = (1, -2, 1.01)$. As previously, the investor has no money to begin with, so the only way to pay the second day's installment on a unit of investment is by initiating a second investment at level two. It is straightforward to show that in order to get a zero payoff, the investor must subscribe at time t to $l_t = -(l_{t-2} m_2 + l_{t-1} m_1)$ investments. A simple calculus leads to a positive payoff after 32 periods. So, with this investment, it is possible to become arbitrarily rich after 32 periods (assuming one can buy an arbitrarily large number of investment m).

As we saw before a strategy will be defined as follows :

Definition 5.1 *A strategy is defined by the choice of :*
  - a finite number of investments indexed on a finite subset J included in I,
  - an investment horizon n,
  - a buying strategy for the set of investments J modeled by a family of positive Radon measures l_j which support is included in $[0, n - T_j]$, for all j in J.

We now want to define the absence of arbitrage opportunities. In fact, we will consider a general notion of no arbitrage, which has been developed by

Kreps (1981) : no free lunches. We recall that an arbitrage opportunity is the possibility to get something positive in the future for nothing or less today. The no free lunches concept allows us to eliminate the possibility of getting arbitrarily close to something positive at an arbitrarily small cost. In fact, Back and Pliska (1990) provide an example of a securities market where there is no arbitrage and where the classical theorem of asset pricing does not hold, essentially where there is no linear pricing rule. More recently, Schachermayer (1994) has introduced a more precise version of the no free lunches concept : no free lunches with bounded risk, which makes more sense from an economic point of view. This concept is also equivalent to the existence of a martingale measure for discrete time process. In this work, we will say that the set of investments $(\mu_i)_{i \in I}$ admits no free lunches if it is possible to get arbitrarily close to a nonnegative payoff in a certain way. To define the type of convergence that we use, we have to recall some properties of the Radon measure (see for example Bourbaki (1987)). We denote by E_n the space of continuous functions with support in $[0, n]$, and we attribute to E_n the topology $\mathcal{T}_n$ of the uniform convergence on $[0, n]$ (E_n is a classical Banach space). Recall that the strict inductive limit topology $\mathcal{T}$ is defined such as, for all n, the topology induced by $\mathcal{T}$ on E_n is the same as $\mathcal{T}_n$. More precisely, a sequence (φ_j) in E is said to be converging to φ in the sens of the topology $\mathcal{T}$ if there exists n, such that all the considered functions have their support in $[0, n]$ and such that the considered sequence converges to φ in the sens of the topology $\mathcal{T}_n$, i.e. in the sens of the uniform convergence on $[0, n]$. The completeness of E is shown in Bourbaki (1987), and we recall that with this topology on E, the space E^* of continuous linear forms on E is the Radon space measure. Notice that, using one of the Riesz representation theorem, a positive Radon measure is uniquely associated to a Borel-Radon measure, and we will use the same notation for both of them. We will now consider the weak-* topology on the space E^* of the Radon measures, which is called the vague topology. This means that the sequence (π_n) of Radon measures converges vaguely to π if for all continuous function φ with compact support $\pi_n(\varphi)$ converges to $\pi(\varphi)$. In fact as in Schachermayer, we will consider only the limit of weak-* sequences and not all the weak-* closure as in the classical definition of free lunches. Our definition of a free lunch will be :

Definition 5.2 *There is a free lunch if and only if there exists an investment horizon n and a sequence of strategies $(l_j^p)_{j \in J}$ with the same investment horizon n such that the corresponding payoff sequence $(\sum_{j \in J} l_j^p * \mu_j)$ converges vaguely to a nonnegative and nonzero measure π.*

Note that the "limit payoff" π also has its support in $[0, n]$ and with this definition we do not include free lunches which occur in an infinite time. We will see, that we can use a weaker definition of free lunch in the case of investments having discrete or continuous cash flows, and also in the case of an investment set reduced to a single investment.

We want to show that the absence of free lunches is equivalent to the existence of a discount rate, such that the net present value of all projects is nonpositive.

To prove this, we will assume that there exists at least one investment which is positive at the beginning, and another, at the end. Note that if we consider a discrete time model or even a continuous time model, this condition seems to be quite natural. If all the investments are negative at the beginning, it is straightforward to see that the payoff associated to a nonnegative strategy is necessarily negative at the beginning and then there is no free lunches. The same can be applied at the end and our condition seems therefore to be redundant. In fact, some particular situations are excluded by such a reasoning : the case of investments with oscillations in the neighborhood of the initial or final date such as we can not define a sign to the investment at these dates. Nevertheless, our condition is justified if we admit that such situations are pathological.

We say that a measure μ_k (resp. μ_l) is positive in zero (resp. T_l) if there exists a positive real ε_k (resp. ε_l) such that for all function φ with support contained in $[0, \varepsilon_k]$ (resp. $[T_l - \varepsilon_l, T_l]$), continuous and nonnegative on its support and positive in zero (resp. in T_l), the integral $\int \varphi d\mu_k$ (resp. $\int \varphi d\mu_l$) is positive.

Assumption 5.1 *There exist at least two investments k and l, and a positive real number ε, such that the measure μ_k is positive in zero , and the measure μ_l is positive in T_l.*

We will denote by ε the infimum of ε_k and ε_l. Under this assumption, our main result states as follows.

Theorem 5.1 *Under assumption 5.1, the absence of free lunches is equivalent to the existence of a discount rate r such that for all i in I, the net present value $\int e^{-rt} d\mu_i(t)$ is nonpositive.*

Another way to say the same thing is that there is a free lunch if and only if there exists a finite subset J of I, such that $\sup_{j \in J} \int e^{-rt} d\mu_j(t)$ is positive for all rate r. Furthermore, if we add for all investment μ_k the investment $-\mu_k$ in the model we obtain the situation where all investments can be sold, and the proof of the following result becomes straightforward.

Corollary 5.2 *If all investments can either be bought or sold, under assumption 5.1, the absence of free lunches is equivalent to the existence of a discount rate r, such that for all i in I, the net present value $\int e^{-rt} d\mu_i(t)$ is equal to zero.*

We recall that the lending rate r_0 (resp. the borrowing rate r_1) is the rate at which the investor is allowed to save (resp. to borrow). A lending rate is modeled by the following family of investments $\mu_t = -\delta_0 + e^{r_0 t} \delta_t$ (you lend one dollar at time zero and you will get back $e^{r_0 t}$ at all the possible repayment dates t), and similarly a borrowing rate can be represented as the family $\mu'_t = \delta_0 - e^{r_1 t} \delta_t$.

Corollary 5.3 *If there exists a lending rate r_0 and a borrowing rate r_1, under assumption 6.1, the absence of free lunches is equivalent to the existence of a discount rate r included in $[r_0, r_1]$ such that for all i in I, the net present value $\int e^{-rt} d\mu_i(t)$ is nonpositive.*

The proof of this result is given in Carassus and Jouini (1996).

In the case of a single investment or in the discrete case we can prove that it is sufficient to consider instead of the free lunch concept the classical arbitrage opportunity concept. Furthermore, in the discrete case we can prove that assumption 6.1 is meaningless. The main result of Adler and Gale (1993) appears then as a consequence of our results.

Before to end this section we give some situations where we can apply the previous results.

First, consider the case of a "plan d'épargne logement". In this case, and if we simplify, the product is divided in two stages. During the first stage, the investor saves at a fixed rate r. In the second stage, he can obtain a loan at a special rate r' near to r. More precisely, the bank receives 1^F today. After one period it returns $(1 + r)^F$, and lends 1^F. Finally, at the last period the bank receives $(1 + r')^F$. We denote this investment by $m = (1, -2 - r, 1 + r')$. Our main result is that there is an arbitrage opportunity if, for all positive real number x, $1 - (2 + r)x + (1 + r')x^2$ is positive. A simple computation leads to the following condition $r' - r > \frac{r^2}{4}$. Considering a rate r of $5, 25\%$, it is possible for the bank to construct an arbitrage opportunity if $r' > 5.32\%$.

Other examples are provided in Adler and Gale (1993).

References

Adler, I. and Gale, D. (1993): On getting something for nothing, or how to make an infinite amount of money in a finite time, To appear in *Math. Finance*.

Aiyagari, R., and M. Gertler (1991), "Asset returns with transaction costs and uninsured individual risk," *Journal of Monetary Economics*, 27, p. 311-331.

Amihud, Y., and H. Mendelson (1991), "Liquidity, maturity, and the yields on U.S. Treasury securites," *Journal of Finance*, 46, p. 1411-25.

Back, K., and S. Pliska (1990), "On the fundamental theorem of asset pricing with an infinite state space," *Journal of Mathematical Economics*, 20, p. 1-33.

Bensaid, B., Lesne, J.P., Pagès, H., and J. Scheinkman (1992), "Derivative asset pricing with transaction costs," *Mathematical Finance*, 2, p. 63-86.

Billingsley, P. (1986), *Probability and Measure*, Wiley: New York.

Black, F., and M. Scholes (1973), "The pricing of options and corporate liabilities," *Journal of Political Economy*, 81, p. 637-54.

Bourbaki, N. (1981), *Espaces Vectoriels Topologiques*, Masson: Paris.

Cannon, J. (1984), "The one-dimensional heat equation," in G-C. Rota, ed., *Encyclopedia of Mathematics and its Applications*, volume 23, Cambridge University Press: Cambridge.

Cantor, D.G. and Lippman, S.A. (1983): Investment selection with imperfect capital markets, *Econometrica*, 51, 1121-1144.

Cantor, D.G. and Lippman, S.A. (1995): Optimal investment selection with a multitude of projects, *Econometrica*, 63/5, 1231-1241.

Carassus, I. and E. Jouini (1996), "Investment and arbitrage opportunities with short-sales constraints," *Mimeo*.

Clark, S. (1993), "The valuation problem in arbitrage price theory," *Journal of Mathematical Economics*, 22, p. 463-78.

Clarke, F. (1983), *Optimization and Nonsmooth Analysis*, Wiley: New York.

Cochrane, J., and L. Hansen (1992), "Asset pricing explorations for macroeconomics," *NBER Macroeconomics Annual 1992*, 7, p. 115-165.

Constantinides, G. (1986), "Capital market equilibrium with transaction costs," *Journal of Political Economy*, 94, p. 842-62.

Cox, J., and S. Ross (1976), "The valuation of options for alternative stochastic processes," *Journal of Financial Economics*, 3, p. 145-166.

Cox, J., Ross, S., and M. Rubinstein (1979), "Option pricing: a simplified approach," *Journal of Financial Economics*, 7, p. 229-264.

Cox, J. and M. Rubinstein (1985), *Options Markets*, Prentice-Hall: New Jersey.

Cvitanic, J., and I. Karatzas (1993), "Hedging contingent claims with constrained portfolios," *Annals of Applied Probability*, 3, p. 652-681.

Davis, M., and A. Norman (1990), "Portfolio selection with transaction costs," *Mathematics of Operations Research*, 15, p. 676-713.

Debreu, G. (1959), *Theory of Value*, Wiley: New York.

Delbaen, F. (1992): Representing martingale measures when asset prices are continuous and bounded, *Math. Finance*, 2, 107-130.

Dellacherie, C. and P-A. Meyer (1975), *Probabilités et potentiel*, Hermann: Paris.

Dermody, J.C. and Rockafellar, R.T. (1991): Cash stream valuation in the face of transaction costs and taxes, *Math. Finance*, 1, 31-54.

Dermody, J.C. and Rockafellar, R.T. (1995): Tax basis and nonlinearity in cash stream valuation, *Math. Finance*, 5, 97-119.

Duffie, D., and C.-F. Huang (1985), "Implementing Arrow-Debreu equilibria by continuous trading of few long-lived securities," *Econometrica*, 53, p. 1337-56.

Duffie, D., and C.-F. Huang (1986), "Multiperiod security market with differential information," *Journal of Mathematical Economics*, 15, p. 283-303.

Duffie, D., and T. Sun (1990), "Transaction costs and portfolio choice in a discrete-continuous time setting," *Journal of Economic Dynamics and Control*, 14, p. 35-51.

Dumas, B., and E. Luciano (1991), "An exact solution to a dynamic portfolio choice problem under transactions costs," *Journal of Finance*, 46, p. 577-95.

Dybvig, P. (1988a), "Distributional analysis of portfolio choice," *Journal of Business*, 61, p. 369-93.

Dybvig, P. (1988b), "Inefficient dynamic portfolio strategies or how to throw away a million dollars in the stock market," *Review of Financial Studies*, 1, p. 67-88.

Dybvig, P., and J. Ingersoll (1982), "Mean-variance theory in complete markets," *Journal of Business*, 55, p. 233-51.

Dybvig, P., and S. Ross (1982), "Portfolio efficient sets," *Econometrica*, 50, p. 1525-46.

Dybvig, P., and S. Ross (1985a), "Performance measurement using excess returns relative to a market line : differential information," *Journal of Finance*, 40, p. 384-99.

Dybvig, P., and S. Ross (1985b), "The analytics of performance measurement using a security market line," *Journal of Finance*, 40, p. 401-16.

Dybvig, P., and S. Ross (1986), "Tax clienteles and asset pricing," *Journal of Finance*, 41, p. 751-62.

Dybvig, P., and C.-F. Huang (1988), "Nonnegative wealth, absence of arbitrage, and feasible consumption plans," *Review of Financial Studies*, 1, p. 377-401.

Fan, K. (1952), "Fixed point and minimax theorems in locally convex topological linear spaces," *Proceedings of the National Academy of Sciences of the USA*, 38, p. 121-126.

Figlewski, S. (1989), "Options arbitrage in imperfect markets," *Journal of Finance*, 44, p. 1289-1311.

Fleming, W., and H. Soner (1993), *Controlled Markov Processes and Viscosity Solutions*, Springer-Verlag: New York.

Gale, D. (1965): *Optimal programs for sequential investment.* Patterns of market behavior, ed. by M. J. Brennan. Providence, R.I.: Brown University Press.

Gihman, I., and A. Skorohod (1972), *Stochastic Differential Equations*, Springer-Verlag: New York.

Gilster, J. and W. Lee (1984), "The effects of transaction costs and different borrowing and lending rates on the option pricing model: a note" *Journal of Finance*, 39, p. 1215-22.

Girsanov, I. (1960), "On transforming a certain class of stochastic processes by absolutely continuous substitution of measures," *Theory Probability Appl.*, 5, p. 285-301.

Grossman, S., and G. Laroque (1990), "Asset pricing and optimal portfolio choice in the presence of illiquid durable consumption goods," *Econometrica*, 58, p. 25-51.

Hadar, J., and W. Russell (1969), "Rules for ordering uncertain prospects," *American Economic Review*, 59, p. 25-34.

Hansen, L.P., and R. Jagannathan (1991), "Implications of security market data for models of dynamic economies," *Journal of Political Economy*, 99, p. 225-262.

Harrison, J., and D. Kreps (1979), "Martingales and arbitrage in multiperiod securities markets," *Journal of Economic Theory*, 20, p. 381-408.

Harrison, J., and S. Pliska (1979), "Martingales and stochastic integrals in the theory of continuous trading," *Stochastic Processes and their Applications*, 11, p. 215-60.

Harrison, J., and S. Pliska (1983), "A stochastic calculus model of continuous trading: complete markets," *Stochastic Processes and their Applications*, 15, p. 313-16.

Ho, H., and C.-F. Huang (1992), "Consumption-portfolio policies: an inverse optimal problem," *Journal of Economic Theory*, forthcoming.

He, H., and N. Pearson (1991), "Consumption and portfolio policies with incomplete markets and shortsale constraints: the infinite dimensional case," *Journal of Economic Theory*, 54, p. 259-304.

Heaton, J., and D. Lucas (1992), "The effects of incomplete insurance markets and trading costs in a consumption-based asset pricing model," *Journal of Economic Dynamics and Control*, 16, p. 601-620.

Hindy, A. (1991), "Viable prices in financial markets with solvency constraints", Stanford University mimeo.

Jacka, S. (1992), "A martingale representation result and an application to incomplete financial markets," *Mathematical Finance*, 2, p. 239-50.

Jameson, G. (1974), *Topology and Normed Spaces*, Chapman and Hall: London.

Jouini, E., and H. Kallal (1995a), "Martingales and arbitrage in securities markets with transaction costs," *Journal of Economic Theory*, 66(1), p. 178-197.

Jouini, E., and H. Kallal (1995b), "Arbitrage in securities markets with short-sales constraints," *Mathematical finance*, 5(3), p. 197-232.

Jouini, E., and H. Kallal (1996a), "Efficient trading strategies with market frictions," To appear in *Review of Financial Studies*.

Jouini, E., and H. Kallal (1996b), "Equilibrium in securities markets with bid-ask spreads" *Mimeo*.

Jouini, E., P.-F. Koehl and N. Touzi (1995b), "Incomplete markets, transaction costs and liquidity effects," *Mimeo* .

Karatzas, I., and S. Shreve (1988), *Brownian Motion and Stochastic Calculus*, Springer-Verlag: New York.

Kreps, D. (1981), "Arbitrage and equilibrium in economies with infinitely many commodities," *Journal of Mathematical Economics*, 8, p. 15-35.

Kreps, D. (1982), "Multiperiod securities and the efficient allocation of risk: a comment on the Black and Scholes option pricing model," in J. Mc Call, ed., *The Economics of Uncertainty and Information*, University of Chicago Press: Chicago.

Kunita, H. and S. Watanabe, (1967), "On square integrable martingales," *Nagoya Math. J.*, 30, p. 209-245.

Leland, H. (1985), "Option pricing and replication with transaction costs," *Journal of Finance*, 40, p. 1283-1301.

Levy, H., and Y. Kroll (1978), "Ordering uncertain options with borrowing and lending," *Journal of Finance*, 33, p. 553-74.

Luenberger, D., (1969), *Optimization by vector space methods*, John Wiley : New York.

Luttmer, E. (1991), "Asset pricing in economies with frictions," University of Chicago mimeo.

Machina, M. (1982), "Expected utility analysis without the independence axiom," *Econometrica*, 50, p. 277-323.

Magill, M., and G. Constantinides (1976), "Portfolio selection with transaction costs," *Journal of Economic Theory*, 13, p. 245-63.

Mehra, R., and E. Prescott (1985), "The equity premium: a puzzle," *Journal of Monetary Economics*, 15, p. 145-161.

Peleg, B. (1975), "Efficient random variables," *Journal of Mathematical Economics*, 2, p. 243-61.

Peleg, B., and M. E. Yaari (1975), "A price characterization of efficient random variables," *Econometrica*, 43, p. 283-92.

Prisman, E. (1986) "Valuation of risky assets in arbitrage-free economies with frictions," *Journal of Finance*, 41, p. 545-60.

Quirk, J., and R. Saposnik (1962), "Admissibility and measurable utility functions," *Review of Economic Studies*, 29, p. 140-46.

Ross, S. (1987), "Arbitrage and martingales with taxation," *Journal of Political Economy*, 95, p. 371-393.

Rockafellar, R. T., (1970), *Convex Analysis*, Princeton University Press : Princeton.

Schachermayer, W. (1994): Martingale measures for discrete-time processes with infinite horizon, *Math. Finance*, 4, 25-56.

Sharpe, W. (1990), *Investments*, fourth edition, Prentice Hall: Englewood Cliffs.

Shorack, G. and J. Wellner (1986), *Empirical Processes with Applications to Statistics*, Wiley: New York.

Soner, M., S. Shreve et J. Cvitanic (1995), "There is no nontrivial hedging portfolio for option pricing with transaction costs", *Annals of applied probability*, 5, 327-355.

Stigum, M. (1983), *The Money Market*, Dow Jones-Irwin: Homewood.

Taksar, M., Klass, M., and D. Assaf (1988), "A diffusion model for optimal portfolio selection in the presence of brokerage fees," *Mathematics of Operations Research*, 13, p. 277-94.

Tuckman, B., and J.-L. Vila (1992), "Arbitrage with holding costs: a utility based approach," *Journal of Finance*, 47, p. 1283-1302.

C.I.M.E. Session on "Financial Mathematics"

List of participants

A. AGLIARI, Istituto di Matematica, Fac. di Ingegneria, Via Kennedy 14/b, 43100 Parma, Italy
J. AMENDINGER, TU Berlin, Fachbereich 3, MA7-4, Str. des 17. Juni 136, 10623 Berlin, Germany
F. ANTONELLI, Dip.to di Mat., Univ; "La Sapienza", P.le A. Moro 2, 00185 Roma, Italy
J. APPLEBY, 354 Griffith Av., White Hall, Dublin 9, Ireland
A. R. BACINELLO, Dip.to di Mat. Appl., P.le Europa 1, 34127 Trieste, Italy
E. BARUCCI, DIMADEFAS, Via Lombroso 6/17, 50134 Firenze, Italy
L. BARZANTI, Facoltà di Economia, P.le della Vittoria 15, 47100 Forlì, Italy
B. BASSAN, Dip.to di Mat. del Politecnico, P.zza L. da Vinci, 32, 20133 Milano, Italy
G. BECCHERE, Dip.to di Mat., Via Buonarroti 2, 56127 Pisa, Italy
A. BENDA, Via Volontari del Sangue 12, 20066 Melzo (MI), Italy
S. BOMBELLI, Dip.to di Scienze Statistiche, Univ. di Perugia, 06100 Perugia, Italy
M. BORKOVEC, Univ. Mainz, FB 17 (Mathematik), Staudingerweg 9, 55099 Mainz, Germany
W. BRANNATH, Inst. of Statistics, Brunnestr. 72, 1210 Wien, Austria
E. BUFFET, School of Math. Sci., Dublin City Univ., Dublin 9, Ireland
B. BUSNELLO, Dip.to di Mat., Via Buonarroti 2, 56127 Pisa, Italy
E. CAPOBIANCO, Dip.to di Scienze Statistiche, Via San Francesco 33, 35121 Padova, Italy
L. CARASSUS, CREST Lab. Finance Assurance, 15 blvd G. Péri, 92245 Malakoff Cedex, France
A. CARBONE, Dip.to di Mat., Univ. della Calabria, 87036 Arcavacata di Rende (CS), Italy
G. CARCANO, Dip.to di Metodi Quantitativi, Contrada S. Chiara 48/B, 25122 Brescia, Italy
C. CASCIATI, Viale Gorizia 48, 61100 Pesaro, Italy
R. CASTELLANO, Via Casilina 1616, 00133 Roma, Italy
D. CHEVANCE, INRIA, 2004 route des Lucioles, BP 93, 06902 Sophia Antipolis Cedex, France
K.-H. CHO, Lab. de Probabilités, 4 place Jussieu, 75252 Paris cedex 05, France
A. CONGEDO, Facoltà di Economia, Univ. di Lecce, 73100 Lecce, Italy
M. CORAZZA, Dip.to di Mat. Appl. ed Inf., Ca' Dolfin, Dorsoduro 3825/E, 30123 Venezia, Italy
V. COSTA, Via E. Scarfoglio 20/7, 00159 Roma, Italy
R.-A. DANA, 9 square Port Royal, 75013 Paris, France
J.-P. DECAMPS, GRMAQ, Univ. de Toulouse I, 21 Allée de Brienne, 31000 Toulouse, France
R. L. D'ECCLESIA, Ist. di Scienze Economiche, Fac. di Economia, 61029 Urbino, Italy
L. DE CESARE, IRMA-CNR, II Fac. di Economia, Via Amendola 122/I, 70125 Bari, Italy
A. DI CESARE, Dip.to di Scienze, Univ. di Chieti, V.le Pindaro 42, 65127 Pescara, Italy
R. DIECI, Istituto di Matematica, Fac. di Economia, Via Kennedy 14/b, 43100 Parma, Italy
F. DÖBERLEIN, TU Berlin, Fachbereich Mathematik, Str. des 17 Juni 135, 10623 Berlin, Germany
N. ELHASSAN, School of Finance and Economics, Univ. of Technology Sydney, PO Box 123,
        Broadway NSW 2007, Australia
L. FAINA, Dip.to di Mat., Via Vanvitelli 1, 06123 Perugia, Italy
G. FIGA' TALAMANCA, Dip.to di Statistica, Univ. di Perugia, 06100 Perugia, Italy
D. FILIPOVIC, HG G 36.1, ETH Zentrum, Ramistr. 101, 8092 Zürich, Switzerland
S. FLORIO, Via Don Luigi Rizzo 38, 35042 Este (PD), Italy
M. FRITTELLI, Ist. di Metodi Quantitativi, Fac.di Economia, Via Sigieri 6, 20135 Milano, Italy
B. FUGLSBJERG, Inst. of Math., Bldg 530, Ny Munkegade, 8000 Aarhus C, Denmark
G. FUSAI, Ist. di Metodi Quantitativi, Univ. L. Bocconi, Via Sarfatti 25, 20136 Milano, Italy
M. GALEOTTI, DIMADEFAS, Via Lombroso 6/17, 50134 Firenze, Italy
A. GAMBA, Dip.to di Mat. Appl. ed Inf., Ca' Dolfin, Dorsoduro 3815/E, 30123 Venezia, Italy
R. GIACOMETTI, Dip.to di Mat., Piazza Rosate 2, 24129 Bergamo, Italy

A. GOMBANI, LADSEB-CNR, Corso Stati Uniti 4, 35127 Padova, Italy

P. GRANDITS, Inst. of Statistics, Univ. of Wien, Brunnerstr. 72, 1210 Wien, Austria

S. GRECO, Istituto di Matematica, Fac. di Economia, Corso Italia 55, 95129 Catania, Italy

A. GUALTIEROTTI, IDHEAP, 21 route de la Maladière, 1022 Chavannes-près-Renens,Switzerland

M. L. GUERRA, Ist. di Scienze Economiche, Fac. di Economia, Via Saffi 2, 61029 Urbino, Italy

S. HERZEL, Ist. di Mat. Gen. e Finanz., Fac. di Ec. e Comm., Via Pascoli 1, 06100 Perugia, Italy

M. HLUSEK, CERGE-EI, Politickych Veznu 7, 111 21 Prague 1, Czech Republic

F. HUBALEK, Inst. of Statistics, Univ. of Wien, Brunnerstr. 72, 1210 Wien, Austria

M. G. IOVINO, Ist. di Mat. Gen. e Finanz., Fac. di Economia, Via Pascoli 1, 06100 Perugia, Italy

U. KELLER, Inst. f. Math. Stoch., Univ. Freiburg, Hebelstr. 27, 79104 Freiburg, Germany

I. KLEIN, Inst. of Statistics, Univ. of Wien, Brunnerstr. 72, 1210 Wien, Austria

V. LACOSTE, Dept. of Finance ESSEC, Av. B. Hirsch, BP 105, 95021 Cergy Pointoise, France

C. LANDEN, Royal Inst. of Techn., Dept. of Math., 100 44 Stockholm, Sweden

A. LAZRAK, GREMAQ, Univ. des Sci. Sociales, 21 Allée de Brienne, 31000 Toulouse, France

D. P. J. LEISEN, Dept. of Stat., Univ. of Bonn, Adenauerallée 14-22, 53119 Bonn, Germany

T. LIEBIG, Univ. Ulm, Abteilung f. Math., Helmholtzstr. 18, 89069 Ulm, Germany

E. LUCIANO, Dip.to di Stat., P.zza Arbarello 8, 10122 Torino, Italy

B. LUDERER, TU Chemnitz-Zwickau, Fac. of Math., 09107 Chemnitz, Germany

C. MANCINI, Ist. di Mat. Gen. e Fin., Univ. di Perugia, Via Pascoli, 06100 Perugia, Italy

M. E. MANCINO, DIMADEFAS, Via Lombroso 6/17, 50134 Firenze, Italy

R. N. MANTEGNA, Dip.to di Energetica, Viale delle Scienze, 90128 Palermo, Italy

I. MASSABO', Dip.to di Organizzazione Aziendale, Univ. della Calabria,
    87036 Arcavacata di Rende (CS), Italy

L. MASTROENI, Dip.to di Studi Economico-Finanziari, Univ. di Roma "Tor Vergata",
    Via di Tor Vergata, 00133 Roma, Italy

F. MERCURIO, Tinbergen Inst., Erasmus Univ. Rotterdam, 3062 PA Rotterdam, The Netherlands

E. MORETTO, Dip.to di Metodi Quantitativi, Contrada S. Chiara 48b, 25122 Brescia, Italy

M. MOTOCZYNSKI, Univ. of Warsawa, Fac. of Math., ul. Banacha 2, 02 097 Warszawa, Poland

S. MULINACCI, Dip.to di Mat., Via Buonarroti 2, 56127 Pisa, Italy

F. NIEDDU, Ist. di Metodi Quantitativi, Univ. L. Bocconi, Via U. Gobbi 5, 20136 Milano, Italy

C. PACATI, Ist. di Mat. Gen., e Fin., Univ. di Perugia., Via A. Pascoli 1, 06100 Perugia, Italy

M. PAGLIACCI, Dip.to di Organizzazione Aziendale, Univ. della Calabria,
    87036 Arcavacata di Rende (CS), Italy

L. PAPPALARDO, Ist. di Metodi Quantitativi, Via Sigieri 6, 20135 Milano, Italy

J. E. PARNELL, Maths Dept., Dublin City Univ, Glasnevin, Dublin 9, Ireland

C. PICHET, Dép. de Math.-UQAM, C.P. 888 Succ. Centre-Ville, Montréal, Québec, Canada

M. PRATELLI, Dip.to di Mat., Via Buonarroti 2, 56127 Pisa, Italy

S. RICCARELLI, Dip.to di Metodi Quantitativi, Contrada S. Chiara 48b, 25122 Brescia, Italy

S. ROMAGNOLI, Ist. di Mat. Gen. e Fin., P.zza Scaravilli 2, 40126 Bologna, Italy

F. ROSSI, Ist. di Mat., Fac. di Economia, Via dell'Artigliere 19, 37129 Verona, Italy

J.-M. ROSSIGNOL, GREMAQ, Univ. des Sci. Sociales, 21 Allée de Brienne,
    31000 Toulouse, France

M. RUTKOWSKI, Inst. of Math., Politechnika Warszawska, 00 661 Warszawa, Poland

W. SCHACHERMAYER, Inst. of Stat., Univ. of Wien, Brunnerstr. 72, 1210 Wien, Austria

W. SCHACHINGER, Inst. of Stat. Univ. of Wien, Brunnerstr. 72, 1210 Wien, Austria

D. SCOLOZZI, Fac. di Economia, Univ. di Lecce, Via per Monteroni, 73100 Lecce, Italy

S. SMIRNOV, Fac. of Comput. Math. and Cyb., Moscow State Univ., Vorobievy Gory V-234,
    Moscow GSP 119899, Russia

R. SMITH, Dept. of Math., Purdue Univ., West Lafayette, Indiana 47907-1395, USA

G. TESSITORE, Dip.to di Mat. Appl., Fac. di Ing., Via S. Marta 3, 50139 Firenze, Italy

M. C. UBERTI, Dip.to di Stat. e Mat. Appl., P.zza Arbarello 8, 10122 Torino, Italy
M. VANMAELE, Dept. of Quant. Techn., Univ. Gent, Hoveniersberg 4, 9000 Gent, Belgium
T. VARGIOLU, Scuola Normale Superiore, P.zza dei Cavalieri 7, 56126 Pisa, Italy
P. VARIN, Dip.to di Mat. Appl., P.le Europa 1, 34127 Trieste, Italy
V. VESPRI, Dip.to di Mat. pura ed appl., Via Vetoio, 67100 L'Aquila, Italy
N. WELCH, Mathematics, Univ. of Kansas, Lawrence, Kansas 66045-2142 USA
J. ZABCZYK, Inst. of Math., Polish Acad. of Sci., Sniadeckich 8, 00-950 Warszaw, Poland
P. A. ZANZOTTO, Dip.to di Mat., Via Buonarroti 2, 56127 Pisa, Italy
P. ZIMMER, Mathematics, Univ. of Kansas, Lawrence, Kansas 66045-2142 USA
M. ZUANON, Ist. di Econometria e Matematica, Univ. Cattolica del Sacro Cuore,
         Largo Gemelli 1, 20123 Milano, Italy
C. ZUHLSDORFF, Dept. of Statistics, Univ. of Bonn, Adenauerallée 24-42, 53113 Bonn, Germany

312

314

1983 - 90. Complete Intersections   (LNM 1092) Springer-Verlag
       91. Bifurcation Theory and Applications   (LNM 1057)   "
       92. Numerical Methods in Fluid Dynamics   (LNM 1127)   "

1984 - 93. Harmonic Mappings and Minimal Immersions   (LNM 1161)   "
       94. Schrödinger Operators   (LNM 1159)   "
       95. Buildings and the Geometry of Diagrams   (LNM 1181)   "

1985 - 96. Probability and Analysis   (LNM 1206)   "
       97. Some Problems in Nonlinear Diffusion   (LNM 1224)   "
       98. Theory of Moduli   (LNM 1337)   "

1986 - 99. Inverse Problems   (LNM 1225)   "
      100. Mathematical Economics   (LNM 1330)   "
      101. Combinatorial Optimization   (LNM 1403)   "

1987 - 102. Relativistic Fluid Dynamics   (LNM 1385)   "
       103. Topics in Calculus of Variations   (LNM 1365)   "

1988 - 104. Logic and Computer Science   (LNM 1429)   "
       105. Global Geometry and Mathematical Physics   (LNM 1451)   "

1989 - 106. Methods of nonconvex analysis   (LNM 1446)   "
       107. Microlocal Analysis and Applications   (LNM 1495)   "

1990 - 108. Geoemtric Topology: Recent Developments   (LNM 1504)   "
       109. H Control Theory   (LNM 1496)   "
       110. Mathematical Modelling of Industrial   (LNM 1521)   "
            Processes

1991 - 111. Topological Methods for Ordinary   (LNM 1537)   "
            Differential Equations
       112. Arithmetic Algebraic Geometry   (LNM 1553)   "
       113. Transition to Chaos in Classical and   (LNM 1589)   "
            Quantum Mechanics

1992 - 114. Dirichlet Forms   (LNM 1563)   "
       115. D-Modules, Representation Theory,   (LNM 1565)   "
            and Quantum Groups
       116. Nonequilibrium Problems in Many-Particle   (LNM 1551)   "
            Systems

316